T0305885

The Science of Water Reuse

The general public's aversion to drinking treated wastewater is evident, rooted in the reluctance to accept a direct pipe-to-pipe connection, for example, from a toilet to a drinking water tap. Despite advancements in treating black water from sewage sources to meet drinking water standards, there persists a general hesitancy to fully accept this possibility or understand the science behind it. *The Science of Water Reuse* explains how technology can sufficiently purify reclaimed water to potable water quality—even surpassing the cleanliness of the water available from conventional taps. It addresses the significant gap in the existing literature on water reuse, focusing particularly on the varied applications of reused or reclaimed water within municipal and agricultural contexts, with a specific emphasis on issues and technologies related to both direct and indirect potable water reuse. It serves as a valuable resource for policymakers, municipal planners, environmental engineering professionals, as well as undergraduate and graduate students.

- Bridges the gap between technical details and public comprehension, making the complex subject of water reuse accessible and relevant.
- Provides a comprehensive understanding of water reuse, including case studies for practical application.
- Contributes to changing public attitudes, making reclaimed water an acceptable source for potable use.

The Science of Water Reuse

Frank R. Spellman

CRC Press
Taylor & Francis Group
Boca Raton London New York

CRC Press is an imprint of the
Taylor & Francis Group, an **informa** business

Designed cover image: Shutterstock

First edition published 2025
by CRC Press
2385 NW Executive Center Drive, Suite 320, Boca Raton FL 33431

and by CRC Press
4 Park Square, Milton Park, Abingdon, Oxon, OX14 4RN

CRC Press is an imprint of Taylor & Francis Group, LLC

© 2025 Frank R. Spellman

Library of Congress Cataloging-in-Publication Data
Names: Spellman, Frank R., author.
Title: The science of water reuse / Frank R. Spellman.
Description: First edition. | Boca Raton, FL : CRC Press, 2024. |
Includes bibliographical references and index. | Identifiers: LCCN 2024013212 |
ISBN 9781032806693 (hbk) | ISBN 9781032806754 (pbk) | ISBN 9781003498049 (ebk)
Subjects: LCSH: Water reuse. | Water—Purification.
Classification: LCC TD429 .S683 2024 | DDC 628.1/62—dc23/eng/20240701
LC record available at https://lccn.loc.gov/2024013212

ISBN: 978-1-032-80669-3 (hbk)
ISBN: 978-1-032-80675-4 (pbk)
ISBN: 978-1-003-49804-9 (ebk)

DOI: 10.1201/9781003498049

Typeset in Times
by codeMantra

Contents

PART 1   The Basics

PART 2   *Toilet to Tap and Nonpotable Reuse: Waste Not, Want Not: Water Reuse*

PART 3   The Real Deal: Purple to Blue PVC Pipe

Preface

This book, *The Science of Water Reuse*, is the thirteenth volume in the acclaimed series that includes *The Science of Land Subsidence* (in production), *The Science of Green Energy, The Science of Carbon Capture and Storage (CCS), The Science of Ocean Pollution* (in production), *The Science of Lithium (Li), The Science of Electric Vehicles (EVs): Concepts and Applications, The Science of Rare Earth Elements: Concepts and Applications, The Science of Water, The Science of Air, The Science of Environmental Pollution, The Science of Renewable Energy, The Science of Waste,* and *The Science of Wind Power* all of which bring this highly successful series fully into the 21st century. This book continues the series mantra based on good science and not on feel-good science. It also continues to be presented in the Author's trademark conversational style—making sure communication is certain—not a failure … and maybe because I just want to converse with the reader. My aim/my goal is to be comprehensive in coverage and comprehensible in what I deliver.

Several water reuse case studies are provided within to make the text more diverse, inclusive, concrete, and contextual, providing in-depth knowledge about the real-world subject matter presented.

Although this book is directed at policymakers, legislators, water planners, and water reuse practitioners including utility staff, undergraduate and graduate students, professionals, scientists, general water practitioners, engineers, and consultants, it is written to be understood by the public.

Following the successful format of the other editions in this series, this no-holds-barred book aims to provide an understanding of the current science underpinning natural and human-made water cycling that enables the reuse of water on a continuous basis, hopefully.

Frank R. Spellman, Norfolk, VA

Prologue: The Yuck Factor

Here are a few questions pertinent to this discussion:

Would you drink water from your toilet?
Would you drink water from anyone's toilet?
Would you drink water from anyone's sink, deep sink, floor drain, car wash, or
 industrial complex?
How about your dishwasher, would you drink the water from its drain?
After bathing, would you drink the used water in your bathtub?
How about the water from your washing machine's drain, would you drink it?
How about stormwater, would you drink it?

When I taught water/wastewater topics in my university environmental health/environmental engineering classes to undergrad and grad students and to water/wastewater operators in the numerous short courses, I always began lecture number one with these questions.

Why? I did so because in my classes the goal was (and always is) designed to facilitate thought and also response. And the questions above accomplished this goal because whether my classes were attended by dozens or hundreds I had no problem soliciting some thought and an overabundance, literally a flood of responses from all in attendance.

I always expected this mass of negative responses. I always received this mass of negative responses spouted out by screwed-up faces of revulsion. However, at the same time, to my surprise, I always had someone (sometimes more than one) who waited for the uproar, the clamor, the din, and the outcries to subside and then would speak out words along lines of: "We are already drinking water from all of those sources."

Amen! There is hope for future generations.

Wow! It was (is) always encouraging to me as an instructor that no matter the size of my classes I always had someone who was knowledgeable and far-thinking to know and understand the truth.

The truth?

Yes.

Yes, because of the natural water cycle and the human-made water cycle, we are drinking water that was and is spent from other uses, including toilet water, sink water, runoff water from fields filled with cow pies (i.e., droppings of cow dung), and basically wastewaters from all other sources.

How can this be?

THE BOTTOM LINE

The 'how' this can be is what this book is all about.

About the Author

Frank R. Spellman, PhD, is a retired full-time adjunct assistant professor of environmental health at Old Dominion University, Norfolk, Virginia, and the author of more than 160 books covering topics ranging from concentrated animal feeding operations (CAFOs) to all areas of environmental science and occupational health. Many of his texts are readily available online at Amazon.com and Barnes and Noble.com, and several have been adopted for classroom use at major universities throughout the United States, Canada, Europe, and Russia; two have been translated into Spanish for South American markets. Dr. Spellman has been cited in more than 1,150 publications. He serves as a professional expert witness for three law groups and as an incident/accident investigator for the U.S. Department of Justice and a northern Virginia law firm. In addition, he consults on homeland security vulnerability assessments for critical infrastructures including water/wastewater facilities nationwide and conducts pre-Occupational Safety and Health Administration (OSHA)/Environmental Protection Agency EPA audits throughout the country. He receives frequent requests to co-author with well-recognized experts in several scientific fields; for example, he is a contributing author of the prestigious text *The Engineering Handbook*, 2nd ed. (CRC Press). He lectures on wastewater treatment, water treatment, homeland security, and safety topics throughout the country and teaches water/wastewater operator short courses at Virginia Tech (Blacksburg, Virginia). In 2011–2012, he traced and documented the ancient water distribution system at Machu Picchu, Peru, and surveyed several drinking water resources in Amazonia-Coco, Ecuador. He continues to collect and analyze contaminated sediments in the major river systems in the world. He also studied and surveyed two separate potable water supplies in the Galapagos Islands. He also studied and researched Darwin's finches while in the Galapagos. He holds a BA in public administration, a BS in business management, an MBA, and an MS and PhD in environmental engineering.

Units and Conversions

Most of the calculations made in the water/wastewater operations involve using *units.* While the number tells us how many, the units tell us what we have. Examples of units include inches, feet, square feet, cubic feet, gallons, pounds, milliliters, milligrams per liter, pounds per square inch, miles per hour, and so on. *Conversions* are a process of changing the units of a number to make the number usable in a specific instance. Multiplying or dividing into another number to change the units of the number accomplishes conversions. Common conversions in water/wastewater operations are:

- Gallons per minute (gpm) to cubic feet second (cfs)
- Million gallons to acre feet
- Cubic feet to acre feet
- Cubic feet of water to weight
- Cubic feet of water to gallons
- Gallons of water to weight
- Gallons per minute (gpm) to million gallons per day (MGD)
- Pounds per square inch (psi) to feet of head (the measure of the pressure of water expressed as the height of water in feet); 1 psi = 2.31 ft of head.

In many instances, the conversion factor cannot be derived—it must be known. Therefore, we use tables such as the one below to determine the common conversions.

Note: Conversion factors are used to change measurements or calculated values from one unit of measure to another. In making the conversion from one unit to another, you must know two things:

1. The exact number that relates the two units
2. Whether to multiply or divide by that number

Most operators memorize some standard conversions. This happens because of using the conversions, not because of attempting to memorize them.

COMMON CONVERSIONS

Linear Measurements

1 inch = 2.54 cm

1 foot = 30.5 cm

1 meter = 100 cm, 3.281 ft = 39.4 in.

1 acre = 43,560 ft^2

1 yard = 3 ft

Volume

1 gal = 3.78 L

1 ft^3 = 7.48 gal

1 L = 1,000 mL

1 acre foot = 43,560 ft^3

1 gal = 32 cups

1 pound = 16 oz dry wt.

Weight

1 ft^3 of water = 62.4 lbs

1 gal = 8.34 lbs

1 lb =453.6 g

1 kg = 1,000 g = 2.2 lbs

1% = 10,000 mg/L

Pressure

1 ft of head = 0.433 psi

1 psi = 2.31 ft of head

Flow

1 cfs = 448 gpm

1 gpm = 1,440 gpd

Water Reuse Term Box

Practitioners of Water Reuse or Water Reuse Practitioners, like the handyperson has a toolbox, have their own term box. To get close, to get familiar, and to connect with the material presented in this book, it is important to have a ready 'term box'; because of this foundational need, the members of Water Reuse Term Box are presented in the following.

ESSENTIAL AND PRACTICAL DETAILS[1]

The basics upon which water reuse science (e.g., reverse osmosis and more) stands (its foundation) depends on the nuts and bolts of foundational material that supports it. Ok, so what are the basics, the foundational materials? I refer to Voltaire who wisely stated: "If you wish to converse with me, please define your terms." To present the principles and operations of water reuse science in an understandable form, the terms associated with the technology are presented in plain English in this chapter. Moreover, to make sure that the essential and practical details and basic aspects or working components of the science of water reuse are easily understandable, the concepts, units of expression, and pertinent nomenclature are also presented.

Concepts

Miscible, Solubility

1. **Miscible**: capable of being mixed in all proportions. Simply, when two or more substances disperse themselves uniformly in all proportions when brought into contact they are said to be completely soluble in one another or completely miscible. The precise chemistry definition is: "homogenous molecular dispersion of two or more substances" (Jost, 1992). Examples are:
   - All gases are completely miscible.
   - Water and alcohol are completely miscible.
   - Water and mercury (in its liquid form) are immiscible liquids.
2. Between the two extremes of miscibility, there is a range of solubility; that is, various substances mix with one another up to a certain proportion. In many environmental situations, a rather small amount of contaminant is soluble in water in contrast to the complete miscibility of water and alcohol. The amounts are measured in parts per million (ppm).

Suspension, Sediment, Particles, Solids

Often, water carries solids or particles in suspension. These dispersed particles are much larger than molecules and may be comprised of millions of molecules. The particles may be suspended in flowing conditions and initially under quiescent conditions, but eventually, gravity causes settling of the particles. The resultant accumulation by settling is often called sediment or biosolids (sludge) or residual solids in wastewater treatment vessels. Between this extreme of readily falling out by gravity and permanent dispersal as a solution at the molecular level, there are intermediate types of dispersion or suspension. Particles can be so finely milled or of such small intrinsic size as to remain in suspension almost indefinitely and in some respects similarly to solutions.

Emulsion

Emulsions represent a special case of suspension. As you know, oil and water do not mix. Oil and other hydrocarbons derived from petroleum generally float on water with negligible solubility in water. In many instances, oils may be dispersed as fine oil droplets (an emulsion) in water and not readily separated by floating because of size and/or the addition of dispersal-promoting additives. Oil and, in particular, emulsions can prove detrimental to many treatment technologies and must be treated in the early steps of a multi-step treatment train.

Ion

An ion is an electrically charged particle. For example, sodium chloride or table salt forms charged particles on dissolution in water; sodium is positively charged (a cation), and chloride is negatively charged (an anion). Many salts similarly form cations and anions on dissolution in water.

Mass Concentration

The concentration of an ion or substance in water is often expressed in terms of parts per million (ppm) or mg/L. Sometimes, parts per thousand or parts per trillion (ppt) or parts per billion (ppb) are also used. These are known as units of expression. A ppm is analogous to a full shot glass of swimming pool water as compared to the entire contents of a standard swimming pool full of water. A ppb is analogous to one drop of water from an eye dropper into the total amount of water in a standard swimming pool full of water.

$$ppm = \frac{Mass\,of\,substance}{Mass\,of\,solutions}$$

Because 1 kg of solution with water as a solvent has a volume of ~1 L,

$$1\,ppm \approx 1\,mg/L$$

Permeate

The portion of the feed stream that passes through an RO membrane.

Concentrate, Reject, Retentate, Brine, Residual Stream

The membrane output stream contains water that has not passed through the membrane barrier and concentrated feedwater constituents that are rejected by the membrane.

Tonicity

Is a measure of the effective osmotic pressure gradient (as defined by the water potential of the two solutions) of two solutions separated by a semipermeable membrane. It is important to point out that, unlike osmotic pressure, tonicity is influenced only by solutes that cannot cross the membrane, as only these exert an effective osmotic pressure. Solutes able to freely cross do not affect tonicity because they will always be in equal concentrations on both sides of the membrane. There are three classifications of tonicity that one solution can have relative to another. They are *hypertonic, hypotonic*, and *isotonic* (Sperelakis, 2011).

- **Hypertonic**: It refers to a greater concentration. In biology, a hypertonic solution is one with a higher concentration of solutes outside the cell than inside the cell; the cell will lose water by osmosis.
- **Hypotonic**: It refers to a lesser concentration. In biology, a hypotonic solution has a lower concentration of solutes outside the cell than inside the cell; the cell will gain water through osmosis.
- **Isotonic**: It refers to a solution in which the solute and solvent are equally distributed. In biology, a cell normally wants to remain in an isotonic solution, where the concentration of the liquid inside it equals the concentration of liquid outside it; there will be no net movement of water across the cell membrane.

"Osmosis"

The naturally occurring transport of water through a membrane from a solution of low salt content to a solution of high salt content in order to equalize salt concentrations.

Osmotic Pressure

A measurement of the potential energy difference between solutions on either side of a semipermeable membrane due to osmosis. Osmotic pressure is a colligative property, meaning that the property depends on the concentration of the solute, but not on its identity.

Osmotic Gradient

The osmotic gradient is the difference in concentration between two solutions on either side of a semipermeable membrane and is used to tell the difference in percentages of the concentration of a specific particle dissolved in a solution. Usually, the osmotic gradient is used while comparing solutions that have a semipermeable membrane allowing water to diffuse between the two solutions, toward the hypertonic solutions. Eventually, the force of the column of water on the hypertonic side of the semipermeable membrane will equal the force of diffusion on the hypotonic side, creating equilibrium. When equilibrium is reached, water continues to flow, but it flows both ways in equal amounts as well as force, therefore stabilizing the solution.

Membrane

A thin layer of material capable of separating materials as a function of their chemical or physical properties when a driving force is applied.

Semipermeable Membrane
A membrane that is permeable only by certain molecules or ions.

RO System Flow Rating
Although the influent and reject flows are usually not indicated, the product flow rate is used to rate an RO System. A 600-gpm RO would yield 600 gpm of permeate; thus, it is rated at 600 gpm.

Recovery Conversion
The ratio of the permeate flow to the feed flow is fixed by the designer and is generally expressed as a percentage. Used to describe what volume percentage of influent water is recovered. Exceeding the design recovery can result in accelerated and increased fouling and scaling of the membranes.

$$\% \text{Recovery} = (\text{Recovery flow/feed flow}) \times 100$$

Concentration Factor (CF)
Concentration factor is the ratio of solute contamination in the concentrate stream to solute concentration in the feed system. It is related to recovery in that at 40% recovery, and the concentrate would be two-fifths that of the influent water.

Rejection
The term rejection is used to describe what percentage of an influent species a membrane retains. For example, 97% rejection of salt means that the membrane will retain 97% of the influent salt. It also means that 3% of influent salt will pass through the membrane into the permeate; this is known as salt passage. The following equation is used to calculate the rejection of a given species.

$$\% \text{Rejection} = \left[(C_i - C_p)/C_i \right] \times 100$$

where:
C_i = influent concentration of a specific component
C_p = permeate concentration of a specific component

The RO processes a semi-permeable to reject a wide variety of impurities. Table 1 is a partial list of the general rejection ability of the most commonly thin film composite (TFC) RO membranes. Note that these are averaged based on experience and are generally accepted within the industry. They are not a guarantee of performance. Actual rejection can vary according to the chemistry of the water, temperature, pressure, pH, and other factors (Pure Water Products, LLC, 2014).

Flux
The word flux comes from Latin: *fluxus* means "flow", and *fluere* is "to flow"; as *fluxion*, this term was introduced into differential calculus by Isaac Newton. With regard to RO systems, flux is the rate of water flow (volumetric flow rate) across a

TABLE 1
Estimated Reverse Osmosis Rejection Percentages of Selected Impurities for Thin Film Composite Membranes

Impurities	Rejection Percentages (%)
Aluminum	97–98
Ammonium	85–95
Arsenic	95–96
Bacteria	99+
Bicarbonate	95–96
Boron	50–70
Bromide	93–96
Cadmium	96–98
Calcium	96–98
Chloride	94–95
Chromium	96–98
Copper	97–99
Cyanide	90–95
Detergents	97
Fluoride	94–96
Herbicides	97
Insecticides	97
Iron	98–99
Lead	96–98
Magnesium	96–98
Manganese	96–98
Mercury	96–98
Nickel	97–99
Nitrate	93–96
Phosphate	99+
Radioactivity	95–98
Radium	97
Selenium	97
Silica	85–90
Silicate	95–97
Silver	95–97
Sodium	92.98
Sulfate	99+
Sulfite	96–98
Virus	99+
Zinc	98–99

Source: Adaptation from Pure Water Products, LLC (2014).

unit surface area (membrane); it is expressed as gallons of water per square foot of membrane area per day, (gpd) or liters per hour per square meter (L/hr-m^2 or Lmh). In general, flux is proportional to the density of flow; it varies by how the boundary faces the direction of flow; and it is proportional within the area of the boundary.

Specific Flux (also Called Permeability)

Refers to membrane flux normalized for temperature and pressure, expressed as gallons per square foot per day per pound per square inch (gfd/psi) or liters per square meter per hour per bar (Lmh/bar). Specific flux is sometimes discussed in comparing the performance of one type of membrane with another. In comparing membranes, the higher the specific flux the lower the driving pressure required to operate the RO system (Kucera, 2010).

Concentration Polarization

Similar to the flow of water through a pipe, Figure 1a and b, concentration polarization is the phenomenon of increased solute (salt) concentration relative to the bulk solution that occurs in a thin boundary layer at the membrane surface on the feed side (Figure 1c, For reference and clearer understanding let's first look at Figure 1a, which shows that flow may be *laminar* (streamline), and then look at Figure 1b where the flow may be or turns *turbulent*. Laminar flow occurs at extremely low velocities. The water moves in straight parallel lines, called streamlines, or laminae, which slide upon each other as they travel, rather than mixing up. Normal pipe flow is turbulent flow, which occurs because of friction encountered on the inside of the pipe. The outside layers of flow are thrown into the inner layers; the result is that all the layers mix and are moving in different directions and at different velocities; However, the direction of flow is forward. Figure 1c shows the hydraulic boundary layer formed by fluid flow through a pipe. Kucera (2010) points out that concentration polarization has a negative effect on the performance of an RO membrane; specifically, it reduces the throughput of the membrane. Flow may be steady or unsteady. For our purposes, we consider steady-state flow only; that is most of the hydraulic calculations in this manual assume steady-state flow.

Membrane Fouling

Fouling is a process where a loss of membrane performance occurs due to the deposition of suspended or dissolved substances on its external surfaces, at its pore openings, or with its pores, forming a fouling layer, or from internal changes of the membrane material. Both forms of fouling can cause membrane permeability to decline.

DID YOU KNOW?

The fouling of a reverse osmosis membrane is almost inevitable. Particulate matter will be retained and is an ideal nutrient for biomass, resulting in biofouling.

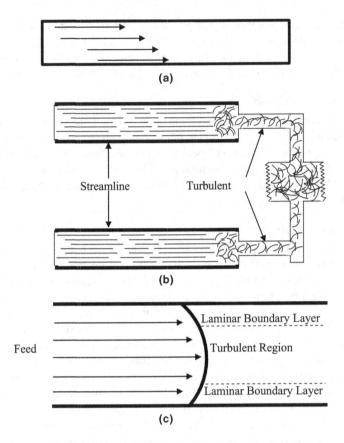

FIGURE 1   (a) Laminar (streamline) flow; (b) turbulent flow; (c) when fluid flows through a pipe, a hydraulic boundary layer is formed.

Membrane Scaling

Scaling is a form of fouling on the feed-concentrate side of the membrane that occurs when dissolved species are concentrated in excess of their solubility limit. Scaling is exacerbated by low cross-flow velocity and high membrane flux (Kucera, 2010).

Silt Density Index (SDI)

A dimensionless value resulting from an empirical test used to measure the level of suspended and colloidal material in water. Calculated from the time it takes to filter 500 mL of the test water through a 0.45-μm-pore-diameter filter at 30 psi pressure at the beginning and at the end of a specified test duration. The lower the SDI, the lower the potential for fouling a membrane with suspended solids. Visually, the deposited foulant on a filter membrane can be identified by its color. For example, a foulant that is yellow could possibly indicate iron or organics; a red foulant indicates iron; and black may indicate manganese (Kucera, 2010).

Langelier Saturation Index (LSI)

A calculated value based on total dissolved solids, calcium concentration, total alkalinity, pH, and solution temperature, indicating the tendency of a water solution to precipitate or dissolve calcium carbonate. The LSI is based on the pH and temperature of the water in question as well as the concentrations of total dissolved solids (TDS), calcium harness, and alkalinity. The LSI generally ranges from <0.0 (no scale, very slight tendency to dissolve scale) to 3.0 (extremely severe scaling). For RO applications, a positive LSI indicates that the influent water has the tendency to form a calcium carbonate scale (Kucera, 2010).

Antiscalants

A chemical sequestering agent added to feedwater inhibits scale formation.

Foundational Terms and Definitions

- **Chemistry**: The science that deals with the composition and changes in composition of substances. Water is an example of this composition; it is composed of two gases, hydrogen and oxygen. Water also changes form from liquid to solid to gas but does not necessarily change composition.
- **Matter**: Anything that has weight (mass) and occupies space. Kinds of matter include elements, compounds, and mixtures.
- **Solids**: Substances that maintain definite size and shape. Solids in water fall into one of the following categories.
  1. Dissolved
  2. Colloidal
  3. Suspended
- **Dissolved solids** are in solution and pass through a filter. The solution consisting of the dissolved components and water forms a single phase (a homogenous solution).
- **Colloidal solids** (sols) are uniformly dispersed in solution but they form a solid phase that is distinct from the water phase.
- **Suspended solids** are also a separate phase from the solution. Some suspended solids are classified as settleable solids. Placing a sample in a cylinder and measuring the amount of solids that have settled after a set amount of time determine settleable solids. The size of solids increases going from dissolved solids to suspended solids.
- **Liquids**: A definite volume, but not shape, liquid will fill containers to certain levels and form free-level surfaces.
- **Gases**: Of neither definite volume nor shape, they completely fill any container in which they are placed.
- **Element**: The simplest form of chemical matter. Each element has chemical and physical characteristics different from all other kinds of matter.
- **Compound**: A substance of two or more chemical elements chemically combined. Examples: water (H_2O) is a compound formed by hydrogen and oxygen. Carbon dioxide (CO_2) is composed of carbon and oxygen.
- **Mixture**: A physical, not chemical, intermingling of two or more substances. Sand and salt stirred together form a mixture.

- **Atom**: The smallest particle of an element that can unite chemically with other elements. All the atoms of an element are the same in chemical behavior, although they may differ slightly in weight. Most atoms can combine chemically with other atoms to form molecules.
- **Molecule**: The smallest particle of matter or a compound that possesses the same composition and characteristics as the rest of the substance. A molecule may consist of a single atom, two or more atoms of the same kind, or two or more atoms of different kinds.
- **Radical**: Two or more atoms that unite in a solution and behave chemically as if a single atom.
- **Solvent**: The component of a solution that does the dissolving (see Figure 2.1).
- **Solute**: The component of a solution that is dissolved by the solvent (see Figure 2.1).
- **Ionization**: The formation of ions by splitting of molecules or electrolytes in solution. Water molecules are in continuous motion, even at lower temperatures. When two water molecules collide, a hydrogen ion is transferred from one molecule to the other. The water molecule that loses the hydrogen ion becomes a negatively charged hydroxide ion. The water molecule that gains the hydrogen ion becomes a positively charged hydronium ion. This process is commonly referred to as the self-ionization of water.
- **Cation**: A positively charged ion.
- **Anion**: A negatively charged ion.
- **Organic**: Chemical substances of animal or vegetable origin made of carbon structure.
- **Inorganic**: Chemical substances of mineral origin.
- **Solids**: As it pertains to water-suspended and dissolved material in water.
- **Dissolved solids**: The material in water that will pass through a glass fiber filter and remain in an evaporating dish after evaporation of the water.
- **Suspended solids**: The quantity of material deposited when a quantity of water, sewage, or other liquid is filtered through a glass fiber filter.
- **Total solids**: The solids in water, sewage, or other liquids; it includes the suspended solids (largely removable by a filter) and filterable solids (those which pass through the filter).
- **Saturated solution**: The physical state in which a solution will no longer dissolve more of the dissolving substance—solute.
- **Colloidal**: Any substance in a certain state of fine division in which the particles are less than one micron in diameter.
- **Turbidity**: A condition in water caused by the presence of suspended matter, resulting in the scattering and absorption of light rays.
- **Precipitate**: A solid substance that can be dissolved but is separated from the solution because of a chemical reaction or change in conditions such as pH or temperature.

NOTE

1 Based on material in Spellman, F.R. (2020). *The Science of Water: Concepts and Applications*, 4th ed. Boca Raton, FL: CRC Press.

REFERENCES

Jost, N.J. (1992). Surface and ground water pollution control technology. In P.C. Knowles (ed.), *Fundamentals of Environmental Science and Technology*. Rockville, MD: Government Institutes, Inc., pp. 261–264.

Kucera, J. (2010). *Reverse Osmosis: Industrial Applications and Processes*. New York: John Wiley & Sons, Inc.

Pure Water Products, LLC. (2014). Reverse osmosis rejection percentages. Denton, TX.

Sperelakis, N. (2011). *Cell Physiology Sources Book: Essentials of Membrane Biophysics*. New York: Academic Press.

Part 1

The Basics

Pollution can corrupt water quality and make surface waters too filthy for recreational, drinking, or wildlife habitat uses. Water reuse collects and treats polluted water so that it can be used again in the community.

DOI: 10.1201/9781003498049-1

1 Used Water Yuck Factor Overstated

SETTING THE RECORD STRAIGHT

That great mythical hero, Hercules, arguably the world's first environmental engineer, was ordered to perform his fifth labor by Eurystheus to clean up King Augeas' stables. Hercules, faced with a mountain of horse and cattle waste piled high in the stable area, had to devise some method to dispose of the waste; he did. He diverted a couple of river streams to the inside of the stable area so that all the animal waste could simply be deposited into the river streams: Out of sight out of mind. The waste simply flowed downstream. Hercules understood the principal point in pollution control technology that is pertinent to this very day and to this discussion, that is, *dilution is the solution to pollution.*

When people say they would never drink toilet water, they have no idea what they are saying. As pointed out in my textbook, *The Science of Water,* 4th edition, the fact is that we drink recycled wastewater every day. In Hampton Roads, Virginia, for example, Hampton Roads Sanitation District's (HRSDs) wastewater treatment plants outfall (discharge) treated water to the major rivers in the region. Many of the region's rivers are sources of local drinking water supplies. Even local groundwater supplies are routinely infiltrated with surface water inputs, which, again, are commonly supplied by treated wastewater (and sometimes infiltrated by raw sewage that is accidentally spilled).

My compliments to Mr. Henifin, General Manager (ret) of Hampton Roads Sanitation District (HRSD), who stated in a recent local newspaper article that he would be the first to drink the treated wastewater effluent from the unit treatment processes at York River Treatment Plant (and he did). My only contention with his statement is that because of Mother Nature's Water Cycle, the one we all learned about in grade school, we have been drinking toilet water all along. I have yet to find anything yucky about it or its taste.

****

PUBLIC PERCEPTION

There is little doubt about public perception when it comes to choosing to drink toilet water or not. "Not" is the verdict. This choice is no surprise. The thought of having a pipe-to-pipe connection from a toilet to a drinking water tap is difficult to accept. Direct connection is simply unacceptable and unimaginable. Even if black water from sewage sources is properly treated to drinking water standards, and then is provided indirectly to the tap, there is still a reluctance to accept by the public—a reluctance to accept that natural and human-made cycles can clean used water

DOI: 10.1201/9781003498049-2

(aka reclaimed water) to potable water quality that has been probably treated and in some cases is cleaner than the water at our taps.

THE REALITY

Have you ever wondered why it is that even though humans, wildlife and domesticated cattle, and Mother Nature constantly do things that contaminate potential potable water sources, we are still able to drink clean, safe, healthy drinking water? Whether it be deposits made in the toilet, or seeps from leaking septic tanks, or wastewater from industrial/manufacturing processes, or from cow pies deposited in an open barnyard or animal pen or from concentrated animal feeding operations (CAFOs), or chemicals from vulcanism, or contaminants from human- or nature-caused forest fires that contaminate local streams, or from fertilizers, pesticides, and other chemicals that have been applied to land near water, from sewer overflows, from stormwater, from rocks and soil that naturally contain chemicals such as arsenic, radon, and uranium, or from mining operations waste, or from cracks in water distribution pipes, or from improper operation of water and wastewater treatment plants, we are still able to find enough safe drinking water to consume?

Why is that?

Why is that even with a human population of billions of people and untold numbers of terrestrial and aquatic wildlife who constantly contaminate water sources on a global scale, we still have drinking water to consume that will refresh us and not make us sick or worse?

NATURE'S WAY

Most of us have access to clean and safe drinking water because of Nature's Way of cleaning up a messy environment and also because of human-made technology.

Let us begin with Nature's Way of cleaning contaminated water.

First, the evaporation of water is ongoing and is critical to our water cycle. This evaporation helps to clean the water to a degree but when water evaporates is it not totally cleaned?

No, not completely.

Second, whenever natural or human-made pollutants are outfalled into flowing waters, nature self-purifies the water? This self-purification process in flowing freshwater bodies works to clean up waste in the water, especially if the waste is organic like sewage. This occurs because, in a flowing river or stream, there are several factors present that can work to purify contaminated water. First, and most important, is the flow itself. When river water flows over river beds and obstructions like rocks and boulders turbulence occurs in the water flow. This is a good thing. This natural aeration of stream water also occurs in waterfalls. Turbulence in the river or stream or waterfall causes the water to be aerated, adding oxygen—the more oxygen the better. Note that natural aeration can also occur through sub-surface aquatic plants. Through the natural process of photosynthesis, water plants release oxygen into the water providing it with the oxygen necessary for aquatic organisms to live and aerobic bacteria to break down excess nutrients (Spellman, 1996). Oxygen is driven into the surface of the water by the wind.

Third, Hercules was correct (to a degree) when he espoused that dilution is the solution to pollution. This depends on the dilution ratio (or dilution factor). When the dilution ratio is high, large quantities of dissolved oxygen become available to reduce the chances of rot and pollution effects. When I stated that Hercules was correct about dilution being the solution to pollution keep in mind that this is dependent on the quantity of sewage or other organics discharged into the stream or river, too large of a quantity of sewage throws a wrench into the entire dilution process.

Fourth, sedimentation on settleable solids contained in waste as they drop down to the bottom of the stream or river where the sediments form sludge in which anaerobic decomposition takes place. This aids the self-purification of streams or rivers polluted by sewage. Note that in lakes cleaning the water is aided by annual turnover.

Finally, bleaching of water by the powerful action of sunlight stabilizes rivers or streams—and adds automatic, natural purification to some surface waters.

THE ARTIFICIAL WAY

Artificial purification of water is conducted using human-designed, made, and operated wastewater and water treatment processes. In a sense and to a degree, we can say that wastewater and water treatment operations are boxes containing streams. Humans clone the self-purification of streams and rivers into artificial purification processes. When humans intervene in the natural water cycle, they generate artificial water cycles or *urban water cycles* (local sub-systems of the water cycle—an integrated water cycle; see Figure 1.1). Although many communities withdraw groundwater for public supply, the majority rely on surface sources. After treatment, water is distributed to households and industries. Water that is wasted (wastewater) is collected in a sewer system and transported to a treatment plant for processing prior to disposal. Current processing technologies provide only partial recovery of the original water quality. The upstream community (the first water user, shown in Figure 1.1) is able to achieve additional quality improvement by dilution into a surface water body and natural purification. However (as also shown in Figure 1.1), the next community downstream is likely to withdraw the water for a drinking water supply before complete restoration. This practice is intensified and further complicated as existing communities continue to grow, and new communities spring up along the same watercourse. Obviously, increases in the number of users bring additional need for increased quantities of water. This withdrawal and return process by successive communities in a river basin results in *indirect water reuse* (use of used toilet water via *de facto* water recycling—this occurs when communities withdraw drinking water from rivers that contain varying amounts of dischargers from upstream cities and industries).

THE BOTTOM LINE

The reluctance to accept reused or reclaimed wastewater for potable purposes is to be expected, especially when there is a perception that no matter what type of natural or artificial treatment is provided the end users do not have a lot of faith in dilution, natural or human-made purification simply because used water's connotation is that the

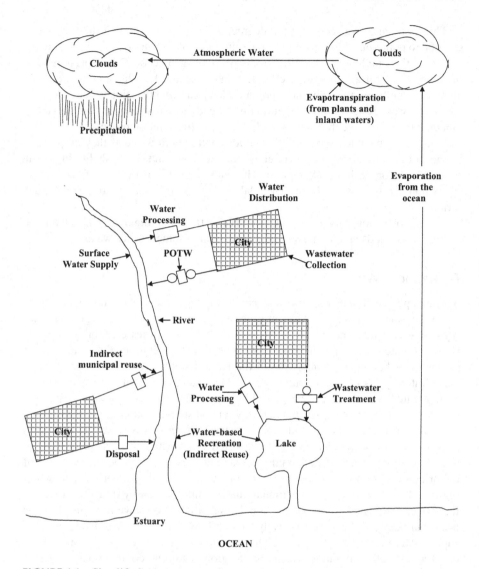

FIGURE 1.1   Simplified urban water cycle.

water is from the toilet and thus will and does change or assuage the feelings of the end user, the customer. Note, however, that based on recent experience in Tidewater (aka Hampton Roads—Norfolk/Va. Beach, South Eastern Virginia region), with proper public outreach and education, support for indirect potable reuse can be achieved. Additional information on public perception can be found in Hampton Roads Sanitation District's (HRSD's) Sustainable Water Initiative for the Future (SWIFT), which is presented later in the text (Spellman, 2021). However, before we get to the SWIFT process foundational materials, the building blocks related to the reuse of reclaimed water are presented.

REFERENCES

Spellman, F.R. (1996). *Self-Purification of Streams*. Lancaster, PA: Technomic Publishing Company.
Spellman, F.R. (2021). *Sustainable Water Initiative for the Future (SWIFT)*. Lanham, MD: Bernan Publishing.

2 Setting the Stage

INTRODUCTION

Note that even though the focus of this book is on treating used/reclaimed/recycled water for indirect potable water reuse, it is important to point out that along with providing clean and safe drinking water via advanced treatment; there are other uses of recycled water that are important (see Figure 2.1).

DID YOU KNOW?

In broad terms, water use pertains to the interaction of humans with and influence on the natural water cycle and individual elements such as water withdrawal, delivery, consumption use, wastewater release, reclaimed wastewater, return flow, and instream use.

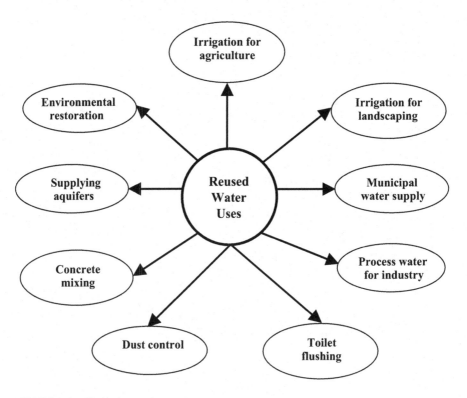

FIGURE 2.1 Uses of reused water.

DOI: 10.1201/9781003498049-3

FIT-FOR-PURPOSE SPECIFICATIONS

Note that planned water reuse refers to water systems designed with the goal of beneficially reusing a recycled water supply. Time and again, communities seek to optimize their overall water use by reusing water to the extent possible within the community before the water is outfalled into the environment. Examples of planned reuse include agricultural and landscape irrigation, industrial process water, potable water supplies, and groundwater supply management (see Figure 2.1).

DID YOU KNOW?

USEPA does not require or restrict any type of reuse. Generally, states maintain primacy (primary) regulator authority in allocating and developing water resources. The Safe Drinking Water Act and the Clean Water Act provides the guidelines and a foundation from which states enable, regulate, and oversee water reuse as they consider appropriate.

So, what are the potential sources of water for potential reuse? Potential sources of water include municipal wastewater, industry process, and cooling water (e.g., used for such purposes as fabricating, processing, washing, cooling [such as for condensers at power plants or factories]), or transporting a product, stormwater, agriculture runoff and return flows, and produced water from natural resource extraction activities (e.g., water produced as a by-product during extraction of oil and gas). Note that these sources of water are adequately treated to meet "fit-for-purpose specifications" for a particular next use. USGS (2018), with regard to reused water, defines "fit-for-purpose specifications" as the treatment requirements to bring water from a particular source to the quality needed, to ensure public health, environmental protection, or specific user needs. Also note that reused water for crop irrigation must be of sufficient quality so as to prevent harm to plants and soils, maintain food safety, and protect the health of farm workers. In uses where there is a greater human exposure water may require more advanced treatment.

TERMINOLOGY

Note that the terminology associated with treating and reusing municipal wastewater varies both within the United States and globally. Also, the terms "water reuse" and "water recycling" and reclaimed water are often used synonymously. This book uses the terms reclaimed water and water reuse. Definitions of terms used herein, except their use in case studies, are provided in the following text (USEPA, 2023).

- **Planned potable reuse**: The publicly acknowledged, intentional use of reclaimed wastewater for drinking water supply. Commonly referred to simply as potable reuse.

- *De facto* **reuse**: A situation where reuse of treated wastewater is practiced but is not officially recognized (e.g., a drinking water supply intake located downstream from a wastewater treatment plant outfall or discharge point).
- **Direct potable reuse**: The introduction of reclaimed water (with or without retention in an engineered storage buffer) directly into a drinking water treatment plant. This includes the treatment of reclaimed water at an Advanced Wastewater Treatment Facility for direct distribution.
- **Indirect potable reuse**: Deliberative augmentation of a drinking water sources (surface water or groundwater aquifer) with treated reclaimed water, which provides an environmental buffer, such as a lake, river, or a groundwater aquifer before treatment at drinking water plant and prior to subsequent use.

THE BOTTOM LINE

Water reuse can involve improving water supply reliability, reuse opportunities, economic considerations, public police, and regulatory factors (Metcalf & Eddy, Inc. et al., 2007). The benefits provided by water reuse are discussed in the following chapter.

REFERENCES

Metcalf & Eddy, Inc., an AECOM Company, Asano, T., Burton, F.L., Leverenz, H.L., Tsuchihashi, R., and George Tchobanoglous. (2007). *Water Reuse: Issues, Technologies, and Applications*. New York: McGraw-Hill.
USEPA. (2023). Case Studies that Demonstrate the benefits of Water Reuse. Accessed 12/22/23 @ https://epa.gov/waterreuse/forms/contact-us--about-water-reuse-and-recycling.
USGS (2018). 2018 United States (Lower 48) Seismic Hazards Long-Term Model. Accessed 12/11/23 @ https://www.usgs.gov/programs/earthquake-hazards/science/2018.

3 Water Reuse
Social Benefits

INTRODUCTION

Water reuse provides a wide range of benefits as shown in Figure 3.1. In this chapter, the social benefits are described with real-world usage in the United States.

SOCIAL BENEFITS

People highly encourage the creation of new greenspaces in cities because they provide social, cultural, and recreational benefits. Providing greenspaces requires the use of water and water reuse projects can help create the new greenspaces. The new greenspaces also enable the preservation of wildlife and work to sustain stream flow in culturally important water bodies. An example of water reuse supporting social benefits is described in Case Study 3.1.

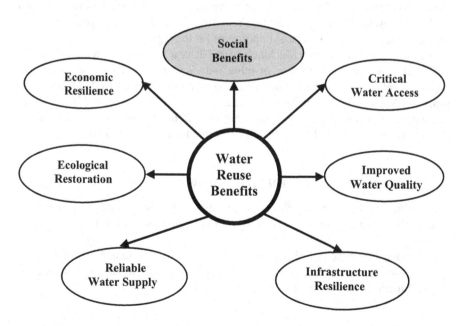

FIGURE 3.1 Water reuse: social benefits.

DOI: 10.1201/9781003498049-4

CASE STUDY 3.1: WATER REUSE: COLLIER COUNTY, FLORIDA (USEPA, 2023)

Operational since 1995, the Corkscrew Swamp Sanctuary in Collier County, Florida, occupies 13,450 acres (5,440 ha) and contains a system—a "Living Machine®"—that treats wastewater generated onsite from the facility restrooms and recycles it for toilet flushing, effectively reducing demand on potable, freshwater supplies, and helping to conserve groundwater.

Because the popularity of the sanctuary increased (more than 100,000 visitors, the amount of wastewater generated onsite also increased to a level beyond the capacity of the old septic system. In response, the sanctuary installed an innovative treatment system, the Living Machine®, that employs plants and microorganisms instead of a traditional onsite wastewater treatment system—the system capacity is not 7,500 gallons per day (28,400 L per day). This innovative setup also provides other benefits, such as educating the sanctuary's visitors about the ability of natural wetlands to improve water quality and be aesthetically appealing.

The Corkscrew Swamp Sanctuary morphed from an Audubon Sanctuary in 1954. National Audubon Society purchased the land to protect the largest virgin bald cypress forest in the world (about 700 acres or 280 ha) from the logging industry and presser the habitat of the largest nesting colony of federally endangered wood storks.

Note that the Sanctury's wetlands recharge local aquifers, rivers, and estuaries, including local rivers. These rivers serve as an important connection from the watershed to coastal habitat. The surrounding communities rely on the aquifers as a source of drinking water and for agriculture. Protecting the quality of the groundwater, as well as surface water is of critical importance for the region.

So, how does the Living Machine® work? It is a wastewater treatment system constructed within a greenhouse or other indoor structure that is composed of a series of tanks and subsurface flow treatment wetlands. Natural wetlands treatment processes such as sedimentation, chemical adsorption (fixation), nitrification and denitrification, volatilization, and anaerobic and aerobic decomposition are included. What is unique about the Living Machine® is that the combined treatment.

DID YOU KNOW?

Aerobic: conditions in which free, elemental oxygen is present. Also used to describe organisms, biological activity, or treatment processes that require free oxygen.

Anaerobic: conditions in which no oxygen (free or combined) is available. Also used to describe organisms, biological activity, or treatment processes that function in the absence of oxygen.

THE LIVING MACHINE®

At Corkscrew Swamp Sanctuary the Living Machine® is used as an onsite wastewater treatment and reuse system, has a capacity of 7,500 gallons (28,400 L) per day, and a retention time of 11 days. Its main components are as follows:

ANAEROBIC TREATMENT PROCESS

The anaerobic treatment process consists of two steps, occurs completely in the absence of oxygen, and produces a useable by-product, methane gas. In the first step of the process, facultative microorganisms use organic matter as food to produce more organisms, volatile(organic) acids, carbon dioxide, hydrogen sulfide, and other gases and some stable solids (see Figure 3.2).

In the second step, anaerobic microorganisms use the volatile acids as their food source. The process produces more organisms, stable solids, and methane gas that can be used to provide energy for various treatment system components (see Figure 3.3).

At the Sanctuary, wastewater from the visitor toilets is pumped to two 10,000-gallon (37,854-L) fiberglass tanks below ground for anaerobic treatment to reduce BOD_5 (5-day biochemical oxygen demand). The BOD_5 test procedure consists of ten steps (for unchlorinated water) as shown in Table 3.1.

Note: BOD_5 is calculated individually for all sample dilutions that meet the criteria. The reported result is the average of the BOD_5 of each valid sample dilution.

Facultative bacteria		More bacteria
Organic matter	$\Rightarrow$	Volatile solids
Nutrients		Settleable solids
		Hydrogen sulfide

FIGURE 3.2 Anaerobic decomposition—first step.

Anaerobic bacteria		More bacteria
Volatile acids	$\Rightarrow$	Stable solids
Nutrients		Settleable solids
		Methane

FIGURE 3.3 Anaerobic decomposition—second step.

TABLE 3.1
BOD_5 Test Procedure

1. Fill two bottles with BOD dilution water; insert stoppers.
2. Place the sample in two BOD bottles; fill with dilution water; insert stoppers.
3. Test for dissolved oxygen (DO).
4. Incubate for 5 days.
5. Test for DO.
6. Add 1 mL $MnSO_4$ below surface.
7. Add 1 mL alkaline KI below the surface.
8. Add 1 mL H_2SO_4.
9. Transfer 203 mL to flask.
10. Titrate with PAO or thiosulfate.

BOD$_5$ Calculation

Unlike the direct reading instrument used in the DO analysis, BOD results require calculation. There are several criteria used in selecting which BOD$_5$ dilutions should be used for calculating test results. Consult a laboratory testing reference manual such as Standard Methods (APHA, 1998), for this information. Currently, there are two basic calculations for BOD$_5$. The first is used for samples that have not been seeded. The second must be used whenever BOD$_5$ samples are seeded. In this section, we illustrate the calculation procedure for unseeded samples.

$$BOD_5(unseeded) = \frac{(DO_{start}(mg/L) - DO_{final}(mg/L)) \times 300\,mL}{Sample\,volume(mL)} \quad (3.1)$$

Example 3.1

Problem: The BOD$_5$ test is completed. Bottle 1 of the test had a DO of 7.1 mg/L at the start of the test. After 5 days, bottle 1 had a DO of 2.9 mg/L. Bottle 1 contained 120 mg/L of sample.

Solution:

$$BOD_5(unseeded) = \frac{(7.1\,mg/L - 2.9\,mg/L) \times 300\,mL}{120\,mL} = 10.5\,mg/L$$

If the BOD$_5$ sample has been exposed to conditions that could reduce the number of healthy, active organisms, the sample must be seeded with organisms. Seeding requires the use of a correction factor to remove the BOD$_5$ contribution of the seed material:

$$Seed\,correction = \frac{Seed\,material\,BOD_5 \times seed\,in\,dilution(mL)}{300\,mL} \quad (3.2)$$

$$BOD_5(seeded) = \frac{\left[(DO_{start}(mg/L) - DO_{final}(mg/L)) - Seed\,corr.\right] \times 300}{Sample\,volume(mL)} \quad (3.3)$$

Anaerobic treatment also works to reduce **Total Suspended Solids** (TSS). TSS are any material suspended or dissolved in water and wastewater. Although normal domestic wastewater contains a very small amount of solids (usually <0.1%), most treatment processes are designed specifically to remove or convert solids to a form that can be removed or discharged without causing environmental harm.

In sampling for TSS, samples may be either grab or composite and can be collected in either glass or plastic containers. TSS samples can be preserved by refrigeration at or below 4°C (not frozen). However, composite samples must be refrigerated during collection. The maximum holding time for preserved samples is 7 days.

TEST PROCEDURE

To conduct a TSS test procedure, a will-mixed measured sample is poured into a filtration apparatus and, with the aid of a vacuum pump or aspirator, is drawn through a pre-weighted glass fiber filter. After filtration, the glass filter is dried at 103°C–105°C, cooled, and reweighed. The increase in weight of the filter and solids compared to the filter alone represents the TSS. An example of the specific test procedure used for TSS is given below:

1. Select a sample volume that will yield between 10 and 200 mg of residue with a filtration time of 10 minutes or less.
   Note: If filtration time exceeds 10 minutes, increase the filter area or decrease volume to reduce filtration time.
   Note: For non-homogenous samples or samples with very high solids concentrations (i.e., raw wastewater or mixed liquor), use a larger filter to ensure a representative sample volume can be filtered.
2. Place the pre-weighed glass fiber filter on the filtration assembly in a filter flask.
3. Mix the sample well and measure the selected volume of the sample.
4. Apply suction to the filter flask, and wet the filter with a small amount of laboratory-grade water to seal it.
5. Pour the selected sample volume into the filtration apparatus.
6. Draw the sample through the filter.
7. Rinse the measuring device into the filtration apparatus with three successive 10 mL portions of laboratory-grade water. Allow complete drainage between rinsings.
8. Continue suction for 3 minutes after filtration of the final rinse is completed.
9. Remove the glass filter from the filtration assembly (membrane filter funnel or clean Gooch crucible). If using the large disks and membrane filter assembly, transfer the glass filter to a support (aluminum pan or evaporating dish) for drying.
10. Place the glass filter with solids and support (pan, dish, or crucible) in a drying oven.
11. Dry filter and solids to constant weight at 103°C–105°C (at least 1 hour).
12. Cool to room temperature in a desiccator.
13. Weigh the filter, support, and record constant weight in the test record.

TSS CALCULATIONS

To determine the TSS concentration in mg/L, we use the following equations:

1. To determine the weight of dry solids in grams

$$\text{Dry solids}(g) = \text{Wt. of dry solids and filter}(g) - \text{Wt. of dry filter}(g) \qquad (3.4)$$

2. To determine the weight of dry solids in milligrams (mg)

$$\text{Dry solids}(mg) = \text{Wt. of solids and filter}(g) - \text{Wt. of dry filter}(g) \qquad (3.5)$$

3. To determine the TSS concentration in mg/L

$$TSS(mg/L) = \frac{\text{Dry solids}(mg) \times 1,000\,mL}{mL\ sample} \qquad (3.6)$$

Example 3.2

Problem: Using the data provided below, calculate TSS:

Sample volume (mL): 250 mL
Weight of dry solids and filter (g): 2.305 g
Weight of dry filter (g): 2.297 g

Solution:

$$\text{Dry solids}(g) = 2.305\,g - 2.297\,g = 0.008\,g$$

$$\text{Dry solids}(mg) = 0.008\,g \times 1,000\,mg/g = 8\,mg$$

$$TSS(mg/L) = \frac{8.0 \times 1,000\,mL/L}{250\,mL} = 32.0\,mg/L$$

Note that in practice and operation BOD_5 and TSS, prior to the other treatment steps. In the anaerobic tanks, retention time is 2 days. The gas produced is released via a vent. The quantity of solids that settle within the tank are removed approximately every 5 years by a third party and disposed of in landfills.

DID YOU KNOW?

The process of nitrification is utilized to convert ammonia to nitrates. *Nitrification* is a biological process that involves the addition of oxygen to the wastewater. If further treatment is necessary, another biological process called *denitrification* is used. In this process, nitrate is converted into nitrogen gas, which is lost to the atmosphere. From the wastewater operator's point of view, nitrogen and phosphorus are both considered limiting factors for productivity. Phosphorus discharged into streams contributes to pollution. Of the two, nitrogen is harder to control but is found in smaller quantities in wastewater.

AEROBIC REACTOR AND AERATION TANKS

The water from the anaerobic treatment tanks goes into one of two series of five tanks (2,500-gallon or 9,464-L, each), arranged in parallel. Seventy-five percent of the maximum daily flow is handled by each one of these series. The aerated tanks contain bacteria, plants (ranging from algae to trees), snails, shrimp, insects, and fish. The plants provide surface area for microbial growth, uptake nutrients, and support beneficial insects and microorganisms. The function of these reactors is to further reduce BOD and complete the nitrification processes, converting ammonia and organic nitrogen to nitrate. Keep in mind that If organic nitrogen and ammonia are being converted to nitrate (nitrification), sufficient alkalinity must be available to support this process, as well. The function of the reactors is to further reduce BOD and complete the nitrification processes, converting ammonia and organic nitrogen to nitrate. The nitrate is either taken up into the plants along with other macro- and micronutrients in the water or is converted into harmless nitrogen gas through denitrification in later portions of the treatment system. At regular intervals, the plants are harvested to prevent overgrowth and remove nutrients permanently from the system.

CLARIFICATION

Clarification is also called *Sedimentation* is also called. Sedimentation removes settleable solids by gravity. In this system, clarification takes place in the next tank, a settling tank that separates any remaining suspended solids from the treated wastewater. The clear water flows onward while the settled solids are pumped back into the anaerobic tanks.

DID YOU KNOW?

Wetlands (aka biological supermarkets or earth's kidneys) are among the most productive ecosystems in the world, comparable to rainforests and coral reefs. They are home to an immense variety of species of microbes, plants, insects, amphibians, reptiles, birds, fish, and mammals. Wetlands provide great volumes of food that attract many animal species. These animals use wetlands for part of or all of their lifecycle. Up to one-half of North American bird species nest or feed in wetlands. Because these systems can improve water quality, engineers and environmental scientists construct systems that replicate the functions of natural wetlands (Kadlec and Knight, 2004; USEPA, 2000; Zedler, 2000; Sharifi et al., 2013).

CONSTRUCTED WETLANDS

The water continues to one of two constructed wetlands (30 by 30 ft, or 9 by 9 m). These artificial marshes are lined with plastic, filled with crushed limestone, and vegetated with typical wetland species from the Corkscrew Swamp, including cypress, slash pine trees, wax myrtle, cabbage palms, saw palmetto, red maple, goldenrod, swamp lily, busy bluestem, wiregrass, tickseed, sawgrass, live oak trees, resurrection fern, pond apple trees, and many more. These plants and the microorganisms that occupy the surfaces on the granular medium, remove remaining nitrogen through the uptake into plants or through denitrification, converting nitrate to harmless nitrogen gas.

DID YOU KNOW?

Chlorination for disinfection follows all other steps in conventional wastewater treatment. The purpose of chlorination is to reduce the population of organisms in the wastewater to levels low enough to ensure that pathogenic organisms will not be present in sufficient quantities to cause disease when discharged.

    Note: Chlorine gas is heavier than (vapor density of 2.5). Therefore, exhaust from a chlorinator room should be taken from the floor level.

    Note: The safest action to take in the event of a major chlorine container leak is to call the fire department.

    Note: You might wonder why it is that chlorination of critical waters such as natural trout streams is not normal practice. This practice is strictly prohibited because chlorine and its by-products (i.e., chloramines) are extremely toxic to aquatic organisms.

    Note: Chlorine may be applied as a gas, a solid, or in liquid hypochlorite form.

DISINFECTION

Florida state permit requirements were complied with when a disinfection step was added. This step works to limit potential pathogenic risks for users at the point of use. The water is disinfected with chlorine in a holding tank, then dechlorinated with sodium sulfite in a subsequent chamber to prevent the chlorine from interfering with the natural microbial processes inherent to the Living Machine® when the water is recycled.

CHLORINATION TERMINOLOGY

- **Chlorine**: a strong oxidizing agent which has strong disinfecting capability. A yellow-green gas that is extremely corrosive and is toxic to humans in extremely low concentrations in the air.
- **Contact time**: the length of time the time the disinfecting agent and the wastewater remain in contact.
- **Demand**: the chemical reactions, which must be satisfied before a residual or excess chemical will appear.
- **Disinfection**: refers to the selective destruction of disease-causing organisms. All the organisms are not destroyed during the process. This differentiates disinfection from sterilization, which is the destruction of all organisms.
- **Dose**: the amount of chemical being added in milligrams/liter.
- **Feed rate**: the amount of chemical being added in pounds per day.
- **Residual**: the amount of disinfecting chemical remaining after the demand has been satisfied.
- **Sterilization**: the removal of all living organisms.

In operation, the Living Machine® (Audubon Society, n.d.) reclaimed almost 90% of the treated wastewater, which is pumped to storage tanks, and used in the toilets for flushing (a separate potable water supply is used for the handwashing sinks and drinking fountains). The remaining 10% of the treated effluent is discharged to a drain field at the visitor parking lot as a preventative measure to reduce the buildup of minerals in the Living Machine®.

The vegetated tanks are monitored for excessive growth of biomass and it is routinely trimmed to prevent plant roots from damaging the tanks. These trimmings are composted onsite. A 0.9-hp (0.7-kW) engine connected to a cast-iron blower motor works to bubble air into the aeration tanks.

THE BOTTOM LINE

Experience has shown that the Living Machine® has proven to be a social benefit to those who visit the Corkscrew Swamp Sanctuary—it has created an aesthetically pleasing environment containing plants, insects, and animals with the greenhouse— a benefit that most conventional wastewater treatment systems do not provide.

REFERENCES

APHA. (1998). *Standard Methods for the Evaluation of Water and Wastewater.* Washinton, DC: American Public Health Association.

Audubon Society. (n.d.). The Living Machine. Accessed 12/24/23 @ https://corkscrew. audubon.org/aboutlivingmachine.

Kadlec, R.H., and Knight, R.L. (2004). *Treatment Wetlands.* Boca Raton, FL: Lewis Publishers.

Sharifi, A., L. Kalin, M. Hantush, and S. Isik. (2013). Wetlands: Earth's Kidneys. In J. Lamar, and B.G. Lockaby (eds.), *Auburn Speaks.* Auburn, AL: Auburn University, pp. 140–143.

USEPA. (2000). Guiding Principles for Constructed Treatment Wetlands: Providing for Water for Water Quality and Wildlife Habitat. Accessed 12/23/23 @ www.epa.gog.owow/ wetlands/constructed/guide.html.

USEPA. (2023). Case Studies that Demonstrate the Benefits of Water Reuse. Accessed 12/22/23 @ https://epa.gov/waterreuse/forms/contact-us--about-water-reuse-and-recycling.

Zedler, J.B. (2000). *Handbook for Restoring Tidal Wetlands.* Boca Raton, FL: CRC Press.

4 Water Reuse
Economic Resilience

INTRODUCTION

Water reuse provides a wide range of benefits as shown in Figure 4.1. In this chapter, the economic benefits are described with real-world usage in the United States.

A local community can benefit from employing water reuse projects to lower the overall cost of infrastructure over its lifetime. For example, water reuse projects that capture and treat stormwater before infiltration it into the ground can help prevent sewers from being swamped with too much water. What this does is, it lowers the cost to clean up sewer overflows and to treat the smaller volume of sewage flowing into the downstream wastewater treatment plant. Water reuse investments can be a real plus because can create green jobs in the public and private sectors and lower the cost of fining, buying, and transporting conventional water supplies from further away.

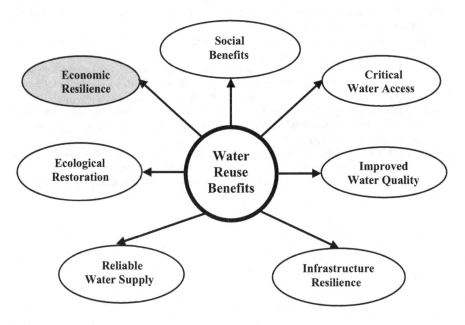

FIGURE 4.1 Water reuse: economic resilience.

DOI: 10.1201/9781003498049-5

CASE STUDY 4.1: NATIVE AMERICAN TRIBE
USES WATER REUSE (USEPA, 2023)

In Scott County, Minnesota, the Shakopee Mdewakanton Sioux Community (SMSC) put their water reclamation treatment facility into operation in 2006. The tribe utilizes treated municipal wastewater from tribal residences and community business enterprises to be reused for landscape irrigation and environmental restoration. What the tribal leaders wanted to do was to put the reused water into beneficial reuse projects—projects with benefits. For example, almost 20% of the total average treated wastewater is reclaimed to irrigate The Meadows at Mystic Lake, an 18-hole golf course, as well as other landscaped areas. The benefit is derived from making it unnecessary to pump any irrigation water (35 million gallons, or 132 million liters) from the already over-drafted aquifer. Wetland restoration and preservation is another advantage of water reuse. The wetlands are very important to the SMSC as part of its stewardship goals. The wetlands are utilized year-round by Canadian geese, muskrats, and mallards. This is notable during the cold Minnesota winters when most other bodies of water are frozen over. The biosolids (aka sludge) from the Water Reclamation Facility (WRF) are dried and composted at the SMCS Organics Recycling Facility. This nutrient recovery practice closes the recycle loop as the material is brought out to local fields and gardens in place of fertilizer sources from outside of the reservation. Another benefit for the local organic economy is recycling biosolids from the WRF for financial gain is the marketing of fertilizers and mulch to businesses and households.

So, how did the tribe accomplish these benefits?

When a new wastewater treatment plant was needed in the early 2000, the SMSC saw an opportunity to develop a facility that could reclaim its municipal wastewater to supplement its freshwater supplies. After a year and a half of construction, the SMSC's WRF became operational in 2006. From the start of operation, the WRF has treated wastewater from tribal residences and local enterprises and reused the water for landscape irrigation and environmental restoration. Approximately 200 million gallons of wastewater (about 757 million liters) are treated by the facility annually. To accommodate future growth, the far-thinkers ensured that the annual design capacity of the WRF is larger than its current use, about 350 million gallons (1.3 billion liters) and 550 million gallons (2 billion liters) during dry and wet weather, respectively.

Note that the WRF's reclaimed water meets high-quality water standards fit for non-potable reuse, following the requirements set by the state of Minnesota as their guide—recall that it is the States that provide guidance and regulation in regard to wastewater reuse practices. The reclaimed water is discharged to natural surface water wetlands which store some of the water for non-potable reuse. Specifically, ~35 million gallons (132 million liters) per year of reclaimed water are used to irrigate their golf course, as well as other landscaped areas within the reservation. This practice of reclaiming water for irrigation has directly offset the use of groundwater resources for irrigation and helps to less groundwater over drafting in the region. The rest of the reclaimed water flows through the surface water wetlands within the Minnesota River watershed. An important side benefit from this procedure is that because the reclaimed water is warmer than the air, it provides a liquid water source to the wetland during the winter which attracts a variety of waterfowl such as Mallards, Canadian Geese, Mergansers, Pintails, and Wood ducks to the wetland when most other waterbodies have frozen over. Maintaining the source water in its wetlands by using reclaimed water is a significant part of the SMSC's responsible stewardship goals.

The SMSC also conducted a groundwater injection and aquifer storage study to investigate whether the reclaimed water could also be used to augment local groundwater supplies. However, the reclaimed water continues to be reused solely for irrigation and wetland preservation because even though a pilot study showed that additional advanced treatment could remove recalcitrant organic compounds from the reclaimed water. Bringing online additional advanced treatment processes for aquifer augmentation is not the problem. Instead, the study indicated that a portion of the water would be conveyed away from where they wanted it because the geology showed a bedrock valley adjacent to the reservation could impede successful injection. So, along with issues and other concerns, full-scale injection has stalled; the wastewater is only used for irrigation and wetland preservation.

A THREE-STEP PROCESS

Wastewater treatment at the WRF, with a tribal population of 325 people and an effective population of 15,000 people, follows a three-step purification process. Again, far-thinkers have designed the WRF to accommodate future growth and higher flow rates. This forward-thinking also includes the processing of biosolids (aka sewage sludge). WRF treatment facility with a treatment capacity of 0.96 million gallons per day (MGD) in dry weather and 1.5 MGD (see Figure 4.2) in wet weather consists of several unit processes in its treatment train including screening to remove refuse to protect the equipment downstream from clogging and damage. Vortex grit removal is next in line to remove small particles like sand. Clarification follows and two parallel plate clarifiers work to remove particulates not removed in the previous two steps. The next unit process is the biologically aerated filtration (BAF) where continuous upward flow aeration takes place through a tightly packed medium. Medium provides filtering and surface area for microorganisms (i.e., *Zoogleal slime*—the biological slime that forms on fixed film treatment devices), it contains a wide variety of organisms essential to the treatment process.

To grow on and consume organic material in the wastewater. Next in the train is membrane filtration (GE ZeeWeed 1,000 membranes) which separates any remaining biomass. UV (ultraviolet) disinfection is next. [*Note* that for UV to be effective in disinfecting wastewater effluent, UV light must be able to penetrate the stream flow. Obviously, stream flow that is turbid works to reduce the effectiveness of irradiation (penetration of light)]. After disinfection, the waste stream is outfalled into the wetlands for environmental restoration where the reclaimed water is pumped and piped for irrigation and 35 million gallons per year is pumped to the golf course and other landscaped areas, the rest of the flow is outfalled into the Minnesota River.

Note that in 2013, the biosolids produced at the WRF amounted to an average of 136 tons per year (123 metric tons). Following treatment, biosolids from the WRF are processed and then sent to the SMSC Organics Recycling Facility to create high-quality compost.

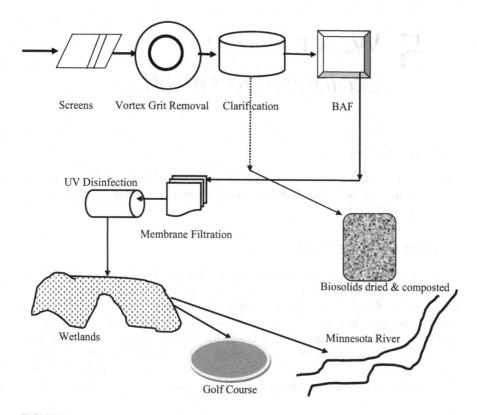

Screens   Vortex Grit Removal   Clarification   BAF

UV Disinfection

Membrane Filtration

Biosolids dried & composted

Wetlands   Minnesota River

Golf Course

FIGURE 4.2   WRF treatment train.

THE BOTTOM LINE

The WRF has been honored with multiple awards for excellence in waste and pollution prevention. Simply, we can say that the Dakota people are not only Smart (with capital S) and far-thinking but also great friends of Mother Nature (and that anthropomorphic celestial deity: the Great Spirit or Grandmother Earth: Unci Maka).

REFERENCE

USEPA. (2023). Case Studies that Demonstrate the Benefits of Water Reuse. Accessed 12./22/23 @ https://epa.gov/waterreuse/forms/contact-us--about-water-reuse-and-recycling.

5 Water Reuse
Ecological Restoration

INTRODUCTION

Water reuse provides a wide range of benefits as shown in Figure 5.1. In this chapter, ecological restoration is described and an example of real-world usage in the United States is provided.

HABITAT LOSS: THE CORRECTION

Correcting habitat loss is the driver for ecological restoration. Habitat loss threatens wildlife and the ability of ecosystems to freely provide humans with food, clean water, and clean air. Simply, water reuse can help restore ecosystems by providing them with a consistent water source. For example, wetlands can be created near wastewater treatment plants, and reused water can be used to help maintain healthy streamflow to support aquatic species. In arid regions, such as the desert southwest, treated wastewater and stormwater may be the sole source of water for important habitats. The bottom line is that wetlands are important because they improve water quality,

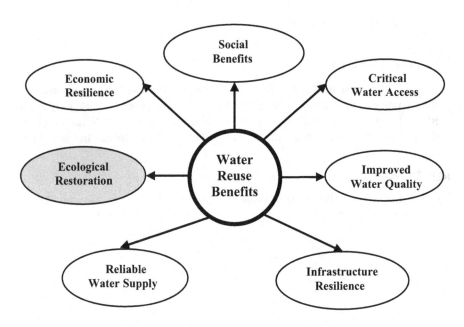

FIGURE 5.1 Water reuse: ecological restoration.

 DOI: 10.1201/9781003498049-6

provide wildlife habitat maintain ecosystem productivity, reduce coastal storm damage, provide recreation opportunities, improve the water supply, and provide opportunities for education.

CASE STUDY 5.1: BROOKLYN REDUCES POTABLE WATER DEMAND & IMPROVES SEWER CAPACITY (USEPA, 2023)

The rapid population growth in New York City (NYC) has caused an increase in the demand for potable water. Along with this increasing stream of population growth flows an expanding generation of more wastewater, straining the present wastewater treatment infrastructure's capacity. Complicating this growth in population and its associated increasing waste load is that NYC has a 150-year-old combined wastewater system, a system that transports stormwater and sewage in the same piles, and is often overwhelmed by rainfall, causing combined sewer overflows (CSOs) in various parts of the city.

Obviously, storm-caused sewage overflows within NYC are a pressing problem. In order to mitigate this issue a new redevelopment project which is currently under construction in Brooklyn, NYC, NY, has incorporated water reuse into its design. Specifically, this reuse project is part of the revitalization of the historic Domino Factory—a mixed development and former sugar refinery—and consists of a non-potable water reuse system that will collect and treat wastewater from five newly constructed buildings and reuse it for toilet flushing, cooling tower make-up and irrigation in three of those same buildings. The treatment system is able to process 400,000 gallons per day (gpd) (~1,500,000 L/day) of wastewater. The project's 3-year timeline with a completion date of 2025/2026 includes the goal of reusing half of the treated wastewater onsite and the remainder to be discharged directly to the East River after treatment. Non-potable reuse will significantly reduce the demand on the water supply in the rapidly growing neighborhood by offsetting the use of potable water. Onsite treatment and reuse also lessen the impact of CSOs by diverting wastewater away from combined sewers and lowering pressure on downstream wastewater treatment facilities. Through an innovative private–public parentship, the project additionally integrates green infrastructure and public park space which lowers stormwater flows to the combined sewer system and enhances public access to the waterfront.

THE DOMINO DISTRICT

The Domino District Non-Potable Water Reuse Project is being constructed as part of an effort to revitalize the historic Domino Factory in Brooklyn, NYC, NY (see Figure 5.2). The redevelopment project includes 600,00 ft^2 of office space and 3,000 apartments, 700 of which are subsidized for low- and middle-income tenants.

Key Definition: A district-scale water reuse system is an onsite water reuse system for a defined service area that covers two or more properties and may cross public rights-of-way.

When completed and in operation the planned district-scale water-reuse system will function as follows:

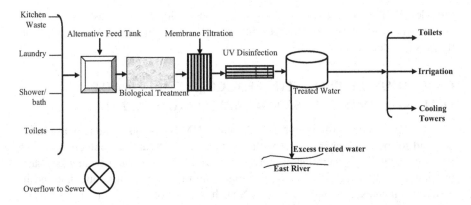

FIGURE 5.2   Schematic of Domino district non-potable water reuse project flow.

1. Wastewater is collected from five newly constructed buildings, serving over 8,000 people, with a capacity of up to 400,000 gallons per day (~1,500,000 L/day);
2. Wastewater is treated using an innovative treatment process that uses only 10,000 ft^2 of low value below ground space. The treatment processes remove trash and fine particles, filter and purify the water using a membrane bioreactor, and disinfect the water with ultraviolet light and ozone; then
3. High-quality recycled water is pumped back to three of the buildings through non-potable-designated pipe (painted purple) for non-potable use (e.g., toilet flushing, cooling tower makeup, and irrigation; see Figure 5.2).

THE BOTTOM LINE

Using recycled water for non-potable uses will be a true benefit in that it reduces the need to use portable. The reuse system is being designed to reduce the discharge of untreated wastewater into NYC's combined sewer system. The decreased burden on the combined sewer system can prevent untreated wastewater from flowing into local waterways through CSOs. Other important aspects of the redevelopment project will work to lower stormwater flows to NYC's sewer system which can help reduce the frequency of CSOs. Through an innovative private–public partnership, for example, the project integrates green infrastructure and public park space with absorbs a portion of the stormwater generated onsite. Even though the project does not treat or reuse the stormwater, the use of green infrastructure lowers the flow of water into the area's sewers. The project does provide a drought-resilient water supply to the buildings for non-potable uses. It also provides financial incentives for funding district-scale onsite reuse projects.

REFERENCE

USEPA. (2023). Case Studies that Demonstrate the Benefits of Water Reuse. Accessed 12/22/23 @ https://epa.gov/waterreuse/forms/contact-us--about-water-reuse-and-recycling.

6 Water Reuse
Reliable Water Supply

INTRODUCTION

Water reuse provides a wide range of benefits as shown in Figure 6.1. In this chapter, water reuse for a reliable water supply is described, and an example of real-world usage in the United States is provided.

The current droughts affecting various water supplies globally are usually blamed on climate change and this may be the case. Whatever the causal factor(s) is/are severe droughts have been ongoing in areas where populations have expanded requiring cities and towns to find or procure more water to support new homes and businesses. Water reuse can help cities and towns meet these growing needs—even during droughts—because it creates a sustainable and reliable water supply from water that would otherwise be wasted, or lost. Note that water reuse is also often cheaper than constructing new reservoirs or desalinating seawater.

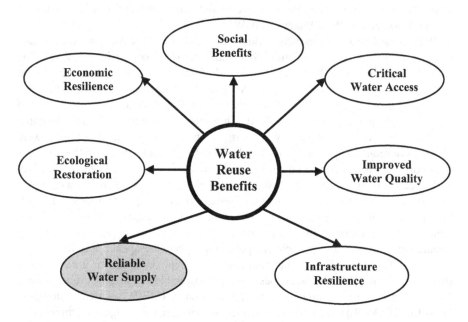

FIGURE 6.1 Water reuse: reliable water supply.

DOI: 10.1201/9781003498049-7

CASE STUDY 6.1: FAIRFAX COUNTY, VIRGINIA (USEPA, 2023)

Due to many factors such as the need to reduce nutrient discharges to the Chesapeake Bay, to lower the demand for potable water supplies, to meet total maximum daily load (TMDL) requirements (i.e., Total Maximum Daily Load is a calculation of the maximum amount of pollutant that a waterbody can receive and still meet EPA water quality standards; aka "pollution diet") and to accommodate a growing population without costly upgrades to its municipal wastewater treatment plant, Fairfax County reuses its treated municipal wastewater for centralized non-potable applications (such as for street cleaning, constructions and commercial car washing).

Fairfax County's wastewater treatment system has a treatment capacity of up to 6.6 million gallons (24.9 million L) per day, an average of 3 million gallons per day. Its Wastewater Management Program consists of sewage collection, treatment, and residual waste facilities for commercial users and more than 268,000 residents. The county's wastewater program is managed by its Department of Public Works and Environmental Services and includes the Norman M. Cole, Jr., Water Pollution Control Plant (NMCPCP). An average of 50 million gallons (189 million L) of wastewater are treated at the NMCPCP and then the treated water is outfalled into the Pohick River, a tributary in the Chesapeake Bay watershed. Note that the need for waste management in the county is expected to grow as the local population and commercial activities increase.

Note: As stated earlier, a TMDL is a "pollution diet" that identifies the maximum amount of a pollutant a waterway can receive and still meets applicable water quality standards under the Clean Water Act (CWA).

With regard to the Chesapeake Bay watershed note that the Chesapeake Bay is the largest estuary in the United States and the third largest in the world. The watershed encompasses the entire District of Columbia, as well as parts of six states: Delaware, Mayland, New York, Pennsylvania, Virginia, and West Virginia. Due to excess nutrients, nitrogen, and phosphorus, the Chesapeake Bay is an impaired waterbody, despite restoration efforts that have taken place for more than 30 years. Prior to substantial human activity in the region, most nitrogen and phosphorus were absorbed or retained by natural forest and wetland vegetation. However, the activities of almost 14 million people in the watershed have overwhelmed the Chesapeake Bay with excess amounts of nutrients. Nitrogen and phosphorus come from a wide range of point and nonpoint sources, including sewage treatment plants, industrial facilities, agricultural fields, lawns, and the atmosphere. Unfortunately, as forests and wetlands have been replaced by cities, farms, and suburbs to accommodate a growing population, nitrogen, and phosphorus pollution in the Chesapeake Bay has greatly increased. These excess nutrients cause seasonal hypoxia zones, or areas without oxygen, in the Bay, which makes these zones inhospitable to fish, blue crabs, oysters, bay grasses, and other aquatic wildlife. In response, USEPA set strict Maximum Daily Load (TMDL) requirements to reduce phosphorus, nitrogen, and sediment discharges throughout its 64,000 mile^2 (166,000 km^2) watershed that includes portions of Delaware, Maryland, New York, Pennsylvania, Virginia, West Virginia, and the District of Columbia.

What this dumping of nutrients and other contaminants into the Chesapeake Bay watershed really points to is that the surface waters in the area are used as potable water sources for water treatment plants in the region—the human-made water cycle feeding the public's water taps in action.

So, what does all this have to do with NMCPCP?

What it has to do with NMCPCP means that in order to restore and to help further reduce nutrient loading to the Chesapeake Bay is that the Commonwealth of Virginia, for example, assigned TMDL limits to the NMCPCP. This enables (mandates) the NMCPCP to provide an extremely advanced level of treatment which has assisted in continuously meeting these strict nutrient discharge requirements. Moreover, the ongoing improvements in treatment technologies are allowing more stringent loading requirements of the Chesapeake Bay TMDL—this is a good thing but also means that the county must look for other ways to reduce nutrient discharges to the Bay.

The bottom line: The Chesapeake Bay TMDL and the associated nutrient discharge limits have been drivers for water reuse across the various wastewater plants that discharge to the Bay.

NEED FOR USERS AND COMMUNICATION

The county's Wastewater Management Leadership Team focuses on long-range planning, strategy, continuous improvement, wastewater capacity issues, and financial management to evaluate options for reducing nutrient discharges. Also, along with the key driver to meeting nutrient TMDLs, there are additional factors driving compliance and the need for reuse. These additional needs include avoiding costly upgrades needed to treat the wastewater of a growing county, the cost savings from using recycled water to offset potable water use, and industry-wide support for the adoption of water reuse.

Ah! Industry-wide support? This is the potential wrench often thrown into reuse projects—it is the old "yuck factor: syndrome." This is a key challenge the county faced when they attempted to find local entities who would commit to long-term reclaimed water use and the costs associated with constructing a pumping station and new pipelines to deliver reclaimed water to each user—along with perceptions of, again, the yuck factor involved.

Anyway, the county realized that it needed to incorporate a project designed to deliver reclaimed water to each user. What it really boils down to is the County is seeking a project that will yield a 20-year payback period.

To pursue water reuse (and to comply with applicable regulations), Fairfax County needed to identify potential end-users who could enter a contract to receive the county's reclaimed water; they did this by searching water records and potential land application sites. This approach makes sense because the selection of local users reduces the cost of pumping and distributing the reclaimed water. The county's search revealed several types of potential users and uses, including providing cooling tower water for industries, spray irrigation for local golf courses and ball fields, and other activities, such as car washing, landscaping, and street cleaning.

Note that each of these end uses requires high-quality water in compliance with Virginia's restrictions on water reuse. So, in addition to fairly small treatment upgrades, the county needed to build a suitable pumping station and pipelines to transport the reclaimed water to the end users.

What about using this reclaimed water for potable reuse? No, the county did not consider including potable reuse in this project because the source water for most of the country's potable water treatment facilities (its waterworks) is far upstream of the NMCPCP.

Along with finding reclaimed water users communicating with customers and the public about the benefits of reclaimed water feasibility studies were conducted, and the county held meetings with the reclaimed water customers to educate them about the benefits of water reuse for their facilities. Note that a ball field owned by another branch of the county was very supportive of the reuse concept and the reduced maintenance it would bring to the ballfields. In addition, the local golf course owners were very supportive and worked with the county to determine the best route for the pipeline to reach their facility.

THE BOTTOM LINE

In order to put reclaimed water, reused water, and/or treated wastewater (call it what you may) to work for beneficial purposes, there are a few hurdles that political figures, public works officials, planners, and ratepayers must overcome. First, is the Yuck Factor. When some ratepayers are informed that reclaimed water is toilet water, there is a tendency for these individuals to reject wastewater for any kind of reuse-period. However, Fairfax County Wastewater Management has worked to lessen ratepayers' concerns, to lessen their so-called Yuck Factor, by careful, well-thought-out plans for beneficial reuse for the area. First, it is all about communication. Ratepayers are informed that reclaimed water is wastewater that has been thoroughly treated to remove harmful organisms and substances, such as bacteria, viruses, and heavy metals, so it can be reused for beneficial purposes—this is when honest communication paves the way. Second, public works officials in the Fairfax County region communicated other benefits including one of the major goals of conserving valuable drinking water. Simply, informing the ratepayers that there is an increasing demand for water, and reclaimed water allows for the use of drinking water for its intended use—to be used to drink. Third, communicating to the ratepayers about using reclaimed water to reduce the amount of nutrients—phosphorus and nitrogen—that reach the Chesapeake Bay, and that the county surpasses state and federal clean water regulations is a positive message, especially to those who value their environment and compliance with laws. Lastly, the real deal maker, so to speak, is communicating the beneficial cost benefits of using reclaimed water. When ratepayers are told that they will save money by using reclaimed water instead of potable water and they informed that the county will make money from selling reclaimed water which in turn will offset overall costs for taxpayers this is a winning argument for most people. In one word, it all comes down to communication, communication, and communication.

REFERENCE

USEPA. (2023). Case Studies that Demonstrate the Benefits of Water Reuse. Accessed 12/22/23 @ https://epa.gov/waterreuse/forms/contact-us--about-water-reuse-and-recycling.

7 Water Reuse
Infrastructure Reliance

INTRODUCTION

Water reuse provides a wide range of benefits as shown in Figure 7.1. In this chapter, water reuse for infrastructure reliance is described and an example of real-world usage in the United States is provided.

Generally, when we think about Alaska what comes to many of our minds is a remote wilderness encapsulated in ice and snow, with rivers that freeze over in winter and rage when ice-free, home on the range for Grizzly Bears, Brown Bears, Black Bears, Moose, Caribou, Wolverine, Bald Eagles, and others all encamped in dark, dense, forbidding forests. All of this is true, of course.

So, one might think that with the ice, snow, and rivers that accessing these sources of water would be a piece of cake; and it is. However, for the thousands of residents living in the rural areas lack access to running water or flush toilets due to their remote locations and the local climate. To deal with this matter, the Alaska Department of Environmental Conservation (DEC) funded the Alaska Water and Sewer Challenge—a research and technology development competition intended to single out a proper decentralized water reuse solution for use in rural Alaska homes.

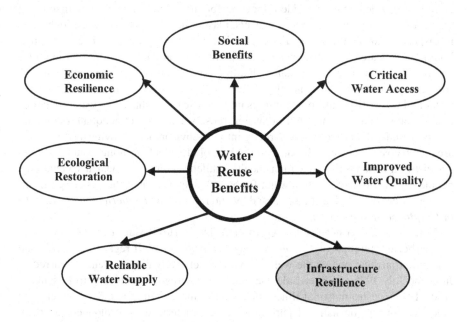

FIGURE 7.1   Water reuse: infrastructure reliance.

DOI: 10.1201/9781003498049-8

What the DEC was looking for was designs that could safely and affordably supply homes with clean water but were also acceptable for the end users and feasible for implementation in rural Alaska. DEC found what they were looking for in the winning design they picked which consisted of an onsite greywater recycling system coupled with urine-diverting dry toilets, which could significantly reduce the reliance on imported water for nonpotable uses. The big plus of the winning design is that the system avoids the use of water for flushing toilets, it limited the need to export sewage for treatment and disposal. As of June 2023, the selected design was progressing toward the implementation stage.

CASE STUDY 7.1: ON-SITE WATER REUSE SOLUTION (USEPA, 2023; ALASKA DEC, 2022)

In the rural areas/villages of Alaska, more than 3,000 homes do not have access to indoor portable water or flush toilets. Many of these homes are in rural villages where access is only possible by plane or boat. Along with distances between homes and community centers and permafrost (i.e., frozen subsurface conditions), it is not reasonable and in many cases not feasible to construct conventional wastewater plants and conveyance systems in these communities. Previously, various agencies funded expensive community-wide water supply and sanitation efforts, typically focused on truck-haul water programs, to meet the needs of the communities amid the challenging conditions. In parts of rural Alaska, residents import an average of 420 gallons (1,590 L) of potable water per week from truck-haul access points in their communities. On the positive side, this approach provides safe drinking water to citizens by giving them access to potable water in a central location. Additionally, community-wide centers provide a location for rural Alaskans to safely dispose of waste from dry toilets, (i.e., honey buckets, plastic lined buckets that are widely used to dispose of urine and feces; aka bucket latrines; see Figure 7.2). The problem is both practices are expensive because of the high operational costs for communities and the inconvenience for residents—hauling large loads of water and waste to and from central access points is not that convenient. Because of the high cost of potable water and waste disposal, some families may choose to purchase less water than they need, capture and use untreated rainwater, reuse water without adequate treatment, and potentially dispose of waste directly into the environment. Obviously, there also are health costs to consider because of the lack of affordable in-home water and sewage disposal which is linked to many health challenges for rural Alaskans. Generally, people who do not have enough water for basic hygiene are at higher risk of gastrointestinal infections, such as those caused by enterotoxigenic *Escherichia coli* (*E. coli*) or *Cryptosporidium parvum*.

Note that you do not need to go to or reside in rural Alaska to be exposed to a waterborne outbreak of *E. coli* or *Cryptosporidium*. With regard to *E. coli* and epidemic diarrhea, one of the well-known and relatively recent events occurred in June and July 1975 at Crater Lake National Park, Oregon. According to Rosenberg et al. (1977), gastrointestinal illness occurred in more than 200 staff members and 2,000 visitors to the national park and was characterized by prolonged diarrhea, cramps, nausea, and vomiting. The malady lasted a median duration of 8 days and

FIGURE 7.2   A honey bucket.

was associated with the consumption of park water (for those familiar with statistics, $P < 0.001$), which had been contaminated with raw sewage.

Let us take a look at water-borne *Cryptosporidium*.

Ernest E. Tyzzer first described the protozoan parasite *Cryptosporidium* in 1907. Tyzzer frequently found a parasite in the gastric glands of laboratory mice. Tyzzer identified the parasite as a sporozoan, but of uncertain taxonomic status; he named it *Cryptosporidium muris*. Later, in 1910, after a more detailed study, he proposed *Cryptosporidium* as a new genus and *C. muris* as the type of species. Amazingly, except for developmental stages, Tyzzer's original description of the life cycle (see Figure 7.3) was later confirmed by electron microscopy. Later, in 1912, Tyzzer described a new species, *Cryptosporidium parvum* (Tyzzer, 1912).

For almost 50 years, Tyzzer's discovery of the genus *Cryptosporidium* (because it appeared to be of no medical or economic importance) remained (like himself) relatively obscure because it appeared to be of no medical or economic importance. Slight rumblings of the genus' importance were felt in the medical community when Slavin (1955) wrote about a new species, *Cryptosporidium melagridis*, associated with illness and death in turkeys. Interest remained slight even when *Cryptosporidium* was found to be associated with bovine diarrhea (Panciera et al., 1971).

Not until 1982 did worldwide interest focus on the study of organisms in the genus *Cryptosporidium*. During this period, the medical community and other interested parties were beginning to attempt a full-scale, frantic effort to find out as much as

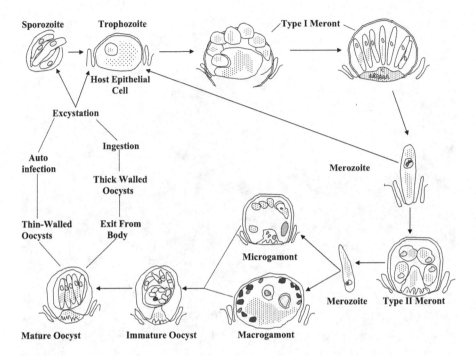

FIGURE 7.3   Life cycle of *Cryptosporidium parvum.*

possible about Acquired Immune Deficiency Syndrome (AIDS). The CDC reported that 21 AIDS-infected males from six large cities in the U.S. had severe protracted diarrhea caused by *Cryptosporidium*. It was in 1993, though, that when the "bug—the pernicious parasite *Cryptosporidium*—made [itself and] Milwaukee famous (Mayo Foundation, 1996)."

Note: The *Cryptosporidium* outbreak in Milwaukee caused the deaths of 100 people—the largest episode of waterborne disease in the U.S. in the 70 years since health officials began tracking such outbreaks.

The massive waterborne outbreak in Milwaukee (more than 400,000 persons developed acute and often prolonged diarrhea or other gastrointestinal symptoms) increased interest in *Cryptosporidium* at an exponential level. The Milwaukee Incident spurred both public interest and the interest of public health agencies, agricultural agencies and groups, environmental agencies and groups, and suppliers of drinking water. This increase in interest level and concern has spurred new studies of *Cryptosporidium* with an emphasis on developing methods for recovery, detection, prevention, and treatment (Fayer et al., 1997).

The USEPA has become particularly interested in this "new" pathogen. For example, in the reexamination of regulations on water treatment and disinfection, the USEPA issued MCLG and CCL for *Cryptosporidium*. The similarity to *Giardia lamblia* and the necessity to provide an efficient conventional water treatment capable of eliminating viruses at the same time forced the USEPA to regulate the surface water supplies in particular. The proposed "Enhanced Surface Water Treatment

Rule" (ESWTR) included regulations from watershed protection to the specialized operation of treatment plants (certification of operators and state overview) and effective chlorination. Protection against *Cryptosporidium* included control of waterborne pathogens such as *Giardia* and viruses (De Zuane, 1997).

THE BASICS OF *CRYPTOSPORIDIUM*

Cryptosporidium (crip-toe-spor-ID-ee-um) is one of several single-celled protozoan genera in the phylum Apircomplexa (all referred to as coccidian). *Cryptosporidium* along with other genera in the phylum Apircomplexa develop in the gastrointestinal tract of vertebrates through all of their life cycle—in short, they live in the intestines of animals and people. This microscopic pathogen causes a disease called *Cryptosporidiosis* (crip-toe-spor-id-ee-O-sis). The dormant (inactive) form of *Cryptosporidium* called an oocyst (O-o-sist) is excreted in the feces (stool) of infected humans and animals. The tough-walled oocysts survive under a wide range of environmental conditions.

Several species of *Cryptosporidium* were incorrectly named after the host in which they were found; subsequent studies have invalidated many species. Now, eight valid species of Cryptosporidium (see Table 7.1) have been named. Upton (1997) reports that *C. muris* infects the gastric glands of laboratory rodents and several other mammalian species, but (even though several texts state otherwise) is not known to infect humans. However, *C. parvum* infects the small intestine of an unusually wide range of mammals, including humans, and is the zoonotic species responsible for human Cryptosporidiosis. In most mammals, *C. parvum* is predominately a parasite of neonate (newborn) animals. He points out that even though exceptions occur, older animals generally develop poor infections, even when unexposed previously to the parasite. Humans are the one host that can be seriously infected at any time in their lives, and only previous exposure to the parasite results in either full or partial immunity to challenge infections.

Oocysts are present in most surface bodies of water across the U.S., many of which supply public drinking water. Oocysts are more prevalent in surface waters when

TABLE 7.1
Valid Named Species of *Cryptosporidium*
(Fayer et al., 1997)

Species	Host
C. baileyi	Chicken
C. felis	Domestic cat
C. meleagridis	Turkey
C. murishouse	House mouse
C. nasorium	Fish
C. parvum	House mouse
C. serpentis	Corn snake
C. wrairi	Guinea pig

heavy rains increase runoff of wild and domestic animal wastes from the land, or when sewage treatment plants are overloaded or break down. Only laboratories with specialized capabilities can detect the presence of *Cryptosporidium* oocysts in water. Unfortunately, present sampling and detection methods are unreliable. Recovering oocysts trapped in the material used to filter water samples is difficult. Once a sample is obtained, however, determining whether the oocyst is alive or whether it is the species *C. parvum* that can infect humans is easily accomplished by looking at the sample under a microscope.

The number of oocysts detected in raw (untreated) water varies with location, sampling time, and laboratory methods. Water treatment plants remove most but not always all, oocysts. Low numbers of oocysts are sufficient to cause Cryptosporidiosis, but the low numbers of oocysts sometimes present in drinking water are not considered cause for alarm in the public.

Protecting water supplies from *Cryptosporidium* demands multiple barriers, why? Because *Cryptosporidium* oocysts have tough walls that can withstand many environmental stresses and are resistant to chemical disinfectants such as chlorine that are traditionally used in municipal drinking water systems.

Physical removal of particles, including oocysts, from water by filtration is an important step in the water treatment process. Typically, water pumped from rivers or lakes into a treatment plant is mixed with coagulants, which help settle out particles suspended in the water. If sand filtration is used, even more particles are removed. Finally, the clarified water is disinfected and piped to customers. Filtration is the only conventional method now in use in the U.S. for controlling Cryptosporidium.

Ozone is a strong disinfectant that kills protozoa if sufficient doses and contact times are used, but ozone leaves no residual for killing microorganisms in the distribution system, as does chlorine. The high costs of new filtration or ozone treatment plants must be weighed against the benefits of additional treatment. Even well-operated water treatment plants cannot ensure that drinking water will be completely free of *Cryptosporidium* oocysts. Water treatment methods alone cannot solve the problem; watershed protection and monitoring of water quality are critical. As mentioned, watershed protection is another barrier to *Cryptosporidium* in drinking water. Land use controls such as septic systems regulations and best management practices to control runoff can help keep human and animal wastes out of water.

Under the Surface Water Treatment Rule of 1989, public water systems must filter surface water sources unless water quality and disinfection requirements are met and a watershed control program is maintained. This rule, however, did not address *Cryptosporidium*. The USEPA has now set standards for turbidity (cloudiness) and coliform bacteria (which indicate that pathogens are probably present) in drinking water. Frequent monitoring must occur to provide officials with early warning of potential problems to enable them to take steps to protect public health. Unfortunately, no water quality indicators can reliably predict the occurrence of Cryptosporidiosis. More accurate and rapid assays of oocysts will make it possible to notify residents promptly if their water supply is contaminated with *Cryptosporidium* and thus avert outbreaks.

The bottom line: The collaborative efforts of water utilities, government agencies, health care providers, and individuals are needed to prevent outbreaks of Cryptosporidiosis.

Cryptosporidiosis

Juranek (1995) wrote in the journal *Clinical Infectious Diseases*:

> Cryptosporidium *parvum* is an important emerging pathogen in the U.S. and a cause of severe, life-threatening disease in patients with AIDS. No safe and effective form of specific treatment for Cryptosporidiosis has been identified to date. The parasite is transmitted by ingestion of oocysts excreted in the feces of infected humans or animals. The infection can therefore be transmitted from person to person through ingestion of contaminated water (drinking water and water used for recreational purposes) or food, from animal to person, or by contact with fecally contaminated environmental surfaces. Outbreaks associated with all of these modes of transmission have been documented. Patients with human immunodeficiency virus infection should be made more aware of the many ways that *Cryptosporidium* species are transmitted, and they should be given guidance on how to reduce their risk of exposure.

Since the Milwaukee outbreak, concern about the safety of drinking water in the U.S. has increased, and new attention has been focused on determining and reducing the risk of Cryptosporidiosis from the community and municipal water supplies. Cryptosporidiosis is spread by putting something in the mouth that has been contaminated with the stool of an infected person or animal. In this way, people swallow the *Cryptosporidium* parasite. As previously mentioned, a person can become infected by drinking contaminated water or eating raw or undercooked food contaminated with *Cryptosporidium* oocysts; direct contact with the droppings of infected animals or stools of infected humans; or hand-to-mouth transfer of oocysts from surfaces that may have become contaminated with microscopic amounts of stool from an infected person or animal (Spellman, 2020).

The symptoms may appear 2–10 days after infection by the parasite. Although some persons may not have symptoms, others have watery diarrhea, headache, abdominal cramps, nausea, vomiting, and low-grade fever. These symptoms may lead to weight loss and dehydration. In otherwise healthy persons, these symptoms usually last 1–2 weeks, at which time the immune system is able to stop the infection. In persons with suppressed immune systems, such as persons who have AIDS or who recently have had an organ or bone marrow transplant, the infection may continue and become life threatening.

Currently, no safe and effective cure for Cryptosporidiosis exists. People with normal immune systems improve without taking antibiotic or antiparasitic medications. The treatment recommended for this diarrheal illness is to drink plenty of fluids and to get extra rest. Physicians may prescribe medication to slow the diarrhea during recovery.

The best way to prevent Cryptosporidiosis is to:

- Avoid water or food that may be contaminated.
- Wash hands after using the toilet and before handling food.
- If you work in a childcare center where you change diapers, be sure to wash your hands thoroughly with plenty of soap and warm water after every diaper change, even if you wear gloves.

During community-wide outbreaks caused by contaminated drinking water, drinking water practitioners should inform the public to boil drinking water for 1 min to kill the *Cryptosporidium* parasite (Spellman, 2020).

In 2013, the Alaska Department of Environmental Conservation (DEC), in coordination with tribal, state, and federal agencies, initiated a research and development competition to find more efficient and affordable ways to deliver drinking water and sewage disposal services to residents in rural Alaska. This competition solicited innovative and cost-effective designs to meet the water supply needs of rural Alaska households. Applications from close to 20 teams were received. The following ten performance targets/questions/requirements were specified in evaluating the submitted designs:

- Capital cost ($$ are always an issue or key parameter)
- Can it be built and made durable?
- Must have a minimum capacity of 15 gallons, or 57 L, per day per person for water use for health benefits
- Keep the monthly cost for operation and maintenance cost less than $135 per household
- Must have a freeze-and-thaw capability
- Modularity of household system (configuration options are important)
- Can it withstand extreme temperatures, permafrost, remote accessibility, operability, and regulator; is it feasible?
- Are the parts available?
- Must comply with the plumbing code
- Finally, and most importantly acceptance and use by end-users.

For this project, a steering committee was formed and during the prototype development stage of the challenge all teams were guided by this committee made up of tribal, state, and federal agencies with knowledge relevant to the technical aspects of the projects. The committee met regularly to review the status of work, make decisions about the progression of each team, and evaluate team proposals and protypes. The most important function of the committee during the development stage and even in the later stages of the challenge was that the public and potential users were informed on the feasibility and usability of possible designs.

Acceptance of reclaimed water reuse depends on communication, communication, and communication with the public and the end-users—this critical step cannot be overlooked or underrated.

For the Prototype Development and Pilot Testing phase of the challenge, three teams were selected as semi-finalists; this phase involved monitoring the pilot-scale version of each team's design over 9 months. Note that because many rural Alaska households lack flush toilets, all three teams incorporated reuse into their design by treating greywater separately from non-potable reuse. The three teams incorporated reuse into their design by treating greywater separately for non-potable reuse; this was necessary because many rural Alaska households lacked flush toilets.

In 2022, one of the three semi-finalist designs was selected as the winning design and moved into the field system development and testing phase. The winning design team, made up of academics from the University of Alaska Anchorage (UAA), engineering consultants, and health professionals, partnered with the rural communities of Kipnuk, and Koyukuk to gather feedback and inform their team's design.

According to Alaska DEC (2022), the winning prototype design consists of a modular system that is composed of three separate systems. In the wash water system greywater is collected from the kitchen and bathroom sinks, shower, and laundry and treats the water to a level suitable for human contact, but not for drinking. The reused water is pumped to fulfill the same non-potable demands (kitchen sink, bathroom sink, shower, and laundry). A separate drinking water source usually provided by a truck-haul supplier is used to supply individual taps in the kitchen and bathroom dedicated to potable uses. Disposal of a small volume of concentrated greywater generated by the greywater recycling system must be disposed of on a weekly basis.

Note that using urine-diverting dry toilets reduces odors, as compared to conventional dry toilets in which urine and feces (aka po) are combined (i.e., honey buckets, see Figure 7.2). Also, note that waste from the urine-diverting toilet must be transported from the home by the residents to the disposal facility.

Alaska DEC (2022) reports that the UAA team points out that by using the greywater recycling system the user can generate 406 gallons (1,537 L) of wash water from an initial input of only 35 gallons (132 L) of water weekly. Depending on availability to the user, this input can consist of non-potable sources, such as rain, river, or lake water. These water sources supply the wash water system and are intended to be used only for non-potable demands. In this system process, there is non-reusable waste produced. This waste results from the reverse osmosis filter when it processes greywater; it generates a concentrated greywater waste stream which must be collected and removed from the treatment system periodically. The UAA team also estimates that 35 gallons (132 L) of concentrated greywater and nine gallons (34 L) of urine will be removed from the home weekly—44 gallons (166 L) total. Households are estimated to need 14 gallons (53 L) of potable water to be hauled in for potable uses.

The selected treatment system is modularized in a 10-ft (3-m) shipping container to allow for relatively easy deployment and cost-saving. Within the container, the treatment process includes the following steps:

- Settling
- Air-assisted soap removal
- Nanofiltration and reverse osmosis filtration
- UV disinfection
- Ozone disinfection

During the pilot testing phase, it was demonstrated that the treatment system can achieve over 7 $\log_{10}$ removal of both bacteria and viruses (i.e., 99.99999% removal), lower turbidity to <0.1 nephelometric turbidity units (NTU), reduce total organic carbon to concentrations <0.5 mg/L, and maintain a pH between 6 and 8. Note that the system is self-contained and energy requirements are met with a 12-V battery, as connection to the electrical grid is challenging for many of these communities. Keep in mind that some means are necessary to recharge the battery or a fully charged standby battery must be close by.

THE BOTTOM LINE

The system selected for onsite reuse system has many hurdles to jump, including identifying funding mechanisms (funding is always a challenge), technical challenges associated with deployment in remote areas, and operation and maintenance of the system. However, the real bottom line is user acceptance and avoidance of the yuck factor associated with water reuse.

REFERENCES

Alaska DEC. (2022). The Alaska Water and Sewer Challenge. Accessed 1/1/24 @ https://dec. alaska.gov/water/water-sewer-chanllenge/&gt.

De Zuane, J. (1997). *Handbook of Drinking Water Quality.* New York: John Wiley & Sons, Inc.

Fayer, R., Speer, C.A., and Dudley, J.P. (1997). The general biology *Cryptosporidium.* In Fayer, R. (ed.), *Cryptosporidium and Cryptosporidiosis.* Boca Raton, FL: CRC Press, pp. 117–123.

Juranek, D.D. (1995). *Cryptosporidiosis*: Sources of infection and guidelines for prevention. *Clinical Infectious Diseases* 21: S57–S61.

Mayo Foundation. (1996). *The "Bug" that Made Milwaukee Famous.* Rochester, MN: Mayo Foundation.

Panciera, R.J., Thomassen, R.W., and Garner, R.M. (1971). Cryptosporidial infection in a calf. *Veterinary Pathology* 8: 479.

Rosenberg, M.L., Koplan, J.P., Wachsmuth, I.K., Wells, J.G., Gnagarosa, R.L., and Sack, D.A. (1977). Epidemic diarrhea at Crater Lake from enterotoxigenic *Escherichia coli*: A large waterborne outbreak. *Annals of Internal Medicine* 86(6): 714–718.

Slavin, D. (1955). *Cryptosporidium* melagridis. *Journal of Comparative Pathology* 65: 262.

Spellman, F.R. (2020). *Handbook of Water/Wastewater Treatment Plant Operations*, 4th ed. Boca Raton, FL: CRC Press.

Tyzzer, E.E. (1912). *Cryptosporidium parvum* sp.: A Coccidium found in the small intestine of the common mouse. *Archiv für Protistenkunde* 26: 394 m.

Upton, S.J. (1997). *Basic Biology of Cryptosporidium.* Manhattan, KS: Kansas State University.

USEPA. (2023). Case Studies that Demonstrate the Benefits of Water Reuse. Accessed 12/22/23 @ https://epa.gov/waterreuse/forms/contact-us--about-water-reuse-and-recycling.

8 Water Reuse
Improved Water Quality

INTRODUCTION

Water reuse provides a wide range of benefits as shown in Figure 8.1. In this chapter, water reuse for improved water quality is described and an example of real-world usage in the United States is provided.

THE 411 ON POLLUTED DRINKING WATER[1]

Is drinking water contamination really a problem—a serious problem? In answer to the first part of the question, we can say it depends upon where your water comes from. As to the second part of the question, we refer you to a book (or the film based upon the book) that concerns a case of toxic contamination—one you might be familiar with—*A Civil Action,* written by Jonathan Harr (1995). The book and film portray the legal repercussions connected with polluted water supplies in Woburn, Massachusetts. Two wells became polluted with industrial solvents, in all apparent likelihood causing 24 of the town's children, who lived in neighborhoods supplied by those wells, to contract leukemia and die.

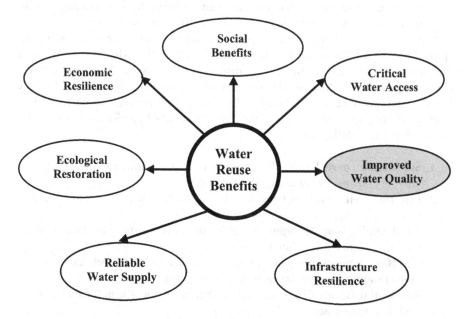

FIGURE 8.1 Water reuse: improved water quality.

DOI: 10.1201/9781003498049-9

Many who have read the book or have seen the movie may mistakenly get the notion that Woburn, a toxic "hot spot," is a rare occurrence. Nothing could be further from the truth. Toxic "hot spots" abound. Most striking is areas of cancer clusters—a short list includes:

- Woburn, where about two dozen children were stricken with leukemia over 12 years, a rate several times the national average for a community of its size.
- Storrs, Connecticut, where wells polluted by a landfill are suspected of sickening and killing residents in nearby homes.
- Bellingham, Washington, where pesticide-contaminated drinking water is thought to be linked to a sixfold increase in childhood cancers.

As Schlichtmann[2] points out, these are only a few examples of an underlying pathology that threatens many other communities. Meanwhile, cancer is now the primary cause of childhood death from disease.

Drinking water contamination is a problem—a very serious problem. In this chapter, we discuss a wide range of water contaminants, the contaminant sources—and their impact on drinking water supplies from both surface water and groundwater sources.

SOURCES OF CONTAMINANTS

If we were to list all the sources of contaminants and the contaminants themselves (the ones that can and do foul our water supply systems), along with a brief description of each contaminant, we could easily fill a book (maybe consisting of several volumes). To give you some idea of the magnitude of the problem we condensed a list of selected sources and contaminants (our "short list"), which includes:

Note: Keep in mind that when we specify "water pollutants" we are in most cases speaking about pollutants that somehow get into the water (by whatever means) from the interactions of the other two environmental mediums: air and soil. Probably the best example of this is the acid rain phenomenon—pollutants originally emitted only into the atmosphere land on Earth and affect both soil and water. Consider that 69% of the anthropogenic lead and 73% of the mercury in Lake Superior reach it by atmospheric deposition (Hill, 1997).

1. **Subsurface percolation**: hydrocarbons, metals, nitrates, phosphates, microorganisms, and cleaning agents (TCE).
2. **Injection wells**: hydrocarbons, metals, non-metals, organics, organic and inorganic acids, microorganisms, and radionuclides.
3. **Land application**: Nitrogen, phosphorous, heavy metals, hydrocarbons, microorganisms, and radionuclides.
4. **Landfills**: organics, inorganics, microorganisms, and radionuclides.
5. **Open dumps**: organics, inorganics, and microorganisms.
6. **Residential (local) disposal**: organic chemicals, metals, non-metal inorganics, inorganic acids, and microorganisms.

7. **Surface impoundments**: organic chemicals, metals, non-metal inorganics, inorganic acids, microorganisms, and radionuclides.
8. **Waste tailings**: Arsenic, sulfuric acid, copper, selenium, molybdenum, uranium, thorium, radium, lead, manganese, and vanadium.
9. **Waste piles**: arsenic, sulfuric acid, copper, selenium, molybdenum, uranium, thorium, radium, lead, manganese, and vanadium.
10. **Materials stockpiles**: coal pile: aluminum, iron, calcium, manganese, sulfur, and traces of arsenic, cadmium, mercury, lead, zinc, uranium, and copper. Other materials piles: metals/non-metals, and microorganisms.
11. **Graveyards**: metals, non-metals, and microorganisms.
12. **Animal burial**: contamination is site-specific—depending on disposal practices, surface and subsurface, hydrology, proximity of the site to water sources, type and amount of disposed material, and cause of death.
13. **Above-ground storage tanks**: organics, metal/non-metal inorganics, inorganic acids, microorganisms, and radionuclides.
14. **Underground storage tanks**: organics, metal, inorganic acids, microorganisms, and radionuclides.
15. **Containers**: organics, metal/non-metal inorganics, inorganic acids, microorganisms, and radionuclides.
16. **Open burning and detonating sites**: inorganics, including heavy metals; organics, including TNT.
17. **Radioactive disposal sites**: radioactive cesium, plutonium, strontium, cobalt, radium, thorium, and uranium.
18. **Pipelines**: organics, metals, inorganic acids, and microorganisms.
19. **Material transport and transfer operations**: organics, metals, inorganic acids, microorganisms, and radionuclides.
20. **Irrigation practices**: fertilizers, pesticides, naturally occurring contamination, and sediments.
21. **Pesticide applications**: 1,200–1,400 active ingredients. Contamination has already been detected: alachlor, aldicarb, atrazine, bromacil, carbofuran, cyanazine, DBCP, DCPA, 1,2-dichloropropane, dyfonate, EDB, metolachlor, metribyzen, oxalyl, siazine, and 1,2,3-trichloropropane. The extent of groundwater contamination cannot be determined with current data.
22. **Animal feeding operations**: Nitrogen, bacteria, viruses, and phosphates.
23. **De-icing salts applications**: chromate, phosphate, ferric ferrocyanide, na-ferrocyan, chlorine.
24. **Urban runoff**: suspended solids and toxic substances, especially heavy metals and hydrocarbons, bacteria, nutrients, and petroleum residues.
25. **Percolation of atmospheric pollutants**: sulfur and nitrogen compounds, asbestos, and heavy metals.
26. **Mining and mine drainage:**
   • **Coal**: acids, toxic inorganics (heavy metals), and nutrients. Phosphate: Radium, uranium, and fluorides.
   • **Metallic ores**: sulfuric acid, lead, cadmium, arsenic, sulfur, and cyan.

27. **Production wells**:
    - **Oil**: 1.2 million abandoned production wells
    - **Irrigation**: farms
    - **All**: potential to contaminate installation, operation, and plugging techniques.
28. **Construction excavation**: pesticides, diesel fuel, oil, salt, and a variety of others.
29. **PPCPs**: any product used by an individual for personal health or cosmetic reasons or used by agribusiness to enhance the growth or health of livestock.

DID YOU KNOW?

PPCPs are the primary ingredient in the conglomeration of products we use each and every day. Why the concern about PPVPs, you might ask? Well, in reality, there is no one concern but instead, there are many:

- Large quantities of PPCPs can enter the environment after use by individuals or domestic animals.
- Wastewater systems are not equipped for PPCP removal. Currently, there are no municipal wastewater treatment plants that are engineered specifically for PPCP removal or other unregulated contaminants. Effective removal of PPCPs from treatment plants varies based on the type of chemical and the individual wastewater treatment facilities.
- The risks are uncertain. The risks posed to aquatic organisms, and to humans are unknown, largely because the concentrations are so low.
- The number of PPCPs is growing. In addition to antibiotics and steroids, over 100 individual PPCPs have been identified (as of 2007) in environmental samples and drinking water (USEPA, 2011).

Note: To better understand specific water pollutants, we must examine several terms important to the understanding of water pollution. One of these is *point source*. The USEPA defines a *point source* as "any single identifiable source of pollution from which pollutants are discharged, e.g., a pipe, ditch, ship, or factory smokestack." For example, the outlet pipes of an industrial facility or a municipal wastewater treatment plant are point sources. In contrast, *non-point sources* are widely dispersed sources and are a major cause of stream pollution. An example of a nonpoint source of pollution is rainwater carrying topsoil and chemical contaminants into a river or stream. Some of the major sources of nonpoint pollution include water runoff from farming, urban areas, forestry, and construction activities. The word *runoff* signals a nonpoint source that originated on land. Runoff may carry a variety of toxic substances and nutrients, as well as bacteria and viruses with it. Nonpoint sources now comprise the largest source of water pollution, contributing ~65% of the contamination in quality-impaired streams and lakes.

CASE STUDY 8.1: MICROSOFT AND THE CITY OF QUINCY, WASHINGTON (USEPA, 2023)

In a nutshell, the Quincy Water Reuse Utility (QWRU) was constructed in a partnership between Microsoft and the city of Qunicy, Washington. The Reuse Utility treats cooling water from a Microsoft data center and recirculates it to the data center for the same purpose, reducing reliance on local potable groundwater. The groundwater also has high levels of total dissolved solids, which can cause problems when used in the data center's cooling equipment. The water supplied from the QWRU is of more suitable quality for the data center's cooling equipment than the conventionally treated, mineral-rich groundwater.

In context, the City of Quincy is in the arid eastern part of the State of Washington and receives only 7.8 inches (199 mm) of annual rainfall. The potable water required to service residential, commercial, and industrial needs within the city is provided by five groundwater wells. These wells in basaltic rocks reach depths up to 1,000 ft (305 m) into the Columbia Plateau regional aquifer system. Although the residential part of Quincy is relatively small with about 8,100 people its large industrial sector means its water demand is akin to that of 30,000 inhabitants. Yahoo, Sabey, Intuit, Vantage, and Microsoft have data centers along with a large food processing industry within the Quincy region.

So, why Quincy, Washington? Why are major companies building their data centers in Quincy? The attraction is one word: affordability, affordability, affordability. Simply, Quincy is an attractive area to bid these data centers because of the area's affordable land (as compared to land in the Silicon Valley region to the south of Washington State in California), cool temperatures, and the availability of renewable, abundant, and cheap hydropower. The surrounding area is also a hub of agricultural activity.

Let us look at the city's water needs.

Typically, on a yearly basis, the city uses ~2.1 billion gallons (8 million L) of groundwater. Until recently, food processors used 57% of the pumped groundwater; 33% went to residents and public and commercial services such as schools, hotels, or restaurants; and the remaining 10% was used by data centers for cooling purposes.

Note that since 2007, Microsoft has operated one of the biggest data centers in Quincy on a 270-acre (110-ha) campus. The large volume of water needed to cool high-density cloud computing server farms is making water management a growing priority for data center operators, including Microsoft. In the past, Microsoft received potable groundwater from Quincy for cooling; they would discharge the water to the sewer system for treatment at the city's municipal wastewater reclamation facility. However, after four to five cooling cycles this water was rich in minerals including calcium, magnesium, and silica. This was not a good situation because the high mineral content meant that long cycles in the data centers could damage heat transfer equipment, or generally impair heat transfer, as water evaporated during cooling and the concentration of total dissolved solids (TDS) increased. To mitigate this mineral buildup water, it had to be "blowdown" or "bled" (i.e., drained) more often from the cooling equipment to remove mineral build-up.

Complicating the TDS problem in the cooling water was the original wastewater treatment approach. The city possessed and operated a municipal reclamation facility (MWRF) that uses reclaimed water to irrigate public areas and to recharge shallow aquifers in the area through groundwater recharge basins. However, the treatment plant could not handle the high TDS concentrations in the discharges from the regional data to the MMRF. Thus, the data centers were charged an extra fee to accommodate adjustments to the facility's process and increased maintenance needs. Another fly in the ointment or icing on the "cake," so to speak, is the high levels of TDS, calcium, magnesium, and silica that together formed a visible cake on the surface of the groundwater recharge basins in the summer, shih was causing it to exceed discharge and indirect potable reuse requirements, for example, for aquifer recharge, set by the Washington State Department of Ecology.

The City of Quincy and Microsoft both recognized the technical difficulty of using mineral-rich municipal groundwater for cooling data center equipment; moreover, they also recognized the subsequent challenge of treating the data center's industrial wastewater with high levels of TDS at the city's municipal wastewater reclamation facility. So what the City and Microsoft did was to form a partnership to create the QWRU, basically an entirely new facility specifically designed to treat the cooling water from Microsoft's data center. Ten years of planning and construction were finally completed when QWRU officially became operational on June 30, 2001. Pipes more than 30 miles (48 km) long run across Quincy's industrial area connected to ten distinct treatment systems. To date, only Microsoft's data center campus is connected to the QWRU.

The QWRU is a closed-loop system and it neither discharges industrial wastewater with high mineral content to the environment nor does it discharge industrial wastewater to the city's MWRF. Note that in order to prevent mineral buildup in the cooling water, salts are removed from the water prior to ruse at the data center and concentrated in lined brine ponds to form a solid which is disposed of in a manner that prevents entry into the state ground or surface waters. What this does is ensure that the cooling water quality does not adversely affect cooling equipment or heat transfer within the data center. Just as important (maybe even more so) the QWRU substantially reduces the data center's reliance on portable groundwater for data center cooling, saving millions of gallons of water per year.

As shown in Figure 8.2 for its treatment system, the QWRU relies on a combination of:

* New online systems, including a lime softening facility (like alum treatment using calcium hydroxide), an ultrafiltration facility, a permeate tank facility [used in reverse osmosis (RO)], a blend tank facility storage, and brine ponds.
* Previously existing infrastructure which was part of the city's industrial wastewater treatment systems, including a high-efficiency softening (HES) facility and reverse osmosis (RO) plant.

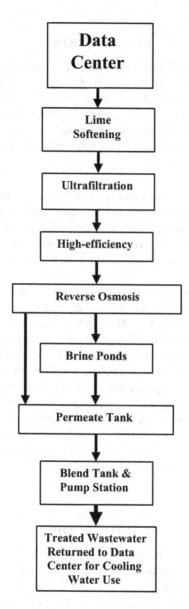

FIGURE 8.2 QWRU's treatment train.

The water is cycled four to five times at the data center and then the resulting blowdown water is transferred to a lime softening facility (see Figure 8.2) that adds limewater (calcium hydroxide) to remove hardness (deposits of calcium and magnesium salts)—next in the treatment train is the ultrafiltration facility.

DID YOU KNOW?

Ultrafiltration (UF) has many applications, some the applications of UF include, the treatment of oily wastewater, processing of whey, recovery of electrocoat paint, concentration of later emulsion, pulp and paper industry, textile industry, clarification of fruit juice, cold sterilization, and water treatment. In water treatment, UF is effective for the removal of colloids, proteins, bacteria, pyrogens, and other organic molecules larger than the membrane pore.

After ultrafiltration, some of the water is transferred to a blend tank, while the rest is further softened in the HES facility and filtered at the RO plant (see Figure 8.2). A pipe transfers the HES calcium and magnesium-laden brin and the RO brine silica to brine ponds. The nature takes over and the brine undergoes maximum evaporation, and the residuals are disposed of separately. Note that Qunicy expects it will need to do this every 2 or 3 years.

The treated wastewater continues to flow from the RO system and is stored in a permeate tank (sce Figure 8.2) for ~8 hours and acts as a buffer whenever the system is down for maintenance of the HES or RO systems. The ultrafiltration and RO water are blended at various ratios throughout the day depending on the conductivity levels (ideally: 300–350 microsiemens). Finally, the treated industrial wastewater is returned to the data center for use as cooling water (see Figure 8.2).

In operation, the QWRU system must be topped off or topped up often because water evaporates during the cooling process and a small volume of highly concentrated brines needs disposal. Because the data centers do not need potable water for cooling, the City of Quincy and Microsoft have been able to use make-up water from untreated surface water sources by accessing rights to water drawn from the Columbia Basin Project in the U.S. Bureau of Reclamation and the Quincy Irrigation District canal.

DID YOU KNOW?

According to the U.S. Bureau of Reclamation (USBR, 2023), more than 670,000 acres in east central Washington serve the Columbia Basin Project. The main facilities of the project include the Grand Coulee Dam, Franklin D. Roosevelt Lake, three power plants, four switchyards, and a pump-generating plant. Primary irrigation facilities include the Feeder Canal, Banks Lake, the Main, West, east High, and East Low canals, O'Sullivan Dam, Potholes Reservoir, and Potholes Canal. There are over 300 miles of main canals, 2,000 miles of laterals, and 3,500 miles of drains and wasteways on the project. In addition to supplying water for irrigation, producing electricity, controlling floods, providing recreation, and regulating streamflow, the Columbia Basin Project also provides water for cities, industries, navigation, and endangered species.

During the summer, when cooling demand in data centers is highest. The QWRU uses almost 300 million gallons per year of Columbia Basin Project water from the irrigation canal. This is a positive practice because it substantially lowers the need to use groundwater for cooling and reduces the city's potable water demand and treatment costs because there is no need to soften the water before sending it back to the data center. Note that the irrigation canal, where the water is extracted, runs dry in winter—but this is not significant tissue, since cooling demand is comparatively lower in the winter months.

Five percent of the make-up water, on an annual basis, is still potable groundwater, which is mixed and softened before entering the cooling systems. Forging ahead, the City of Qunicy aims to purchase more water from the Bureau of Reclamation and get permission from the Quincy Irrigation District to withdraw more water from the canal because it requires minimum treatment for the data centers and could eliminate the data center's reliance on potable groundwater supplies.

THE BOTTOM LINE

Pollution can degrade water quality and make surface waters too dirty for recreational, drinking, or wildlife habitat uses. Water reuse collects and treats polluted water so that it can be used again within a community. This practice can reduce pollutant discharges, such as nutrients while increasing the reliability of community water supplies. The benefits derived from the Quincy, Washington reuse project are reduced reliance on the surface and groundwater supply, improved water supply for data center cooling operations, and fewer issues associated with the use of high mineral-content groundwater for cooling. So, water recycling helps to lower the use of potable groundwater supplies in the region while supplying cooling water of a higher quality to the Microsoft data center.

NOTES

1 Based on information in Spellman, F. (2024). *The Drinking Water Handbook*, 5th ed. (In production). Boca Raton, FL: CRC Press.
2 This account, written by Jan Schlichtmann, appeared in *USA Today*, February 4, 1999.

REFERENCES

Harr, J. (1995). *A Civil Action*. New York: Vintage Books.
Hill, M.K. (1997). *Understanding Environmental Pollution*. Cambridge, UK: Cambridge University Press.
USBR. (2023). Columbia-Pacific Northwest Region. Accessed 1/5/2024 @ https://www.usbr.gov/pn/grandcoulee/cbp/index.html.
USEPA. (2011). Ground Water Tule (GWR). Accessed 10/17/2011 @ https://water.epa.gov/lawsreg/rulesregs/sdwa/gwr/index.cfm.
USEPA. (2023). Case Studies that Demonstrate the Benefits of Water Reuse. Accessed 12/22/23 @ https://epa.gov/waterreuse/forms/contact-us--about-water-reuse-and-recycling.

9 Water Reuse
Critical Water Access

INTRODUCTION

Water reuse provides a wide range of benefits as shown in Figure 9.1. In this chapter, water reuse for improved water accessibility quality is described, and an example of real-world usage in the United States is provided.

At the present time, there is plenty of fresh water available for access by users. The keywords here are "available for access"—meaning the real keyword that should be substituted for access is location. Also, if easy access to fresh water is a given because of one's location, next in importance is the quality of the water; it does the thirsty consumer little good and a whole bunch of potential harm if he or she is drinking unhealthy ("sick water") water. On the other side of the coin is the accessibility of 'healthy' freshwater; it may be beyond reach, or not that accessible due to its location versus the location of the potential user.

At the small community level located in remote areas, it may be extremely difficult to access adequate, reliable, and safe drinking water and treatment services for wastewater. One way in which to improve water supplies and sanitation is to incorporate water reuse practices. Note that onsite water reuse is especially attractive in areas where it is impractical to build large, centralized treatment plants.

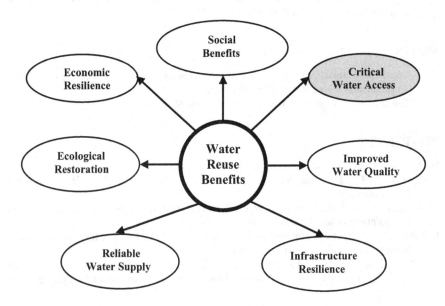

FIGURE 9.1 Water reuse: critical water access.

DOI: 10.1201/9781003498049-10

DID YOU KNOW?

A wastewater treatment plant is just like a natural stream or river in a box. What the plant 4does is to imitate nature's way, but in a much faster manner (Spellman, 2024).

CASE STUDY 9.1: LOS ANGELES COUNTY, CALIFORNIA (USEPA, 2023)

Los Angeles (LA) County, California, is not exactly a remote location. However, without water, a large part of LA County would be desert. The region is located in a semi-arid region and receives only 15″ of water per year. Water travels a long distance (about 215 miles) to serve the needs of the City of LA and locations within LA County. This water stream is fed to the County and City via the LA Aqueduct from Owens Lake (tapped via aqueduct in 1913 and 13 years later became a dry lake), Owens River, Mono Lake Basin, and reservoirs on the east slopes of the southern Sierra. In normal conditions, water is at a premium within the region and what makes matters worse is that the area has been suffering a severe drought in recent years.

Because LA County experiences extreme water stress, it has implemented numerous reuse projects focused on non-potable and indirect potable reuse and treated wastewater. The problem is that dues to increased water efficiency and lowered residential use during drought, less and less wastewater has been available for existing reuse applications. In LA County, wastewater and stormwater are conveyed through separate sewer networks. This system design feature prevents stormwater from overwhelming the sanitary sewer and wastewater treatment plants' design capacity. Note that separate sewers also prevent stormwater from being available for water use at conventional wastewater reclamation plants. Because of this the LA County Sanitation Districts (LACSD) started building stormwater-sewer diversion projects—engineered structures that bring stormwater (i.e., the volume of water following a storm that runs off or travels over the ground surface to a drainage area or channel) into the sanitary sewer (i.e., domestic wastewater) at a controlled rate.

The Los Angeles County Sanitation Districts (LACSD) owns and operates one of the largest wastewater recycling programs in the world, providing affordable, high-quality recycled water to public and private water suppliers. The provisions of Section 305 of the Wastewater Ordinance as amended in 1998 specify Sanitation Districts' policies on rainwater, groundwater, and other water discharges. Note that Section 305 specifies that no rainwater stormwater, groundwater, artesian well water, street drainage, yard drainage, water from yard fountains, pons, or lawn sprays shall be discharged to the Sanitation Districts' sewerage system, except where prior approval has been given by the Chief Engineer and General Manager (LACSD, 2023).

Since the 1960s, recycled water produced by the LACSD has augmented surface water and groundwater supplies to help meet the water supply needs of more than 5 million people within the LACSD service area. This service area includes 78 municipalities surrounding the City of LA and adjacent unincorporated areas.

Roughly 40% of the population served by the LACSD is considered disadvantaged based on income, thus recycled water affordability is a priority.

Within the LACSD service area, the wastewater generated is collected in small local sewers managed by various municipalities and subsequently conveyed to a regional system of 1,400 miles (2,250 km) of large trunk sewers and 49 pumping stations owned and operated by the LACSD. The wastewater stream flows through this sewer network to the 11 wastewater treatment plants located throughout LA County. Stormwater generated within the service area is managed as a separate system by other local municipalities, the LA County Department of Public Works, and the LA County Flood Control District, with limited involvement from the LACSD.

Note that 10 of 11 LACSD wastewater treatment plants also serve as water reclamation plants that produce around 150 million gallons per day (570 million L/day) of disinfected recycled water. For a variety of non-potable purposes, the recycled water is used at more than 900 sites for street cleaning; seawater intrusion barrier injection; irrigation for parks, schools, golf course, commercial buildings, and agriculture. It is to be observed that the recycled water is also used for indirect potable reuse at the Montebello Forebay Groundwater Recharge Project (begun in 1962, it is the oldest planned potable reuse groundwater recharge project in California), where it is injected into groundwater supplies for later drinking water use. The eleventh facility, the A. K. Warren Water Resource Facility (formerly known as the Joint Water Pollution Control Plant; see Figure 9.2), discharges 250 million gallons per day (984 million L/day) of treated wastewater effluent to the ocean. At present, after implementing a water purification project, the Warren Facility provides an additional source of water for indirect potable reuse via groundwater recharge or for another beneficial use.

The Warren Facility is not an isolated wastewater treatment plant; it is a central component of a network of seven treatment plants and over 1,200 miles of trunk sewers known as the Joint Outfall System (JOS), which provides regional wastewater treatment for LA County, covering a broad area that includes 73 cities and unincorporated county domain. The six water reclamation plants in the JOS provide an important level of treatment, churning out a recycled water that is used at hundreds of sites all over the county.

The Warren Facility (Figure 9.2) is the heart of the JOS. It is the largest facility of the system. What it does is to provide centralized processing of solids removed during wastewater treatment for all the JOS plants, producing electricity and reuseable biosolids in the process. The treated water from the Waren Facility is outfalled to the Pacific Ocean through a network of tunnels and outfall pipes that eventually extend ~2 miles off the Paos Verdes Peninsula to a depth of ~200 ft.

In operation, the Warren Facility provides both primary and secondary treatment for ~260 million gallon of water per day (mgd) and has a total permitted capacity of 400 mgd. Solids collected in primary treatment and secondary treatment are processed in anaerobic digestion tanks where bacteria break down organic matter and produce methane gas.

Anaerobic digestion is the traditional method of sludge stabilization. It involves using bacteria that thrive in the absence of oxygen and is slower than aerobic digestion but has the advantage that only a small percentage of the wastes are converted into new bacterial cells. Instead, most of the organics are converted into carbon dioxide and methane gas.

Note: In an anaerobic digester, the entrance of air should be prevented because of the potential for air to mix with the gas produced in the digester which could create an explosive mixture.

Equipment used in anaerobic digestion includes a sealed digestion tank with either a fixed or a floating cover, heating and mixing equipment, gas storage tanks, solids and supernatant withdrawal equipment, and safety equipment (e.g., vacuum relief, pressure relief, flame traps, explosion proof electrical equipment).

In operation, process residual (thickened or unthickened sludge; aka biosolids) is pumped into the sealed digester. The organic matter digests anaerobically by a two-stage process. Sugars, starches and carbohydrates are converted to volatile acids, carbon dioxide and hydrogen sulfide. The volatile acids are then converted to methane gas. This operation can occur in a single tank (single stage) or in two tanks (two stages). In a single stage system, supernatant and/or digested solids must be removed whenever flow is added. In a two-stage operation, solids and liquids from the first stage flow into the second stage each time fresh solids are added. The supernatant is withdrawn from the second stage to provide additional treatment space. Periodically, solids are withdrawn for dewatering or disposal. The methane gas produced in the process may be used for many plant activities.

Note: The primary purpose of a secondary digester is to allow for solids separation.

Various performance factors affect the operation of the anaerobic digester. For example, % Volatile Matter in raw sludge, digester temperature, mixing, volatile acids/alkalinity ratio, feed rate, % solids in raw sludge and pH are all important operational parameters that the operator must monitor.

Along with being able to recognize normal/abnormal anaerobic digester performance parameters, wastewater operators must also know and understand normal operating procedures. Normal operating procedures include sludge additions, supernatant withdrawal, sludge withdrawal, pH control, temperature control, mixing, and safety requirements. Important performance parameters are listed in Table 9.1.

TABLE 9.1
Anaerobic Digester—Sludge Parameters

Raw Sludge Solids	Impact
<4% solids	Loss of alkalinity
	Decreased sludge retention time
	Increased heating requirements
	Decreased volatile acid:alkalinity ratio
4%–8% solids	Normal operation
>8% solids	Poor mixing
	Organic overloading
	Decreased volatile acid:alkalinity ratio

SLUDGE ADDITIONS

Sludge must be pumped (in small amounts) several times each day to achieve the desired organic loading and optimum performance.

Note: Keep in mind that in fixed cover operations additions must be balanced by withdrawals. If not, structural damage occurs.

SUPERNATANT WITHDRAWAL

Supernatant withdrawal must be controlled for maximum sludge retention time. When sampling, sample all draw off points and select level with the best quality.

SLUDGE WITHDRAWAL

Digested sludge is withdrawn only when necessary—always leave at least 25% seed.

pH CONTROL

pH should be adjusted to maintain 6.8–7.2 pH by adjusting feed rate, sludge withdrawal, or alkalinity additions.

Note: The buffer capacity of an anaerobic digester is indicated by the volatile acid/alkalinity relationship. Decreases in alkalinity cause a corresponding increase in ratio.

TEMPERATURE CONTROL

If the digester is heated, the temperature must be controlled to a normal temperature range of 90°F–95°F. Never adjust the temperature by more than 1°F/day.

MIXING

If the digester is equipped with mixers, mixing should be accomplished to ensure organisms are exposed to food materials.

SAFETY

Anaerobic digesters are inherently dangerous—several catastrophic failures have been recorded. To prevent such failures, safety equipment such as pressure relief and vacuum relief valves, flame traps, condensate traps, and gas collection safety devices are installed. It is important that these critical safety devices be checked and maintained for proper operation.

Note: Because of the inherent danger involved with working inside anaerobic digesters, they are automatically classified as permit-required confined spaces. Therefore, all operations involving internal entry must be made in accordance with OSHA's confined space entry standard.

Process Control Monitoring/Testing/Troubleshooting

During operation, anaerobic digesters must be monitored and tested to ensure proper operation. Testing should be accomplished to determine supernatant pH, volatile acids, alkalinity, BOD or COD, total solids, and temperature. Sludge (in and out) should be routinely tested for % solids and % volatile matter. Normal operating parameters are listed in Table 9.2.

After digestion, the solids are dewatered at solids processing and hauled off-site to composting, land application, and landfill disposal. Methane gas generated in the anaerobic digestion process is used to produce power and digester heating steam in a total energy facility that utilizes gas turbines and waste-heat recovery steam generators. The on-site generation of electricity permits the Warren Facility (see Figure 9.2) to produce most of its electricity.

Note: Warren Facility employs odor control technology and methods to eliminate the migration of fugitive odors from the plant to the surrounding communities.

Rainwater and Stormwater Practice

As a normal practice, in accordance with the Wastewater Ordinance as amended in 1998, Section 305 specific requirements to meet rainwater and stormwater management requirements the Sanitation District requires roofing and/or grading of open

TABLE 9.2
Anaerobic Digester: Normal Operating Ranges

Parameter	Normal Range
Sludge Retention Time	
Heated	30–60 days
Unheated	180+ days
Volatile solids loading	0.04–0.1 lbs V.M/day/ft^3
Operating Temperature	
Heated	90°F–95°F
Unheated	Varies with season
Mixing	
Heated—primary	Yes
Unheated—secondary	No
% Methane in gas	60%–72%
% Carbon dioxide in gas	28%–40%
pH	6.8%–7.2
Volatile acids:alkalinity ratio	≤0.1
Volatile solids reduction	40%–60%
Moisture reduction	40%–60%

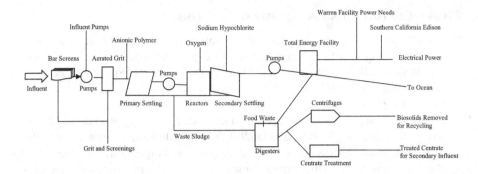

FIGURE 9.2 A.K. Warren water resource facility.

areas with exposed drains that discharge to the public sewer. This ensures the direct conveyance of rainfall, stormwater, and other runoff to the storm sewer. This is a positive practice because it protects the Sanitation Districts' sewerage system from excessive hydraulic loads that can be created by stormwater runoff. In the case where roofing and/or grading of exposed areas may be impossible or prohibited by local regulations the Sanitation Districts recognize and fully accept. Under these conditions, the Sanitation Districts, on a case-by-case basis, may accept the controlled discharge of rainwater or stormwater to the sewerage system—BTW, this occurs only after all other alternatives have been demonstrated to be impractical. Note that sufficient documentation demonstrating that no other alternatives are practical must be submitted via applications for discharge of rainwater or stormwater to the sewerage system. Other possibilities that must be considered include treatment and discharge to the storm sewer, reuse, on-site storage/evaporation and relocation of the processing or treatment areas exposed to rainwater intrusion.

TOWARD A CIRCULAR ECONOMY[1]

The LACSD stormwater-sewer diversion program contributes to the circular economy by facilitating the conversion of pollutant-laden stormwater into recycled water that can be beneficially reused, thus potentially reducing reliance on other potable water supplies in the region.

Okay, so what exactly is a circular economy? What is it all about?

Well, for the purposes of this book "Circular Economy" is defined as an economy that keeps materials, products, and services in circulation for as long as possible (USAID, 2023). The *Save Our Seas 2.0 Act* for example (addresses the increasing quantities of plastic waste in the seas) refers to an economy that uses a systems-focused tactic and involves industrial processes and economic actions that are restorative or regenerative by design, enabling resources used in such processes and activities to maintain their highest value for as long as possible, and points toward for the elimination waste through the superior design of materials, products, and systems.

DID YOU KNOW?

In the circular economy, an arbitrage opportunity entails the benefits in terms of material costs, labor, and energy that circular setups provide over linear models (WEF, 2014).v

Simply, a circular economy shrinks material use, redesigns materials, products, services, and processes to be less resource intensive, and recaptures "waste" as a resource to manufacture new materials and products and to provide clean, safe reuseable water.

What we are talking about here is a paradigm shift whereby a change to the model in which resources are mined, made into products, and then become waste. Let us state this one again, differently, to enhance understanding of a circular economy— the goal of the circular economy is to transition from today's take-make-waste linear pattern of production (see Figure 9.3) and consumption to a circular system in which the societal value of products, materials, and resources is maximized over time. However, make a note of the fact that circularity in and of itself does not ensure social, economic and environmental performance (i.e., sustainability). Sustainability of circular economy strategies needs to be evaluated against their linear (direct) counterparts to identify and avoid strategies that increase circularity yet lead to unintended outwardness. While it is true that the proliferation of circularity metrics has received considerable attention, at the present time, there is no critical review of the methods and combinations of methods that underlie those metrics and that specifically quantify sustainability impacts of circular strategies (EERE, 2023).

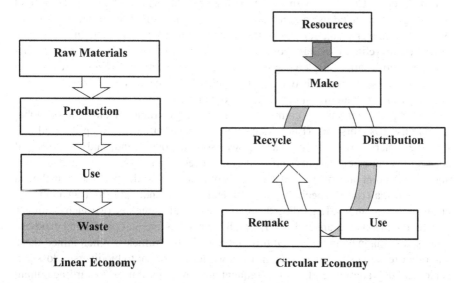

FIGURE 9.3   Linear and circular economy. (Adaptation from C. Westman (2016). Accessed 10/27/23 @ https://commons.wikimedia.org/wiki/file:linear.com/.)

DID YOU KNOW?

End-of-use refers to materials/products at the end of their primary use, that are collected and returned to the same usage, or cascaded to a new one (WEF, 2014).

DID YOU KNOW?

It is important to note that the circular economy is not the linear economy that most of us are familiar with or have been living with and has become used to many without even knowing or understanding the approach that has been practiced since the Industrial Revolution in the 18th century. The linear economy involves taking raw materials from the environment and turning them into finished product. Then the finished products are used and then discarded into the environment. And in many cases the old "I do not want this product anymore so just throw it into the river, lake, ocean or landfill—out of sight out of mind syndrome." The problem with this linear economy should be rather obvious because this system has a beginning and an end where limited raw material, critical minerals run out in the long run. So, what we end up with in this linear economy is a pile of waste, growing by the minute, bring about more waste-disposal issues and expenses.

What the shift from a linear to a circular economy requires is for us to make products differently. Considering this shift the life of materials must be extended beyond their original intended use and not subject to being landfilled when damaged, worn out, and no longer viable for any further use. Note that this is more than redoubling efforts on recycling. Circular economy strategies that reduce the use of raw materials can significantly cut global greenhouse gas emissions and relieve pressure on raw materials such as metals (Circularity Gap, 2020). The idea is to strive to reach a pathway to zero economy and to sustain economic growth.

Shifting to green energy complemented by energy efficiency cut 55% of global greenhouse gas emissions (Ellen MacArthur Foundation, 2019). Keep in mind that for green energy to be a truly clean power source, solar, wind, and batteries and accessories must be fashioned, utilized, and retired in a net zero, safe, and sustainable way. A point of concern is that at the current rates of production, the annual need for critical materials, especially critical metals, could endanger the transition to a clean energy economy. Regarding the reference to critical metals, two such examples include the metals indium and silver, which are needed for solar panels and neodymium used in magnets in both wind turbines and electric vehicles. When mined these and other resources produce contaminated waste, contribute to biodiversity loss and ecological degradation, and involve frequent human rights abuses. Regarding human rights abuses and their connection to mining activities they include increased road traffic, access to water and impacts of poor-quality water, increases in noise and air pollution, child labor involved in the global mining for lithium, cobalt, manganese, copper, nickel, and zinc—all used in green energy technologies like wind turbines, electric vehicles, and solar panels.

The other fly in the ointment is that a significant amount of green energy equipment is approaching the end of its useful life; the quantity of obsolete green energy equipment is expected to grow exponentially over the next three decades. For example, consider that in 2019 the world generated 18,000 tons of solar photovoltaic (PV) panel waste. It is estimated that by 2050, PV panel waste could increase to 10 million tons annually (BNEF, 2020). Wind Europe (2019) estimates that around 14,000 wind turbine blades could be mothballed by 2023, equivalent to between 40,000 and 60,000 tons. In addition there is a concern with and over what we call and classify as e-waste (electronic waste) or separately, given concerns about unsafe handling of waste that results in harm to human health and the environment. By returning this equipment to be reused or recycled, circular economy strategies both keep waste out of landfills and moderate demand for the materials that make energy equipment to begin with.

According to Accenture (2015) research, the circular economy could generate trillions of dollars in economic output by 2030; moreover, circular business models were identified that will help uncouple economic growth and natural resource consumption while driving competitiveness. The circular economy focuses on reuse, repair, remanufacturing, and sharing other significant innovation opportunities and jobs. By 2050, more sustainable use of materials and energy could add more than $1 trillion annually to the global economy.

DID YOU KNOW?

Reshaping the Green Energy Industry into a more circular economy is a paradigm shift that has significant potential to reduce waste and carbon emissions while extending the supply of valuable resources. Moreover, a circular economy also cultivates new business models, innovations, initiatives, sustainable practices, policies, and markets.

LINEAR VERSUS CIRCULAR ECONOMY
The Transition: Linear to Circular Economy

Take + make + dispose − Linear economy

Returning byproducts + circular supplies + waste as a resource

= Circular economy

Repairing, refurbishing, and reusing = Circular economy

In a linear economy where tons of materials are extracted and processed, contributing to at least half of global carbon dioxide emissions, and the resulting waste—including plastics, textiles, food, electronics, and more—is taking its toll on the environment and human health. On the other hand, and in contrast to a linear economy, a circular economy basically circulates materials and products instead of producing new ones (see Figure 9.3). Moreover, whereas the linear economy disposes of waste in landfills at the end of its use (or when we do not it anymore), the circular economy creates several different opportunities for return cycles or what are known as loops that avoid disposal or landfilling. As mentioned, this closed-loop system minimizes the use of resource inputs and the creation of waste, pollution, and carbon emissions by keeping materials, products, equipment, and infrastructure in use for longer periods, thus improving the productivity of these resources.

DID YOU KNOW?

According to USAID (2023) in a circular economy, recycling becomes the last resort, not the first and only option.

THE BOTTOM LINE

"Waste as a resource," is the mantra of any water and waste reuse project. Case Study 9.1 highlights a mature program that has been accepting stormwater diversion to its sanitary sewer system for decades. The LACSD program demonstrates that the diversion of stormwater runoff into the sanitary sewer system can be an effective way to reduce pollutant loads and increase water reuse.

NOTE

1 Based on Spellman, F.R. (2024). *The Science of Green Energy* (in production). Boca Raton, FL: CRC Press.

REFERENCES

Accenture. (2015). Waste to Wealth: Creating Advantage in Circular Economy. Accessed 10/3/23 @ _acnmedia/https://www.accenture.com/Accenture/conversion-assets/dotcom/document/global/pdf/strategy_7/Accenture-wate-wealth-infographic.pdf.
BNEF. (2020). The Afterlife of Solar Panels. Bloomberg New Energy Finance. Accessed 10/3/23 @ https://www.bnef.com/insights/24259.
Circularity Gap Reporting Initiative. (2020). Circularity Gap Report. Accessed 10/3/23 @ https://assets.web.files.com.
EERE. (2023). *The Circular Economy*. Washington, DC: Energy Efficiency & Renewable Energy.
Ellen MacArthur Foundation. (2019). Completing the Picture: How the Circular Economy Tackles Climate Change. Accessed 10/3/23 @ https://emg.thirdlight.com/link/2j2gtyton7ia.

LACSD. (2023). Rainwater, Stormwater, Groundwater and Other Water Discharges. Accessed 1/6/24 @ https://www.lacsd.org/services/wastewater-programs-permits/industrial-waste-pretreatment-program/.

Spellman, F.R. (2024). *The Drinking Water Handbook*, 5th ed. (In production). Boca Raton, FL: CRC Press

USAID. (2023). Clean Energy and the Circular Economy: Opportunities for Increasing the Sustainability of Renewable Energy Value Chains. Accessed 10/29/23 @ usaid.gov/energy/sure.

USEPA. (2023). Case Studies that Demonstrate the Benefits of Water Reuse. Accessed 12/22/23 @ https://epa.gov/waterreuse/forms/contact-us--about-water-reuse-and-recycling.

WEF. (2014). Towards the Circular Economy: Accelerating the Scale-Up Across Global Supply Chains. Accessed 10/31/23 @ www.weformum.org/agenda/2022/06/what-is-a-cicular-economy.

Weetman, C. (2016). *A Linear Economy Handbook*, 2nd ed. London: Koger Page.

Wind Europe (2019). European Wind Study Calls for Landfall Base. Brussels, Belgium: Wind Energy News.

Part 2

Toilet to Tap and Nonpotable Reuse

Waste Not, Want Not: Water Reuse

10 Direct Potable Reuse

INTRODUCTION

Direct potable reuse involves the treatment and distribution of water without an environmental buffer. What is involved in direct potable reuse is the introduction of highly treated reclaimed water either directly into the potable water supply distribution system or downstream of a water treatment plant or into the raw water supply immediately upstream of the water treatment plant. For absolute clarity, it is important to understand that direct potable reuse (as defined and described in this book) is different than indirect potable water reuse in that the latter is the planned incorporation of reclaimed water into a raw water supply, such as in potable water storage reservoir or a groundwater aquifer (discussed in detail in Chapter 11), resulting in mixing and assimilation, thus providing an environmental buffer—"buffer" being the keyword in this usage. The EPA defines an environmental buffer as a natural or engineered system that provides a barrier between the treated wastewater and the point of reuse. It is designed to protect public health by providing a physical separation between the treated wastewater and the point of use (USEPA, 2023). The key to the implementation of direct potable reuse, toilet to tap, is communication. Not only is it important to inform the public and all ratepayers about what is in the water but also what is not in the water.

DID YOU KNOW?

The World Health Organization for Potable Water Reuse recommends a $9.5 \log_{10}$ reduction of viruses and $8.5 \log_{10}$ of enteric bacteria and protozoa, based on a disability-adjust life-year risk approach (WHO, 2017).

RECYCLED WATER FOR DRINKING (USEPA, 2019, 2023)

Case studies about selected direct water reuse activities and water reuse projects based on those provided by USEPA in its *2017 Potable Reuse Compendium* (2019) are presented in this chapter including an ongoing pilot study and actual direct potable reuse projects that are in operation. This discussion begins with the City of Altamonte Springs-Florida (northern suburbs of Orlando-Kissimmee-Sanford Metropolitan area) which has an interesting direct potable reuse (DPR) pilot program and likely a permanent reuse operation for the future.

DOI: 10.1201/9781003498049-12

CASE STUDY 10.1: CITY OF ALTAMONTE SPRINGS, FLORIDA

The City of Altamonte Springs, a population of ~47,000, proactively created pure-Alta® to address their community's future water needs and diversify the City's water portfolio. What goes on is that cutting-edge technology is used to purify reclaimed water to drinking water standards. This makes sense and is practical because of population increases and dwindling Floridian aquifer levels. Truth be told, experts have been predicting for years that the state will not have enough groundwater to satisfy the public's drinking water needs.

So, the City put together a pilot study using the advanced treatment process including the following unit processes in its treatment train (see Figure 10.1): ozonation and biological activated carbon filtration (03/BAF), ultrafiltration (UF), granular activated carbon filtration (GAC), and ultraviolet light with advanced oxidization process (UV AOP) all coupled with advanced system monitoring techniques. Note that the source flow for this process comes directly from the effluent train of the City of Altamont Springs WWTP/DPR.

Findings and Results

The resulting purified water is tested to ensure it meets drinking water standards and removes nonregulated pharmaceuticals and personal care products (PPCPs; see Chapter 8). Approximately 29,000 gallons per day (gpd) will be treated in the potable reuse pilot project. Note that this is <1% of the total water currently produced in the City (6 MGD). If the pilot study is successful, they might build a full-scale treatment system with a capacity of 300,000–500,000 gpd, which is ~5% of the City's future water demand, 9 MGD. This far-thinking plan may provide a purified water supply that supplements the City's drinking water system.

Okay, what is happening currently is that the pilot project is operating in a testing phase. What happens is that the purified water is blended with reclaimed water from the Water Reclamation facility and beneficially reused for irrigation in the City's existing urban reclaimed water system. Based on the success of the pilot, the City might build a full-scaled treatment system in the future to produce purified water to supplement the City's drinking water system by up to 5%.

CASE STUDY 10.2: ORANGE COUNTY'S GROUNDWATER REPLENISHMENT MODEL

Orange County in southern California holds the distinction, as a result of its groundwater replenishment system, of being the world's largest water recycler of its kind. What it does is treat water to a purified state that is of near-distilled quality. Then, the clean water stream is piped to a location where it naturally seeps into a groundwater basin (a buffer) that provides 60% of the potable water needs to 2.4 million residents.

Orange County is a semi-arid region that receives on average 13 inches (330 mm) of rain a year. The population of more than 3 million is projected to grow by more than 10% by 2035; thus a limiting factor comes into play.

A limiting factor?

Yes.

Simply, drinking water availability.

We cannot expect an area's population to grow if drinking water is not available.

Anyway, Orange County has a large groundwater basin that provides 60% of the potable water needs of 2.4 million residents in north and central Orange County. Note that water from the basin is also injected into a barrier to prevent saltwater intrusion from the sea.

In the past, the Santa Ana River was the source of water for replenishing the basin but increasingly unreliable flows meant that the Orange County Water District for forced to choose an expensive option by importing water from other rivers to replenish the basin. By the mid-1990s, demand had increased and there were continued problems with seawater intrusion. Be that as it may, the county's increasing volume of wastewater had become a disposal problem for the Sanitation District. So, the two agencies saw the opportunity to use some of the wastewater to replenish the groundwater basin.

In operation, treated wastewater (sewer water) is purified to drinking water quality standards using the three-step process consisting of microfiltration, reverse osmosis, and ultraviolet light (UV) with hydrogen peroxide (see Figure 10.1). The treated water is stored in the Orange County groundwater basin. Note that half of it is pumped into a string of wells to form a hydraulic barrier that prevents saltwater intrusion from contaminating the county's groundwater supplies. The other half is pumped 21 km through the cities of Fountain Vally, Santa Ana, Orange, and Anaheim, to recharge basins where, as a preventive measure, it is blended with other water in a ratio of 75:25, before it seeps underground. Bear in mind that the quality of the purified water exceeds all state and federal drinking water standards. The system can purify up to 70 million gallons (265 L) of water a day—this is enough to meet the needs of nearly 600,000 residents.

The Orange County program to replenish groundwater is ongoing. In the planning stage, the planners set 2015 as the date whereby the capacity was to be increased to 100 million gallons (378 ML) a day, with the ultimate target of expanding to 130 million gallons (492 ML). Orange County is a role model for groundwater replenishment.

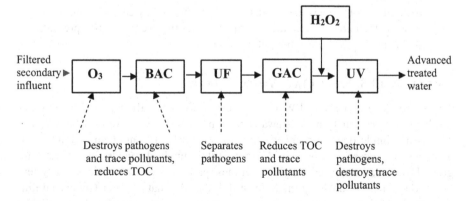

FIGURE 10.1 Treatment train for the pureAlta potable reuse demonstration facility and then returned to the effluent discharge and released. Again, this is only a pilot project.

The system recycles 35% of the Sanitation District's wastewater and contributes about 20% of the water that refills the basin.

With regard to public acceptance of the County's groundwater replenishment system, the far-thinking County administrators employed a creative and proactive approach designed to secure support for the project from USEPA (2023):

- local, state, and federal elected officials
- business and civic leaders
- environmental advocates
- regulatory agencies
- health experts
- media
- (most importantly) the general public—the yuck factor has to be eradicated, or at least lessened

The County's struggle against eliminating and/or reducing the public's perception of the Yuck Factor was eased by establishing objectives including:

1. secure positive media impressions (if you do not have the media on your side, you are dead in the water, so to speak)
2. be prepared to address significant opposition (respond to opposition with the truth and grace—remain calm and no outrage
3. educate people to overcome the negative "toilet-to-tap yuck factor" perception of recycling wastewater
4. start the outreach program ahead of time, prior to the project's start-up, and continue it throughout the project's life to maintain support for future expansions—maintaining support cannot be overstated along with consistency and persistence
5. create a positive perception of recycling wastewater to increase support for indirect and direct potable reuse
6. offer facility tours
7. the real bottom line is to be transparent—transparency can be advanced by offering the public virtual tours of the facilities involved; an active presentation is the key to getting the public on board.

CASE STUDY 10.3: SANTA CLARA VALLEY, CALIFORNIA

The Silicon Valley Advanced Water Purification Center is all about recycling water for drinking; and purifying wastewater for replenishing groundwater in the Valley. This is no simple operation; meeting the water needs of Santa Clara County residents and businesses is complex. Current water supply demands rely on rainfall, groundwater, and imported supplies that must pass through the sensitive ecosystem of the Sacramento-San Joaquin Delta. It did not take that long for County administrators and others to recognize that new approaches to supplying clean, reliable, and safe water supplies had to be found. The problem is that these sources simply are not sufficient to overcome or to balance with climate change, environmental restrictions/regulations, natural disasters (earthquakes, forest fires, and so forth), and population growth—increasing the uncertainty of future water supplies.

Santa Clara County is taking a proactive approach to supplying customers with quality water. The keyword in the County's approach is diversify. In order to meet their objectives the County's goal is to provide water in the long run by using the most appropriate water purification technologies. Complicating the issue is that more than 55% of the water consumed in the Santa Clara Valley is imported from surrounding watersheds and stored in underground aquifers. And informally we can say that this would be all and good except that there has been a reduction in rainfall over the past decade, and very little local water is flowing into the District's reservoirs and groundwater basin. As a result of the California statewide water shortage, and a severe reduction in water available from both federal and state water projects, the Water District has been forced to use its imported water for drinking water, conveying it directly to its drinking water treatment plants (purifying it to ensure safe quality water), instead of storing it underground for future use. Truth be told drawing water from underground is no longer sustainable and another source of water to replenish the groundwater basin is needed. So, the far-thinkers, scientists, technologists, and public administrators are looking at purified water as an option.

So, What Is the Plan?

In 2014, the Water District set out to improve the quality of its recycled wastewater by designing the Silicon Valley Advanced Water Purification Center which produces up to 30 million liters (ML) of purified cycled water a day. The key to the District's plan is to employ advanced water purification processes. At the local level, some of the treated water at the wastewater treatment plant in north San Jose is to be further purified in the new Silicon Valley Advanced Water Purification Center. The facility includes three purification processes (as shown in Figure 10.2): microfiltration, reverse osmosis, and ultraviolet light disinfection. The microfiltration removes particles down to 0.1 µm in size, including all bacteria. In reverse osmosis (RO) solutions of differing ion concentrations are separated by a semipermeable membrane. Typically, water flows from the chamber with a lesser ion concentration into the chamber with a greater ion concentration, resulting in hydrostatic or osmotic pressure. In RO, enough external pressure is applied to overcome this hydrostatic pressure, thus reversing the flow of water. This results in the water on the other side of the membrane becoming depleted in ions and demineralized. Ultraviolet (UV) disinfection systems are used to further purify the water.

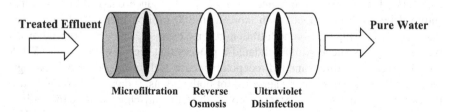

FIGURE 10.2   Silicon Valley advanced water purification treatment train.

The current plan includes investigating the possibility of using this purified recycled wastewater to replenish its groundwater basins. This project is currently in the pre-feasibility phase, and locations for pipelines are being examined. Putting into practice a groundwater replenishment scheme will also create the basis and structure for potential indirect potable reuse. A recycled water and infrastructure master plan was developed by the Water District, which will incorporate individual plans by each of the recycled water producers. Through the employment of recycling and conservation practices the long-term goal of the Water District is to save almost 150 billion liters of water a year by 2030. California's State Water Quality Control Board is working on the regulatory framework for Direct Potable Reuse schemes.

The Santa Clara Valley Water District's groundwater replenishment system gives the growing population of Orange County a reliable source of safe, clean water which reduces the region's dependence on imported water. Moreover, the District insists that the purified water is of the highest quality. The purified water is injected into the basin, not the tap.

CASE STUDY 10.4: SAN DIEGO, CA: RECYCLED WATER FOR DRINKING

By 2035, the City of San Diego's phased, multiyear program called Pure Water San Diego will provide nearly half of San Diego's water supply. This long-range plan has received a momentum push after a successful 5-year wastewater purification trial. The city is currently exploring options of storing the purified water in an existing reservoir or distributing it directly to the city's 1.3 million residents. The proposal, when completed, is expected to supply one-third of the city's water needs.

The Whys and Wherefores

Aqueducts from the California Bay Delta and the Colorado River convey most of the imported water (more than 85%) to the San Deigo region. The problem is that the region's reliance on imported water leaves the City of San Diego's water supply vulnerable to drought, competing demands, and rising costs of imported water. Thus, to avoid outside conveyance systems the region's planners are looking to go local via reuse schemes—instead of outfalling treated wastewater to the ocean, purify it for consumptive uses. Pure Water San Diego' is the city's program to develop this local source of drinking water to reduce its reliance on imported water; keep up with population growth; and combat water supply challenges such as recurring drought.

The Plan in a Nutshell

The City's long-term goal, targeted for 2035, is to produce one-third of the City's future drinking water supply with more than 300 million liters (ML) of purified water per day. Using membrane filtration, reverse osmosis, and UV advanced oxidation the City has successfully validated a process that recycles wastewater. While San Diego explores direct and indirect potable reuse, California is developing a regulatory framework and studying approval for both options. If during its exploration of the two options and pending State approval, the City may choose to use the indirect reuse option. If this option is chosen, the purified water would be conveyed 37 km (23 miles) to the San Vicente Reservoir where it would be blended with imported water supplies in the reservoir before going to a standard drinking water treatment plant. The Reservoir was created by the San Vicente Dam in San Diego County by impounding the waters of San Vicente Creek, and the Colorado River via the First Sanio Aqueduct branch of the Colorado River Aqueduct from Lake Havasu. At present, the Reservoir has a storage capacity of ~250,000 acre-feet. Along with the storage of fresh water, the reservoir is a popular place for recreational activities including boating, waterskiing, waterboarding, and fishing. The Reservoir is currently under construction to increase the storage capacity of the Reservoir. When commissioned it will produce about 315 ML of purified water per day for the city's 1.3 million residents.

Pure Water San Diego has a multi-year program that will provide nearly half of Sandiego's water supply locally by the end of 2035. The far-thinkers at Pure Water San Diego understood the residents' so-called yuck factor involved with the toilet-to-tap process. To assuage the feelings of residents Pure Water San Diego formed a workgroup, the Pure Water Working Group to capture diverse opinions/viewpoints via its public outreach program. The Group put together informational materials and events; tours of the facility; email updates; website content; presentations at the city council meetings and community meetings; press releases for newspaper, radio, and TV; and bog posts.

The real bottom line is that the far-thinkers fully understood that if they could not persuade the public to buy into the program, to overcome the 'yuck factor' the whole concept would be dead in the water—pun intended (City of San Diego, 2023).

SIDEBAR 10.1: JUST LIKE A BEAVER (*CASTOR* SPP.)

Natural infrastructure in dryland streams (NIDS) in a watershed is a climate-smart practice. In arid regions, it is not only important to reuse water but also to capture water in streams so that the streambed can act like a sponge, absorb the water, and percolate it to aquifers—to groundwater supplies. Basically, NIDS is the cloning of Beaver Dams (aka "Beaver Mimicry" or 'Simulated Beaver Structures').

Why?

Well, beavers are an important species in many riparian ecosystems. Beaver dams, along with allowing the water sponge effect in restocking aquifers, also control sediment movement, create wildlife habitat, and control secondary flows and erosion. Thus, beaver mimicry is a fast-growing conservation technique promoted as a low-cost, nature-based strategy with many potential socio-environmental benefits, which Goldfarb (2018) points out has led to an emerging popular culture of 'beaver believers." Actually, there are a number of beaver- and human-made NIDS including:

Note: The practice of straightening streams has changed the impact of NIDS because straightening a stream allows the water to flow freely (unobstructed), quickly, restricted and causes the streambed to deepen in what are called 'incised channels' which have no floodplain (floodplains are desirable because they dissipate flood energies across a wide area, reduce flood peaks as flood flows spread out; Zeedyk, 2009). All of these factors contribute to a waste of the flow with regard to storing the water in groundwater aquifers.

- **Beaver dams**: via their dam-building activities and subsequent water storage, beavers have the potential to restore riparian ecosystems by modulating streamflow (Goldfarb, 2018). (**Note**: Modulating streamflow can be accomplished by restoring secondary streams). They do this by piecing together a structure (dam) perpendicularly in the channel, made of logs, branches, leaves, twigs, sticks, bark, rock grass, leaves, and mud. Observation reveals that beavers often build multiple dams in sequence up or downstream (Fairfax and Whittle, 2020; Wohl, 2021).
- **Beaver dam analogs (BDAs; aka beaver mimicry and simulated beaver structures)**: mimicking Mother Nature, these human-made structures, situated perpendicularly in the channel made of large wood and other materials and constructed in a manner that deliberately mimics the form and function of a naturally occurring beaver dam (Pollock et al., 2018; Silverman et al., 2019; Vanderhoof and Burt, 2018).
- **Check dams**: are relatively small, temporary human-made structures built perpendicularly across a swale or channel, composed of stacking loose rocks ~1 m high, but varying in height and length, depending on channel dimensions (Norman and Niraula, 2016). They are typically constructed out of gravel, rock, sandbags, logs, treated lumber, or straw bales. They are used to slow the velocity of concentrated water flows, a practice that works to reduce erosion. In operation, as stormwater runoff flows through the structure, the check dam catches sediment from the channel itself or the contributing drainage area. Check dams are used where it is impractical to implement other flow-control practices (USEPA, 1993).
- **Gabions**: are human-made chicken wire cages, cylinders, or boxes filled with rocks, situated perpendicularly in the channel (Norman et al., 2010).
- **Leaky weirs**: human-made type of structure constructed of rock cemented into place in areas of exposed bedrock within the channel that allows for water to leak through slowly (Coy et al., 2021).
- **One-rock-dams**: human-made structures, situated perpendicularly in the channel, constructed with a layer of rock on the bed and exactly "one-rock" high but varying in length, depending on channel dimensions (Zeedyk, 2009).
- **Trincheras (Spanish for trench)**: human-made structures, situated on hillslopes perpendicular to downslope flow constructed by one or two more layers of rock (Fish et al., 2013). The term refers to a cultural pattern that existed in northern Sonora, Mexico, and the southernmost portion of the Southwest from about A.D. 800-A.D. 1450 (USNPS, 2018).

When it comes to establishing regenerative water sinks for NIDS, several human-made structures can be built and used in side-stream channels to help store water—the smart move is to mimic the beaver—we can say without question that beavers are ecological engineers (Hammerson, 1994).

DID YOU KNOW?

A thought that might not have occurred to most people as we look at a drinking glass of water is, "Has someone tasted this same water before us?" Certainly, absolutely, without a doubt, for certain. Remember, water is almost a finite entity. What we have now is what we have had in the past. We are drinking the same water consumed by Cleopatra, Aristotle, da Vinci, Napoleon, Joan of Arc, Stonewall Jackson, General Patton (and several billion other folks who preceded us)—because water is dynamic (never at rest), and because water constantly cycles and recycles, as discussed in this book.

CASE STUDY 10.5: EL PASO, TEXAS RECYCLED WATER FOR DRINKING

When you live in the Chihuahuan desert, or any other desert or dryland environment for that matter, water is a limiting factor. Actually, living anywhere in the United States' Southwestern States water can be or is a limiting factor. The population in the Southwest is among the fastest growing in the United States. Before the increased arrival of new residents, the region could be described as vast, open, and undeveloped, with isolated population clusters where water was readily available. However, in the past few decades the populated areas have expanded becoming large metropolitan areas (e.g., El Paso, Texas) and development has not always occurred where water is readily available. Along with water being the limiting factor in the desert Southwest, federal, state, city, and county land is extensive and is also a limiting factor to growth.

Limiting factors pretty much sum up the definition of deserts. El Paso is located in the Chihuahuan desert. The city secures its water supply from groundwater and the Rio Grande River. The problem is water from the Rio Grande is only available during spring, summer, and early autumn and is further limited in dry years. Extreme drought conditions over many years have shown a drying trend which has continuously reduced river flows, leaving less water available for the city.

So, what is the City to do? What has the City done?

The greater metropolitan area of El Paso is home to one of the first wastewater plants to treat the wastewater to drinking water standards. El Paso Water Utilities (EPWU) has met the challenges that come with living in a desert city by diversifying its water supply—water reuse is an important part of the water portfolio.

El Paso Water Utilities Department (EPWU) controls the water systems that convey nearly 90% of all municipal water to more than 800,000 residents of El Paso County (USEPA, 2023). For its potable supply, EPWU uses groundwater and surface water, producing ~34 billion gallons a year of potable water for its customers. At the Fred Hervey Water Reclamation Plant (a model and center of learning for cities facing diminishing supplies of fresh water), EPWU operates an Indirect Potable Reuse

(IPR) facility. This plant treats wastewater to drinking water standards. The treated water is buffered via its injection through a series of wells and infiltration basins to replenish the Hueco Bolson Aquifer. EPWU also treats wastewater at other facilities and discharges it into the Rio Grande.

EPWU intends to send some of the treated water directly to a proposed Advanced Water Purification Facility. This facility will turn the treated water into drinking water and put it directly into the distribution system. This practice is important because it will provide a new source of drinking water to augment the water supply—direct potable reuse.

CASE STUDY 10.6: WICHITA FALLS, TEXAS: A TEMPORARY SOLUTION

Due to a drought-induced water crisis the city of Wichita Falls, Texas, in 2014, as a temporary solution to its water problem became one of the first in the United States to use treated wastewater directly in its drinking water supply. The City's traditional water supply was solely obtained from Lake Chickapoo and Lake Arrowhead. No other water sources including groundwater are available within ~130 km (~81 miles). In the late 1990s, the city experienced a severe multi-year drought, driving the decision to add a reverse osmosis (RO) plant (completed in 2008) to the existing Cypress Water Treatment Facility to treat the brackish water from a third lake—Lake Kemp.

Again in 2010, the area experienced severe drought which, coupled with extreme temperatures of over 38°C (100°F) for more than 100 days at a time, caused reservoir water levels to drop. The City reached a state of emergency in November of 2013 when the water shortage escalated, on a scale of 1-lowest to 5-highest, to a stage-4 drought disaster, lowering production by about 65 million liters (17,171,183 U.S. Gallons) per day.

After evaluating the crisis, the city recognized that it was conveying 26 million liters of wastewater a day from its wastewater treatment plant to other cities downstream and that this treated wastewater could instead be further treated locally at the existing Cypress Water Treatment Plant and used to augment the public drinking water supply.

So, as described earlier in Chapter 1, what the City is really doing here is employing human intervention in the natural water cycle, generating an artificial water cycle or *urban water cycle* (local sub-systems of the water cycle—an integrated water cycle; see Figure 1.1). Although many communities withdraw groundwater for public supply, the majority rely on surface sources. After treatment, water is distributed to households and industries. Water that is wasted (wastewater) is collected in a sewer system and transported to a treatment plant for processing prior to disposal. Current processing technologies provide only partial recovery of the original water quality. The upstream community (the first water user, shown in Figure 1.1) is able to achieve additional quality improvement by dilution into a surface water body and natural purification. However, the next community downstream is likely to withdraw the water for a drinking water supply before complete restoration. This practice is intensified and further complicated as existing communities continue to grow, and new communities spring up along the same watercourse. Obviously, increases in the number of users bring additional need for increased quantities of water.

This withdrawal and return process by successive communities in a river basin results in *indirect water reuse* (use of used toilet water via de facto water recycling).

In order to augment purified wastewater with the City's water supply the treated wastewater is disinfected and pumped to the Cypress Water Treatment Plant where it goes through microfiltration and reverse osmosis before being released into a holding lagoon where it is blended with lake water (50:50). The blended water goes through an eight-step conventional surface water treatment process. The treated water is stored and then pumped to the distribution system. The operation provides 19 million liters a day, satisfying one-third of the city's daily demand. Wichita Falls has a population of about 160,000. The practice is considered a temporary drought response and will be replaced by a permanent indicted portable reuse process whereby high-quality effluent will be stored in Lake Arrowhead. The permanent process will recycle up to 60 million liters (15,850,323 gallons) a day and is scheduled to be completed in a 3-to-5-year period.

THE BOTTOM LINE

In order to accomplish successful augmentation of purified wastewater with normal surface water sources, the community had to be on board with the concept. To accomplish this City planners, politicians and others engaged the community early and were able to speed up the process. A public information officer and staff were hired to create an aggressive public education campaign to inform customers about the water supply situation. It has been said that a picture is worth many thousands of words and the fact that many residents could visually see the water levels in their reservoirs fall helped them to understand the urgency for using alternative water supplies. Convincing rate-payers and others of the need to augment the City's water supply with purified wastewater came down to one winning practice: Communication, communication, communication—further amplified and redirected to "Waste Not, Want Not."

REFERENCES

City of San Diego. (2023). Pure Water San Diego. Accessed 1/14/24 @ https://www.sandiego.gov/sites/default/files/2023.

Coy, H., Wilson, N.R., Bennet, A., Hsieh, D., and Horman, L.M. (2021). Hydrologic Data Collected at Leaky Weirs, Cienega Ranch, Wilcox, AZ (March 2019-Ocotber 2020). Washington, DC: U.S. Geological Survey Data Release.

Fairfax, E., and Whittle, A. (2020). Smokey the beaver: Beaver-dammed riparian corridors stay green during wildfire throughout the western United States. *Ecological Applications* 30(8):e02225.

Fish, S.K., Fish, P.R., Varineau, R., and Villalpando, E.P. (2013). Fight: Adriel Heisey's Images of Trincheras Archeology [WWW Document]. An Exhibition of Arizona State Museum and the Mexican National Institute of Anthropology and History. Accessed 1/17/24 @ https://www.statemuseum.arizona.edu/exhibits/heisey/index.shtml.

Goldfarb, B. (2018). *Let the Rodent Do the Work: Reflections of a Beaver Believer*. New Haven, CT: Yale School of the Environment.

Hammerson, G.A. (1994). Beaver (Casto Canadensis): Ecosystem alteration, management, and monitoring. *Natural Areas Journal* 14:44–57.

Norman, L.M., Levick, L.R., Guertin, D.P., Callegary, J.B., Quintanar Guadarrama, J., Zulena
    Gil Anaya, C., Prichard, A., Gray, F., Castellanos, E., Tepezano, E., P., Rodriguez, S.,
    Nunez, J. Atwood, D., Patricio Olivero Granillo, G., and Octavio Gastelum Ceballos,
    F. (2010). Nogales Flood Detention Study. Washington, DC: U.S. Geological Survey
    Open-File Report 2010-1262, 112.
Norman, L.M., and Niraula, R. (2016). Model analysis of check dam impacts on long-term sed-
    iment and water budgets in Southeast Arizona, USA. *Ecohydrology and Hydrobiology*
    16:125–137.
Pollock, M.M., Lewallen, G., Woodruff, K., Jordan, C., and Castro, J. M. (Eds.) (2018). *The
    Beaver Restoration Guidebook: Working with Beaver to Restore Streams, Wetlands, and
    Floodplains*. Portland, OR: United States Fish and Wildlife Service.
Silverman, N., Allred, B.W., Donelly, J.P., Chapman, T.B., Maestas, J.D., Wheaton, J.M.,
    White, J., and Naugle, D.E. (2019). Low-tech riparian and wet meadow restoration
    increases vegetation productivity and resilience across semiarid rangelands: Low-tech
    restoration increases vegetation productivity. *Restoration Ecology* 27:269–278.
USEPA. (1993). Guidance Specifying Management Measure for Sources of Nonpoint Pollution
    in Coastal Waters. EPA 840-B-92-002. Washington, DC: US. Environmental Protection
    Agency, Office of Water.
USEPA. (2019). 2017 Potable Reuse Compendium. Accessed 1/9/24 @ https://www.epa.gov/
    ground-water-and-drinking-water/2017-potable-reuse-compendium.
USEPA. (2023). Case Studies that Demonstrate the Benefits of Water Reuse. Accessed 12/22/23
    @ https://epa.gov/waterreuse/forms/contact-us--about-water-reuse-and-recycling.
USNPS. (2018). Trincheras. United States National Park Service. Accessed 1/16/24 @ https://
    www.nps.gov/subjects/swscience/trincheras.htm.
Vanderhoof, M., and Burt, C. (2018). Applying high-resolution imagery to evaluate restora-
    tion-induced changes in stream condition, Missouri River Headwaters Basin, Montana,
    Reinduced changes in stream conditions, Missouri River Headwaters Basin, Montana.
    Remote Sensing 10(6):913.
WHO. (2017). Potable Reuse: Guidance for Producing Safe Drinkng Water. Geneva: World
    Health Organization. Accessed 1/9/24 @ https://www.who.int/water_sanitation_health/
    publications/potable-reuse-guidelines/en/.
Wohl, E. (2021). Legacy effects of beavers in the continental United States. *Environmental
    Research Letters* 16:025010.
Zeedyk, B. (2009). An Introduction to Induced Meandering: A Method for Restoring Stability
    to Incised Stream Channels. Accessed 1/17/24 @ https://quiviracoaltion.org/images/
    pdfs/1905-An-introduction_to_induced_meandering.pdf.

11 Indirect Water Reuse

THE SCHEME

Indirect water reuse has been defined as potable reuse by incorporation of reclaimed wastewater into a raw water supply; the wastewater effluent is discharged to the water source, where it mixes, blends with the source water, and is assimilated with it, with the intent of reusing the water instead of as a means of disposal (Tchobanoglous, 1991, pp. 1139–1140). A good example of indirect potable water reuse is the Hampton Roads Sanitation District, Hampton Roads, Virginia SWIFT process (Sustainable Water Initiative for Tomorrow). This process pumps ultra-treated-purified wastewater to drinking water standards into the Potomac Aquifer, which underlies the lower Chesapeake Bay region.

DID YOU KNOW?

There is the same amount of water on Earth as there was when the Earth was formed. The water from your faucet could contain molecules that dinosaurs drank (Spellman, 2024).

CASE STUDY 11.1: PURE WATER MONTEREY (USEPA, 2023)

For those lucky folks who have had the grand pleasure of driving the 17-mile Drive through Pebble Beach and Pacific Grove on the Monterey Peninsula in California, much of which hugs the picturesque Pacific coastline and passes several famous golf courses, mansions, and other scenic attractions it is one of those scenic views never to be forgotten. Simply, the Monterey Peninsula, Carmel, Castro Valley, and the coast with its mild climate are one of the most striking locations on the planet. One thing seems certain; while traversing the Drive, it is not difficult to imagine that you are witnessing one of the grandest landscapes/seascapes bordered by the waves crashing against the shore and the smell of saltwater in the air, and the surroundings make for one of Nature's grandest tranquilizers, a tonic without equal.

With its proximity to the Pacific Ocean, which seems to be everywhere as one passes along the shoreline, one of the last thoughts that probably comes to mind is a water problem—a dry glitch in all the beauty. Well, saltwater is not an issue in this pristine location. No. Plenty of ocean, for sure. The problem is not saltwater but instead the availability of and accessibility of freshwater. Besides the dramatic coastline, Monterey Peninsula's Carmel River Basin includes some of the world's most fertile agricultural lands—the area has been dubbed "America's Salad Bowl" because of the variety of crops grown. Note that besides fertile soil and ideal climate, agriculture requires a source of freshwater—lots of freshwater.

DOI: 10.1201/9781003498049-13

Truth be told the area relies solely on its limited, local water resources; it is a region isolated from state and federal projects. Residents and businesses located in the Monterey Peninsula region have historically relied on obtaining water from two sources: (1) the Carmel River and (2) the Seaside Groundwater Basin. Owing to state and court-ordered reductions, water supplies are trending toward becoming very limited. To help address this challenge, Monterey One Water and its partners have come together to create a new, drought-resistant, and independent water supply: Pure Water Monterey (PWM).

Pure Water Monterey uses a proven, advanced, multi-stage treatment process to turn used water into a safe, reliable, and sustainable water supply that complies with or exceeds strict state and federal drinking water supplies. The purified water is used for groundwater replenishment.

So, where does Monterey's used water come from?

Monterey One Water has helped diversify the local water supply through recycled water production for agriculture—helping irrigate 12,000 acres of freshly edible food crops and prevent seawater intrusion. In order to address both the potable and non-potable water demands, PWM has identified addition used water sources to bring into its existing wastewater-treated system, including roughly 67% from secondary treated wastewater, 16% agriculture drainage water, 17% from agriculture wash water (washing vegetables produces large volumes of water contaminated by soil, vegetable parts, pesticides, and other contaminants) and urban stormwater runoff—roughly 22% of the Monterey Peninsula's water supply will be provided by PWM (USEPA, 2023).

TECHNICAL ASPECTS

At PWM, it is all about using advanced purification technology. Indirect potable reuse is not only used at PWM but also is utilized in many locations throughout the Southwest United States and around the globe. The ultimate goal, of course, is to protect public health and safety. In doing so, PWM uses a four-step advanced purification treatment process shown in Figure 11.1 to meet or exceed all state and federal drinking water regulations.

PWM consists of four major project components: Source Waters, Advanced Water Purification Facility, Conveyance Pipeline, and Basin Injection Wells. As shown in Figure 11.1, PWM utilizes secondary treated water pumping into the Advanced

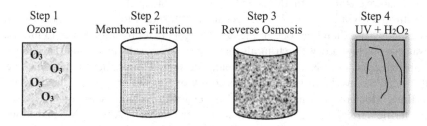

| Step 1 | Step 2 | Step 3 | Step 4 |
| Ozone | Membrane Filtration | Reverse Osmosis | UV + H_2O_2 |

FIGURE 11.1 Monterey groundwater replenishment advanced treatment train.

Treatment plant which is put through the four-step Advanced Water Treatment (AWT) purification process of Ozone (O_3) Pre-Treatment, Membrane Filtration (MF), Reverse Osmosis (RO), and Oxidation with Ultraviolet Light (UV) and Hydrogen Peroxide (H_2O_2). Note that these methods are commonly used in processing strictly regulated items for public consumption like bottled water and baby food. The bottom line: the purified water is near distilled quality and exceeds all drinking water standards.

ADVANCED WATER PURIFICATION PROCESS STEPS

Let us take a look at the Advanced Water Purification stops:

Step 1: Ozone Pre-Treatment

With Ozone (O_3) Pre-Treatment, Ozone, a gas at ordinary temperature and pressures, is a very powerful disinfectant (it disinfects by breaking up molecules in water)—even more effective against some viruses and cysts than chlorine. It has the added advantage of leaving no taste or odor and is unaffected by pH or the ammonia content of the water. When ozone reacts with reduced inorganic compounds and with organic material, an oxygen atom instead of a chloride atom is added to the organics, the end result being an environmentally acceptable compound. But since ozone is unstable and decomposes to elemental oxygen in a short amount of time and cannot be stored, it must be produced on-site. Ozonation usually costs more than chlorination as well but is used in preparing the treated water for Step 2.

Step 2: Membrane Filtration

Simply, membrane filtration involves the flow of wastewater pollutants across a membrane (basically a filter) with pores 300th the size of a human hair. Water permeates through the membrane into a separate channel for recovery. Because of the cross-flow movement of water and the waste constituents, materials left behind do not accumulate at the membrane surface but are carried out of the system for later recovery or disposal. Again, the water passing through the membrane is called the *permeate*, while the water with the more concentrated materials is called the *concentrate* or *retentate*.

Membranes are constructed of cellulose or other polymer material, with a maximum pore size set during the manufacturing process. The requirement is that the membranes prevent the passage of particles the size of microorganisms, or about 1 μm (0.001 mm) so that they remain in the system (a RO system rejects contaminants 0.0001 μm). This means that microfiltration membrane bioreactor (MBR) systems are good for removing solid material but the removal of dissolved wastewater components must be facilitated by using additional treatment steps.

Membranes can be configured in a number of ways. For MBR applications, the two configurations most often used are hollow fibers grouped in bundles or flat plates. The hollow fiber bundles are connected by manifolds in units that are designed for easy changing and servicing.

Designers of MBR systems require only basic information about the wastewater characteristics, (e.g., influent characteristics, effluent requirements, flow data) to design an MBR system. Depending on effluent requirements, certain supplementary options can be included with the MBR system. For example, chemical addition (at various places in the treatment chain, including before the primary settling tank; before the secondary settling tank [clarifier]; and before the MBR or final filters) for phosphorus removal can be included in an MBR system if needed to achieve low phosphorus concentrations in the effluent.

MBR systems historically have been used for small-scale treatment applications when portions of the treatment system were shut down and the wastewater routed around (or bypassed) during maintenance periods.

However, MBR systems are now often used in full-treatment applications. In these instances, it is recommended that the installation include one additional membrane tank/unit beyond what the design would nominally call for. This "N plus 1" concept is a blend between conventional activated sludge and membrane process design. It is especially important to consider both operations and maintenance requirements when selecting the number of units for MBRs. The inclusion of an extra unit gives operators flexibility and ensures that sufficient operating capacity will be available (Wallis-Lage et al., 2006). For example, bioreactor sizing is often limited by oxygen transfer, rather than the volume required to achieve the required SRT—a factor that significantly affects bioreactor numbers and sizing (Crawford et al., 2000).

Although MBR systems provide operational flexibility with respect to flow rates, as well as the ability to readily add or subtract units as conditions dictate, that flexibility has limits. Membranes typically require that the water surface be maintained above a minimum elevation so that the membranes remain wet during operation. Throughput limitations are dictated by the physical properties of the membrane, and the result is that peak design flows should be no more than 1.5–2 times the average design flow. If peak flows exceed that limit, either additional membranes are needed simply to process the peak flow, or equalization should be included in the overall design. The equalization is done by including a separate basin (external equalization) or by maintaining water in the aeration and membrane tanks at depths higher than those required and then removing that water to accommodate high flows when necessary (internal equalization) (USEPA, 2007).

DID YOU KNOW?

Membrane filtration is used regularly in food production such as baby food, bottled water, and beer.

Step 3: Reverse Osmosis

Even though the use of RO is quite common it is difficult for most people to gain an understanding of RO unless they first understand the principles of natural biological osmosis. In the simplest terms, osmosis can be explained as the naturally occurring process whereby water is transported through a membrane from a solution of low salt content to a solution of high salt content in order to equalize salt concentration.

DID YOU KNOW?

RO is not properly a filtration method. In RO, an applied pressure is used to overcome osmotic pressure, a colligative property, that is driven by chemical potential (Chen et al., 2011).

Osmotic Pressure

In a well-practiced experimental demonstration of osmotic pressure as applied to osmosis, water moves spontaneously from an area of high vapor pressure to an area of low vapor pressure (see Figure 11.2). If this experiment were allowed to continue, in the end, all of the water would move to the solution (see Figure 11.3). A similar process will occur when *pure* water is separated from a *concentrated* solution by a **semi-permeable** membrane (i.e., it only allows the passage of water molecules, see Figure 11.4). The *osmotic pressure* is the pressure that is just adequate to prevent osmosis. In dilute solutions, the osmotic pressure is directly proportional to the solute concentration and is *independent* of its identity (see Figure 11.5).

DID YOU KNOW?

The energy consumption of RO is directly related to the salt concentration since a high salt concentration has a high osmotic pressure.

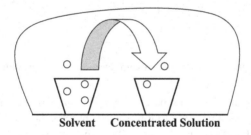

Solvent   Concentrated Solution

FIGURE 11.2 Osmotic pressure.

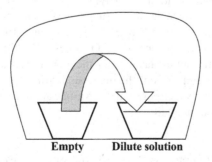

Empty   Dilute solution

FIGURE 11.3 Osmotic pressure (cont'd).

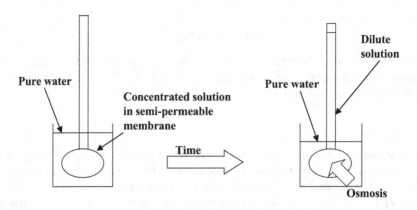

FIGURE 11.4   Colligative properties: passage of water molecules only.

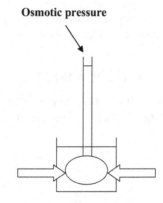

FIGURE 11.5   Osmotic pressure is the pressure just adequate to prevent osmosis.

Okay, let us get to it—RO is a separation or purification process (not properly a filtration process) that uses pressure to force a solvent through a semipermeable membrane that retains the solute one side and allows the pure solvent to pass to the other side, forcing it from a region of high solute concentration through a membrane to a region of low solute concentration by applying a pressure in excess of osmotic pressure. The difference between normal osmosis and RO is shown in Figure 11.6. Although many solvents (liquids) may be used and many applications are described in this book, the primary application of RO discussed in this book is water-based systems. Therefore, after an explanation of the RO process and its many different applications, the major emphasis of the discussion will focus on water as the liquid solvent. That is, drinking water purification, wastewater reuse, and desalination processes will be discussed in detail.

Process Description[1]

As mentioned, in the RO process (Figure 11.6), water passes through a membrane, leaving behind a solution with a smaller volume and a higher concentration of solutes.

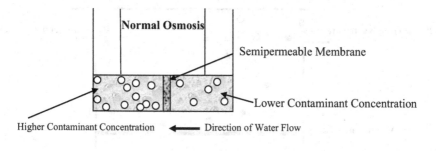

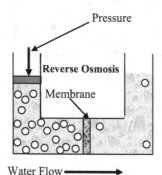

FIGURE 11.6 Normal osmosis and RO.

The solutes can be contaminants or useful chemicals or reagents, such as copper, nickel, and chromium compounds, which can be recycled for further use in metal plating or other metal finishing processes. The recovered water (*permeate*) can be recycled or treated downstream, depending on the quality of the water and the needs of the plant. As shown in Figure 11.7, the water that passes through the membrane is defined as *permeate*, and the concentrated solution left behind is defined as *retentate* (or concentrate).

The RO process does not require thermal energy, only an electrically driven feed pump. RO processes have simple flow sheets and a high energy efficiency. However, RO membranes can be fouled or damaged. This can result in holes in the membrane and passage of the concentrated solution to clean water, and thus a release to the environment. In addition, some membrane materials are susceptible to attack by oxidizing agents, such as free chlorine.

The flux of component A (recall that RO flux is the rate of water flow across a unit surface area) through an RO membrane is given by Equation 11.1:

$$N_A = P_A \left(\frac{\Delta \Phi}{L} \right)$$

(11.1)

where

N_A = Flux of component A through the membrane, mass/time length^2.
P_A = Permeability of A, mass-length/time force

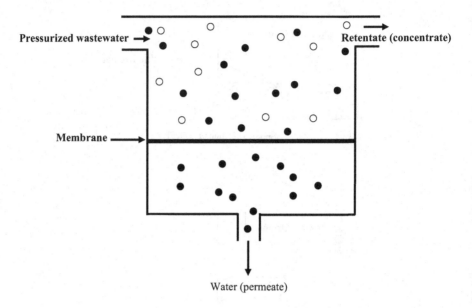

FIGURE 11.7   RO process.

ΔΦ (DF) = Driving force of A across the membrane, either pressure difference or
concentration difference, force/length² or mass/length².
L = Membrane thickness, length.

At equilibrium, the pressure difference between the two sides of the RO membrane
equals the osmotic pressure difference. At low solute concentration, the osmotic
pressure (p) of a solution is given by Equation 11.2:

$$\pi = C_S RT \tag{11.2}$$

where
$\pi(p)$ = Osmotic Pressure, force/length²
C_S = Concentration of solutes in solution, moles/length²
R = Ideal Gas Constant, (force-length)/(mass-temperature)
T = Absolute temperature, K or °R

As a mixture is concentrated by passing water through the membrane, the osmotic
pressure of the solution increases, thereby reducing the driving force of further water
passage. An accurate characterization of the pressure to drive the RO process must
be based on an osmotic pressure computed from the average of the feed and retentate
stream compositions. The water recovery of an RO process may be expressed by
Equation 11.3:

$$REC = \left(Q_p / Q_F \right) \times 100 \tag{11.3}$$

where
   REC = Water recovery, %
   Q_p = Permeate flow rate, length²/time
   Q_F = Feed flow rate, length²/time

Water recovery is determined by temperature, operating pressure, and membrane surface area. Rejection of contaminants determines permeate purity, while water recovery primarily determines the volume reduction of the feed or the amount of permeate produced. Generally, for concentrations of water from the metal finishing industry, greater water recoveries are desirable to obtain overall greater volume reduction.

DID YOU KNOW?

It is common for water professionals (and others) to confuse RO with filtration. However, there are key differences between RO and filtration. The predominant removal mechanism in membrane filtration is straining, or size exclusion, so the process can theoretically achieve perfect exclusion of particles regardless of operational parameters such as influent pressure and concentration. On the other hand, RO involves a diffusive mechanism so that separation efficiency is dependent on solute concentration, pressure, and water flux rate (Wachinski, 2013).

Note: RO technology pushes water through semipermeable membranes under high pressure.

Step 4: UV Disinfection and Hydrogen Peroxide

Ultraviolet (UV) light is electromagnetic radiation just beyond the blue end of the light spectrum, outside the range of visible light. It has a much higher energy level than visible light, and in large doses, it inactivates both bacteria and viruses. UV energy is absorbed by generic material in the microorganisms, interfering with their ability to reproduce and survive, as long as the radiation contacts the microorganisms without interference from turbidity. The big advantage of UV disinfection over chlorine and ozone is that UV does not involve chemical use. Generally, UV light used for disinfecting water is generated by a series of submerged, low-pressure mercury lamps. Continuing advances in UV germicidal lamp technology are making UV disinfection a more reliable and economical option for disinfection in many plants. To ensure the water's purity and definitely remove any molecules that may have slipped through the water is oxidized using hydrogen peroxide (H_2O_2) in the presence of UV light. Together, these break apart any chemical bonds that may be present.

   The Bottom Line: At this stage, the water is already of a high-grade water quality, pure enough for public consumption.

CASE STUDY 11.2: ROSEVILLE MUNICIPAL UTILITY (USEPA, 2023)

The community of Roseville, California, operates a wastewater utility to purify and treat water before it is delivered back to the environment. This inland Northern California community of ~140,000 residents has been recycling tertiary CA Title 22 water (requires stringent filtration and disinfection standards) for nearly 20 years. In 2020, the community operated two wastewater treatment regional facilities that have the capacity to treat up to 30 million gallons of water daily; maintain more than 500 miles of interceptor lines (sewer mainlines); 246 miles of sewer laterals and more than 11,000 manholes. The community's primary source of drinking water is a federally operated surface water reservoir designed and used primarily for flood control. Currently, 20% of all wastewater treated, ~1 billion gallons per year, is recycled for irrigation and industrial needs. The problem with Roseville's surface water supply is attempting to meet the challenges of significant past and projected population growth as well as environmental demands, climate change, and periodic drought. Increasing water recycling is one of the main strategies Roseville plans to meet this water supply challenge. However, all economically viable recycled water needs are being met presently through a traditional purple pipe distribution system completely separated from the system used to deliver potable water. Far-thinking community planners and others understood the need to improve recycled water utilization and in order to meet this challenge significant changes to the recycled water program were needed.

In light of this challenge, Roseville has successfully employed aquifer storage of potable surface water which provides storage and groundwater management capability when surface water is plentiful to ensure that groundwater is always a viable backup water supply. A similar strategy is envisioned for recycled water.

To make the most of its groundwater aquifer for seasonal storage of recycled water and improve distribution options to new areas of the City, Roseville must employ advanced treatment to create indirect potable reuse opportunities. California requires that RO be part of any advanced treatment process unless a proven alternative is capable of providing equivalent water quality. Here is the problem: Roseville is an inland community without access to an ocean outfall, the brine waste generated by RO cannot be disposed of economically thereby eliminating RO as a treatment option for Roseville.

So, to meet this challenge, Roseville is piloting alternative advanced treatment that incorporates ozone biologically active filtration (BAF), in addition to other processes needed to meet water quality-based criteria. Benefits of including ozone-BAF in the treatment train include lower energy requirements, no brine disposal, and improved removal of certain contaminants of emerging concern (CECs) like PCPPs.

The Bottom Line: Roseville's wastewater utility is dealing with reality—drinking water is not always readily accessible and available—thus their ongoing operation to store and recover advanced treated water is an example of turning purple pipes blue (blue for drinking quality water).

CASE STUDY 11.3: EMORY UNIVERSITY: A MODEL FOR WATER REUSE (USEPA, 2023)

In the 2011–2020 timeframe, Emory University reported that Atlanta witnessed numerous water-related stresses, including severe drought, USEPA mandates to put an end to critical infrastructure failures, and long-drawn-out political disputes over water rights in the so-called "Tri-State Water Wars." What this was all about was Georgia, Alabama, and Florida for more than three decades have disputed the use of two shared river basins—the Apalachicola-Chattahoochee-Flint (ACF) and the Alabama-Coosa-Tallapoosa (ACT). These river systems are used to meet multiple needs, including drinking water, power generation agriculture, aquaculture, navigation, and recreation. Because of these challenges, Emory University set out to explore ways to minimize its impact on the community water resources and the environment with a more strategic and impactful water management solution: campus-wide water reclamation and reuse.

Emory's water reclamation system, the WaterHub®, was designed by Sustainable Water to integrate into the existing campus framework using two small parcels near Chappell Park Field. What is going on here is the mining of wastewater directly out of the campus sewer system daily. Supported by solar (PV) energy production. The system has 50,000 gallons of clean water storage capacity, providing $N+1$ redundancy (aka parallel redundancy; to ensure that a constant source is available) for campus district energy systems. A 4,400 linear foot "purple pipe" conveys recycled water to multiple utility plants and select dormitories for toilet flushing. The system reduces Emory University's draw of potable water by up to 146 million gallons annually.

The 411 on WaterHub® is that it is the first system of its kind installed in the United States. It is a decentralized, commercial-scale water reclamation and reuse system serving Emory University's main campus just outside of Atlanta, GA. The WaterHub® mines wastewater directly from the campus sewer system producing up to 400,000 gallons of reclaimed water per day. Beneficial reuse is accomplished using ecological treatment processes. Approximately two-thirds of campus wastewater for nonpotable demands including heating, cooling, and toilet flushing are recycled by the system. The far-thinkers behind WaterHub® have set the standard as the flagship model for commercial-scale sustainable water management in urban areas.

In operation, the WaterHub® enables the University to reduce its draw of potable water by up to 146 gallons annually—displacing nearly 40% of total campus water demand. This system provides a consistent, reliable, and redundant source of water for extensive nonpotable demands and critical heating and air conditions needs—the system enhances campus resiliency. The WaterHub® is designed to safeguard campus operations from potential water service disruptions resulting from drought and aging municipal water infrastructure.

The Bottom Line: WaterHub® is designed to promote research and community outreach, enhancing the concept of the campus as a "living laboratory."

CASE STUDY 11.4: INDIANTOWN, FLORIDA: WATER REUSE (USEPA, 2023)

In 2012, Indiantown Cogeneration Industrial Facility implemented a zero-liquid dis-charge (ZLD) system. In a nutshell, ZLD is a wastewater treatment process intended to reduce wastewater efficiently and produce clean water that is suitable for reuse (e.g., equipment cooling, agriculture). Wastewater treatment technologies including desalination are used in ZLD to purify and recycle virtually all wastewater received (Panagopoulos et al., 2019; Voutchkov and Kaiser, 2020).

At Florida's Indiantown facility, it initiated ZLD in 2012 using two brine concen-trators to process the cooling tower blowdown water. There were drawbacks with this system and managers and operators soon found out that these concentrators were expensive to maintain, used a load of 1.4 Mega Watthour (MWH) electric-ity, and produced a high-volume wastewater stream. Because of expensive mainte-nance, the facility decides to replace concentrators with a ZLD system consisting of Microfiltration (MF) and RO. The new ZLD system is less expensive to maintain and has lower wastewater discharges which returned a higher volume of filtered water into the facility.

The brine concentrators were replaced by MF/RO systems in the ZLD system achieving higher quantities of filter water. This is accomplished in MF when the incoming water passes through several thousand spaghetti-like hollow fiber poly-meric membranes that remove suspended solids and bacteria. For removal of dis-solved solids, the treated water from the MF unit passes through the spiral-would RO membranes (see Figures 11.8a and 11.8b). This technology is in use before the demineralizers. The pores of the RO membrane are only a few angstroms in size and can remove a majority of the dissolved salts.

A typical 500-ton cooling tower, running 24 h/day, 365 days per year will flush over 3.9 million gallons of water each year. By using the ZLD system, electric gen-eration facilities can reuse a bulk of the wastewater stream. The Indiantown facility used the new ZLD system to increase the filtered water volumes, reduce maintenance costs for the facility, and save on 1.4 MWH of electricity used by the old system. The good news is that the new system was more effective in using briny groundwater from aquifers that are not sources of drinking water reducing reliance on the fresh

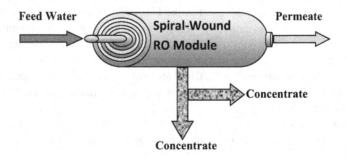

FIGURE 11.8A Cutaway view of a spiral-wound RO module consisting of internal-wound product spacers, RO membranes, feed spacers, and RO membranes.

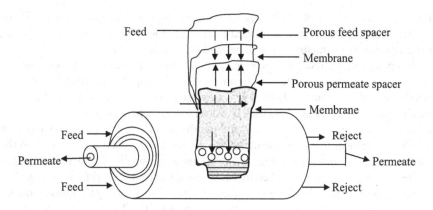

FIGURE 11.8B Internal construction of the spiral-wound module.

stream water. The fly in the ointment, the monkey wrench in the works in using this new system is that the system encountered problems with microbiological fouling and scaling in the second stage RO. However, these problems were resolved by introducing microbicides and lowering the pH of water to 5.

The Bottom Line: The Indiantown Facility was able to use its ZLD system to provide reuse within the facility and have zero liquid discharge with the use of microfiltration and RO.

CASE STUDY 11.5: CITY OF ORLANDO/ORANGE COUNTY, FLORIDA WATER CONSERV II (USEPA 2023)

Parsons (2018) points out that the Water Conserv II project … "was started in 1986 to stop the discharge of treated wastewater from Orlando and Orange County into Lake Tohopekaliga, an important recreation bass fishing lake."

Reclaimed water is pumped from the City's Conserv II Water Reclamation Facility and the County's South Water Reclamation Facility to the Water Conserv II Distribution Center in Western Orange County. The water is then distributed to customers for irrigation or the Rapid Infiltration Basins (RIBs) through a network of distribution pipes.

RIBs are not a new technology. Rapid infiltration (RI) (aka soil aquifer treatment) is a major land treatment technique that uses the soil ecosystem to treat wastewater. The huge advantage of us RI process is that it can treat a much larger area than other land treatment concepts. In RI systems, wastewater is applied to shallow rapid infiltration basins constructed in deep and permeable deposits of highly porous soils. Flooding or sprinkler application of wastewater is used. Treatment, including filtration, adsorption, ion exchange, precipitation, and microbial action, occurs as the wastewater moves through the soil matrix.

The RIBS in the Water Conserv II process are made up of one to five cells, each measuring ~350 ft long by 150 ft wide. They are built over a natural sand ridge ranging in thickness from 30 to 200 ft. Beneath these surficial sands is a dense concentration of semi-permeable clays known as the Hawthorn Formation. This formation is important because it acts as a barrier separating shallow groundwater flow within the surficial sands from deeper, continued flow in the Floridan aquifer, which is comprised primarily of fractured limestones and dolomites.

In operation, wastewater percolates through the soil where it can be collected or it can flow to native surface water or the Floridan aquifer. In West Orange and Southeast Lake Counties, shallow groundwater in the surficial sands follows primarily later flow patterns about the Hawthorn until reaching areas of low resistance (water flow like the flow of electricity follows the path of least resistance) that permit significant vertical flow downward where it is further polished as it flows into the Floridan aquifer, thus replenishing the drinking water supply. The actual operation of the RIB system is accomplished through a computerized management system. The system provides the ability to forecast the impact on the regional groundwater system of loading individual or groups of RIBs at approved rates. The recovered water can be used for beneficial reuse such as irrigating crops or industrial uses. Water not recovered recharges the Floridan aquifer.

Water Conserv ll (2019) indicated that

> Faced with a need to expand wastewater treatment service and a state requirement to eliminate discharge to surface waters, the City of Orlando and Orange Count formed a long-term partnership to develop an innovated water reclamation program. The project is best described as "A cooperative Water Resue Project by the City of Orlando, Ornge county and the Agricultural Community".

The entire process is monitored and carefully controlled by state-of-the-art supervisory control and data acquisition (SCADA) computers housed at the Distribution Center. Simply put, SCADA is a computer-based system that remotely controls processes previously controlled manually. SCADA allows an operator using a central computer to supervise (control and monitor) multiple networked computers at remote locations. Each remote computer can control mechanical processes (pumps, valves, etc.) and collect data from sensors at its remote location. Thus the phrase: Supervisory Control and Data Acquisition, of **SCADA**. The central computer is called the Master Terminal Unit, or MTU. The operator interfaces with the MTU using software called Human Machine Interface, or HMI. The remote computer is called the Program Logic Controller (PLC) or Remote Terminal Unit (RTU). The RTU activates a relay (or switch) that turns mechanical equipment "on" and "off." The RTU also collects data from sensors (Spellman, 2007).

Initially, utilities ran wires, also known as hardwires or landlines, from the central computer (MTU) to the remote computers (RTUs). Because remote locations can be located hundreds of miles from the central location, utilities have begun to use public phone lines and modems, leased telephone company lines, and radio and microwave communication. More recently, they have also begun to use satellite links, the Internet, and newly developed wireless technologies.

Because the SCADA systems' sensors provided valuable information, many utilities established "connections" between their SCADA systems and their business system. This allowed Utility management and other staff access to valuable statistics, such as water usage. When utilities later connected their systems to the Internet, they were able to provide stakeholders with water/wastewater statistics on the Utility web pages.

As stated above, SCADA systems can be designed to measure a variety of equipment operating conditions and parameters or volumes and flow rates or water quality parameters and to respond to changes in those parameters either by alerting operators or by modifying system operation through a feedback loop system without having personnel physically visit each process or piece of equipment on a daily basis to check it and/or ensure that it is functioning properly. SCADA systems can also be used to automate certain functions so that they can be performed without the need to be initiated by an operator (e.g., injecting chlorine in response to periodic low chlorine levels in a distribution system or turning on a pump in response to low water levels in a storage tank). Note that in addition to process equipment, SCADA systems can also integrate specific security alarms and equipment, such as cameras, motion sensors, lights, data from card reading systems, etc., thereby providing a clear picture of what is happening in areas throughout a facility. Finally, SCADA systems also provide constant, real-time data on processes, equipment, location access, etc., the necessary response to be made quickly. This can be extremely useful during emergency conditions, such as when distribution mains break or when potentially disruptive BOD spikes appear in wastewater influent.

Because these systems can monitor multiple processes, equipment, and infrastructure and then provide quick notification of, or response to, problems or upsets. SCADA systems typically provide the first line of detection for atypical or abnormal conditions. For example, a SCADA system connected to sensors that measure specific water quality parameters is measured outside of a specific range. A real-time customized operator interface screen could display and control critical systems monitoring parameters.

The system could transmit warning signals back to the operators, such as by initiating a call to a personal pager. This might allow the operators to initiate actions to prevent contamination and disruption of the water supply. Further automation of the system could ensure that the system-initiated measures to rectify the problem. Preprogrammed control functions (e.g., shutting a valve, controlling flow, increasing chlorination, or adding other chemicals) can be triggered and operated based on SCADA utility.

SCADA Vulnerabilities

In the beginning, SCADA networks were developed with little attention paid to security, making the security of these systems often weak. Studies have found that, while technological advancements introduced vulnerabilities, many water/wastewater utilities have spent little time securing their SCADA networks. As a result, many SCADA networks may be susceptible to attacks and misuse. Remote monitoring and supervisory control of processes began to develop in the early 1960s and adopted many technological advancements. The advent of minicomputers made it possible to automate a vast number of once manually operated switches. Advancements in radio technology reduced the communication costs associated with installing and maintaining buried cables in remote areas. SCADA systems continued to adopt new communication methods including satellite and cellular. As the price of computers and communications dropped, it became economically feasible to distribute operations and expand SCADA networks to include even smaller facilities.

Advances in information technology and the necessity of improved efficiency have resulted in increasingly automated and interlinked infrastructures and created new vulnerabilities due to equipment failure, human error, weather and other natural causes, and physical and cyber-attacks. Some areas and examples of possible SCADA vulnerabilities include:

- **Human**: People can be tricked or corrupted and may commit errors.
- **Communications**: Messages can be fabricated, intercepted, changed, deleted, or blocked.
- **Hardware**: Security features are not easily adapted to small self-contained units with limited power supplies.
- **Physical**: Intruders can break into a facility to steal or damage SCADA equipment.
- **Natural**: Tornadoes, floods, earthquakes, and other natural disasters can damage equipment and connections.
- **Software**: Programs can be poorly written.

A survey found that many water utilities were doing little to secure their SCADA network vulnerabilities (Ezell, 1998); for example, many respondents reported that they had remote access, which can allow an unauthorized person to access the system without being physically present. More than 60% of the respondents believed that their systems were not safe from unauthorized access and use. Twenty percent of the respondents even reported known attempts, successful unauthorized access, or use of their system. Yet 22 of 43 respondents reported that they do not spend any time ensuring their network is safe and 18 of 43 respondents reported that they spend <10% ensuring network safety.

SCADA system computers and their connections are susceptible to different types of information system attacks and misuse such as system penetration and unauthorized access to information. The Computer Security Institute and Federal Bureau of Investigation conduct an annual Computer Crime and Security Survey (FBI, 2007). The survey reported on ten types of attacks or misuse and reported that viruses and denial of service had the greatest negative economic impact. The same study also found that 15% of the respondents reported abuse of wireless networks, which can be a SCADA component. On average, respondents from all sectors did not believe that their organization invested enough in security awareness. Utilities as a group reported a lower average computer security expenditure/investment per employee than many other sectors such as transportation telecommunications, and financial.

Sandia National Laboratories' *Common Vulnerabilities in Critical Infrastructure Control Systems* described some of the common problems it has identified in the following five categories (Stamp et al., 2003):

1. **System data**: Important data attribute for security include availability, authenticity, integrity, and confidentiality. Data should be categorized according to its sensitivity, and ownership and responsibility must be assigned. However, SCADA data is often not classified at all, making it difficult to identify where security precautions are appropriate.

2. **Security administration**: Vulnerabilities emerge because many systems lack a properly structured security policy, equipment and system implementation guides, configuration management, training, and enforcement and compliance auditing.
3. **Architecture**: Many common practices negatively affect SCADA security. For example, while it is convenient to use SCADA capabilities for other purposes such as fire and security systems, these practices create single points of failure. Also, the connection of SCADA networks to other automation systems and business networks introduces multiple entry points for potential adversaries.
4. **Network (including communication links)**: Legacy systems' hardware and software have very limited security capabilities, and the vulnerabilities of contemporary systems (based on modern information technology) are publicized. Wireless and shared links are susceptible to eavesdropping and data manipulation.
5. **Platforms**: Many platform vulnerabilities exist, including default configurations retained, poor password practices, shared accounts, inadequate protection for hardware, and nonexistent security monitoring controls. In most cases, important security patches are not installed, often due to concern about negatively impacting system operation; in some cases, technicians are contractually forbidden from updating systems by their vendor agreements.

The following incident helps to illustrate some of the risks associated with SCADA vulnerabilities.

- During the course of conduction a vulnerability assessment, a contractor stated that personnel from his company penetrated the information system of a utility within minutes. Contractor personnel drove to a remote substation and noticed a wireless network antenna. Without leaving their vehicle, they plugged in their wireless radios and connected to the network within 5 min. Within 20 min, they had mapped the network, including SCADA equipment, and accessed the business network and data.

Note: This illustrates what a cyber security advisor from Sandia National Laboratories specialized in SCADA stated—that utilities are moving to wireless communication without understanding the added risks.

Today, many common communications sector applications for SCADA systems include, but are not limited to those shown below.

- Boiler controls
- Bearing temperature monitor (electric generators and motors)
- Gas processing
- Plant monitoring
- Plant energy management
- Power distribution monitoring
- Electric power monitoring

- Fuel oil handling system
- Hydroelectric load management
- Petroleum pilot plants
- Plant monitoring
- Process controls
- Process stimulators
- Tank controls
- Utility monitoring
- Safety parameter display systems and shutdown systems
- Tank level control and monitoring
- Turbine controls
- Turbine monitoring
- Virtual annunciator panels
- Alarm systems
- Security equipment
- Event logging

Because these systems can monitor multiple processes, equipment, and infrastructure and then provide quick notification of, or response to, problems or upsets, SCADA systems typically provide the first line of detection for atypical or abnormal conditions. For example, a SCADA system connected to sensors that measure specific machining quality parameters is measured outside of a specific range. A real-time customized operator interface screen could display and control critical systems monitoring parameters.

The system could transmit warning signals back to the operators, such as by initiating a call to a personal pager. This might allow the operators to initiate actions to prevent power outages or contamination and disruption of the energy supply. Further automation of the system could ensure that the system-initiated measures to rectify the problem. Preprogrammed control functions (e.g., shutting a valve, controlling flow, throwing a switch, or adding chemicals) can be triggered and operated based on SCADA utility (Spellman, 2007).

The Bottom Line: Let us look at the benefits of Water Conserv II (2017) including the elimination of outfall to surface water; turning a liability into an asset for beneficial use; reduction of the demand on the Floridian aquifer by eliminating the need for well water for irrigation; helps to replenish the Floridian aquifer through the discharge of reclaimed water to the Rapid Infiltration Basins (RIBs); and proven beneficial and cost-effective year-round reclaimed water reuse. With regard to agriculture, most growers in the area understand that they have a good quality resource for year-round use (Parsons, 2018).

CASE STUDY 11.6: LOS ANGELES COUNTY STORMWATER SERVICES PROGRAM (USEPA, 2023)

The Sanitation District of Los Angeles County is helping local jurisdictions with the development of stormwater projects promoting MS4 compliance by improving water quality, and where feasible, achieving co-benefits such as enhancing water supply and resiliency.

So, what is MS4?

Answer: USEPA (2005) points out that small municipal separate storm sewer systems or MS4 often receive polluted stormwater runoff and ultimately discharge into local rivers and streams without treatment. USEPA's Stormwater Phase II Rule establishes an MS4 stormwater management program that is intended to improve the Nation's waterways by reducing the quantity of pollutants that stormwater picks up and carries into the storm sewer systems during storm events. Common pollutants include oil and grease from roadways, pesticides from lawns, sediment from construction sites, and carelessly discarded trash, such as cigarette buttes, paper wrappers, and plastic bottles. When deposited into nearby waterways through MS4 discharges, these pollutants can impair the waterways, in so doing discouraging recreational use of the resource, contaminating drinking water supplies, and interfering with the habitat for fish, other aquatic organisms, and wildlife.

The bottom line: operators of regulated small MS4s are required to design their programs to reduce the discharge of pollutants to the "maximum extent practicable" (MEP); protect water quality; and satisfy the appropriate water quality requirements of the Clean Water Act (USEPA, 2005).

The Sanitation Districts of Los Angeles County's augmentation of municipal recycled water with urban runoff/stormwater includes one project that is a controlled, permitted diversion from storm drains to storage facilities with managed releases to the sanitary sewer system owned by the Sanitation Districts and city and county and satellite sewer systems. Note that in some cases, these projects can supplement the water reclamation facility's recycled water supply and assist the region in meeting its local supply and water resiliency goals.

The real bottom line is that population growth, drought, regulations, and water conservation have increased the demand for recycled water. So what this means is that water administrators and other responsible persons need to be creative, break new ground, generate new ideas, think outside the treatment plants (streams in a box), and simply be far thinkers—this is especially the case with providing sources of safe, healthy, clean freshwater for whatever the need.

The Bottom Line: There can be little doubt that water management today and tomorrow is taking center stage (hopefully) in addressing current and future needs for clean, safe freshwater.

NOTE

1 Material in this section is from USEPA. (1996). Reverse Osmosis Process. EPA/625/R-96/009. Washington, DC: United States Environmental Protection Agency; Spellman, F.R. (2009). *Physics for the Nonphysicist*. Lanham, MD: Government Institutes.

REFERENCES

Chen, J.P., Mou, H., Wang, L.K., Matsuura, T., and Wei, Y. (2011). Membrane separation: Basics and application. In Wang, L.K., Chen, J.P., Hung, T., and Shammas, N.K. (eds.), *Membrane and Desalination Technologies*, Handbook of Environmental Engineering, vol. 13. Totowa, NJ: Human Press, pp. 271–332.

Crawford, G., Thompson, D., Lozier, J., Daigger, G., and Fleischer, E. (2000). Membrane bioreactors: A designer's perspective. In *Proceedings of the Water Environment Federation 73rd Annual Conference & Exposition on Water Quality and Wastewater Treatment,* Anaheim, CA, CD-ROM, October 14–18, 2000.

Ezell, B.C. (1998). *Risks of Cyber Attack to Supervisory Control and Data Acquisition.* Charlottesville, VA: University of Virginia.

FBI. (2007). Protecting Critical Infrastructure and the Importance of Partnerships. Accessed 1/25/24 @ https://www.fib.gov/news/speeches/protecitng-critical-infrastruce-andt-the-importantce-o.

Panagopoulos, A., Haralambous, K.-J., and Loizidou, M. (2019). Desalination brine disposal methods and treatment technologies: A review. *Science of the Total Environment* 693:133545.

Parsons, L.F. (2018). Agricultural use of reclaimed water in Floride: Food for thought. *Journal of Contemporary Water Research & Education* 165:20–27.

Spellman, F.R. (2007). *Water Infrastructure Protection and Homeland Security.* Boca Raton, FL: CRC Press.

Spellman, F.R. (2024). *The Drinking Water Handbook,* 4th ed. (in production). Boca Raton, FL: CRC Press.

Stamp, J., et al. (2003). *Common Vulnerabilities in Critical Infrastructure Control Systems,* 2nd ed. Albuquerque, NM: Sandia National Laboratories.

Tchobanglous, G. (1991). *Wastewater Engineering: Treatment Disposal Reuse.* New York: McGraw-Hill.

USEPA. (2005). *Stormwater Phase II Final Rule.* Washington, DC: United States Environmental Protection Agency, Office of Water.

USEPA. (2007). *Wastewater Management Fact Sheet: Membrane Bioreactors.* Washington, DC: United States Environmental Protection Agency.

USEPA. (2023). Case Studies that Demonstrate the Benefits of Water Reuse. Accessed 12/22/23 @ https://epa.gov/waterreuse/forms/contact-us--about-water-reuse-and-recycling.

Voutchkov, N., and Kaiser, G. (2020). *Management of Concentrate from Desalination Plants.* Amsterdam, Netherlands: Elsevier, pp. 187–203.

Wachinski, A.M. (2013). *Membrane Processes for Water Use.* New York: McGraw-Hill.

Wallis-Lage, C., et al. (2006). MBR plants: Larger and more complicated. *Presented at the Water Reuse Association's 21st Annual Water Reuse Symposium,* Hollywood, CA, September 2006.

Water Conser ll. (2017). Water Conserv ll Tour Materials. Accessed 1/26/24 @ https://www.waterconservii.com/wp-content/themes/divi-child/dis/Tour_Pack.pdf.

Water Conserv ll. (2019). History/Anatomy of a RIB. Accessed 1/25/24 @ https://www.water-conservii.com/.

Part 3

The Real Deal
Purple to Blue PVC Pipe

12 When Purple Becomes Blue

SETTING THE STAGE

In this part of the book, the intention, the goal, the objective is to explain (using a top-notch current example), how toilet water becomes drinking water. This is accomplished via the ingenuity, resourcefulness, initiative, and imagination of humankind (and one organization in particular) when it comes to getting it done; whatever "it" is.

The "it" in this case has to do with my visual cues—beyond my traditional focus points, I look for wastewater treatment plants that turn purple polyvinyl Chloride (PVC) piping to blue PVC piping. The color of piping has been an item that I am not only familiar with but also on the lookout for whenever I conduct pre-OSHA/ EPA environmental health and safety inspections at water and wastewater treatment plants in the United States. During my plant walk-arounds, I find PVC pipes painted white for some drinking water systems; gray for non-potable water; red for electrical and communication lines; blue for drinkable water systems; green for raw sewage; yellow for gas lines (e.g., yellow pipes conveying mostly methane); orange for underground electrical conduit systems; and, finally, purple for reclaimed/recycled water.

With regard to this book and the material contained herein, it is the contents of the purple piping for the most that part has my attention. Wastewater effluent that is treated to drinking water standards and quality is an important step (I think a critical step) forward in mitigating the declining water quantity of aquifers, especially in the Southwestern United States and in the Mid-Atlantic Region.

> **DID YOU KNOW?**
>
> Water for potential reuse can include municipal wastewater, industry process, and cooling water, stormwater, agricultural runoff, and return flows, and produced water from natural resource extraction actives (e.g., from hydraulic fracturing operations).

It is the Mid-Atlantic region, specifically the Tidewater Region, also known as Hampton Roads Virginia, which is the name/area of a body of water that serves a wide channel for the James, Nansemond, and Elizabeth Rivers that flow into the lower Chesapeake Bay and the surrounding metropolitan region. This is composed of the Virginia Beach—Norfolk—Newport News, VA-NC area.

DOI: 10.1201/9781003498049-15

This region is the focus of this part of the book because of Hampton Roads Sanitation District (HRSD) and specifically its Sustainable Water Initiative For Tomorrow (SWIFT) project. SWIFT is an innovative water treatment project in the Hampton Roads Region designed to further protect the region's environment, and enhance the sustainability of the region's long-term groundwater supply. Moreover, the project helps address environmental pressures such as land subsidence, sea level rise, saltwater intrusion, and restoration of Chesapeake Bay. SWIFT takes highly treated water that would otherwise be discharged into the Elizabeth, James, or York rivers and puts it through additional rounds of advanced water treatment to meet drinking water quality standards. The SWIFT Water™ is then added to the Potomac Aquifer, the primary source of groundwater throughout eastern Virginia.

SWIFT's Advanced Water Treatment Process produces clean, safe drinking water quality SWIFT Water™ through a multistep disinfection process that is used throughout the globe. It is a matter of replenishing groundwater that is being used in eastern Virginia at rates faster than it can be naturally replaced. Adding SWIFT Water back into the ground replenishes this natural resource and protects the Potomac Aquifer from further damage caused by overuse. SWIFT is giving local Hampton Roads communities a sustainable source of groundwater. This resource supports the local economy by providing businesses with the water they need to operate. Another issue is sea level rise. The pumping water out of the ground at the current rate has led to the sinking of land, or land subsidence, in some parts of Virginia. This makes us more vulnerable to rising sea levels and the associated impacts. Replenishing the Potomac Aquifer with HRSD's SWIFT Water can help slow or reverse the sinking of land due to withdrawal. Overuse of the aquifer causes about 25% of the sinking of land in parts of eastern Virginia. Replenishing the overdrawn aquifer can improve the environment and help communities adapt to rising seas, contributing to the resiliency of the region. In order to restore the health and productivity of Chesapeake Bay largely depends on reducing the amount of nutrients and sediment that enter Chesapeake Bay waters.

HRSD's SWIFT program is a work in progress. One of its 14 treatment plants is in full operation as an ongoing pilot study site while at the same discharging treated water to drinking water quality standards into the Potomac Aquifer. At the current time, the only absolutely positive result of this practice is the effective discharge of water-containing nutrients to the aquifer and thus not discharging into local waters. In the near future, HRSD projects that when its other plants' effluent becomes treated SWIFT Water and is discharged into the aquifer it will eliminate up to 90% of HRSD's discharge to local waters.

THE BOTTOM LINE

What HRSD's SWIFT Water project accomplishes, as do others using similar processes, is treat wastewater in the Hampton Roads Region to meet "fit-for-purpose specifications" to serve multipurpose objectives: to prevent saltwater intrusion, to restock the Potomac Aquifer with wastewater-treated to drinking water quality. Moreover, HRSD is also attempting to slow land subsidence in the region and also to keep harmful nutrients from being discharged into local surface waters.

Prevention of harmful nutrients from entering the surface waters is ongoing and proving to be effective. The jury is still out on the other hoped-for effects—for example, the question as to will the discharge of HRSD's treated water halt land subsidence and whether will it actually cause a land rebound whereby land subsidence stops and land levels rise? Only time will tell.

13 A Convulsive Event

INTRODUCTION

The *Sustainable Water Initiative for Tomorrow* (SWIFT) program is a forward-looking, innovative solution to tackle one of today's problems.

Today's problems?

Yes.

Okay, is SWIFT technology a one-time application that can be utilized by one specific location only?

No.

First, let us get back to the problems.

At the present time as pointed out earlier, SWIFT is designed to address, mitigate, and help the Lower Chesapeake Bay Region by reducing the amount of nutrients (such as nitrogen and phosphorus) that Hampton Roads Sanitation District (HRSD) treatment plants discharge to the James, Elizabeth, and York Rivers. SWIFT is also designed to replenish dwindling groundwater supplies, allowing this natural resource to remain productive for generations to come. Moreover, it is also designed to fight sea level rise by reducing the rate at which land is sinking in coastal areas. Also, it is designed to protect groundwater from saltwater intrusion due to a shrinking aquifer. Finally, it supports the local economy by providing businesses with the water they need to operate.

The bottom line: SWIFT technology can be applied in different locations where these same issues are present.

SWIFT provides a detailed step-by-step account from pilot project, scientific research, engineering analysis, and comparison of different treatment technologies, to selecting the best treatment technology and receiving the appropriate regulatory approval, through the drilling, construction, equipping, and testing and monitoring stages.

Why did HRSD consider implementing SWIFT?

Well, HRSD arguably is the premier wastewater treatment entity on the globe. HRSD is staffed with forward-thinking innovators. From the former General Manager to the treatment plant's so-called rag persons, they are top-notch professionals who respect our environment and tend to the needs of their ratepayers and others who reside in Hampton Roads. Simply stated: HRSD's mission is to protect public health (is there a more important mission?) and the waters of Hampton Roads by treating wastewater effectively. The residents of HRSD's service area send an estimated 150 million gallons of wastewater to HRSD each day. HRSD cleans and polishes this water to rigorous state-regulated standards and returns it (outfalls it) safely to area waterways— the treated wastewater is cleaner than the waters contained within the receiving bodies. The substantially treated water HRSD currently delivers exceeds the water quality requirements of local waterways; with the addition of advanced treatment processes, HRSD can produce water that meets drinking water standards.

DOI: 10.1201/9781003498049-16

The bottom line: HRSD is exploring ways to use SWIFT Water to provide Hampton Roads and Virginia with added environmental and societal benefits.

This chapter and subsequent chapters provide a real-world template, a gauge, a model, and a guide for any other city, town, township, village, or community that is experiencing the same type of issues as Hampton Roads Sanitation District and its surrounding region. Moreover, this book is accessible and is designed to reach a wide range of diverse technical personnel, readers, and students. The lack of far-thinking and innovation, in regard to the disposal and/or reuse of treated wastewater, is a world-class problem. Thus, it logically follows that we *can* solve this problem because every problem has a solution and the key to finding the solution is achieved through innovation, which is the core message of this text.

Note: In order to obtain the big picture, to gain understanding and insight and to understand what SWIFT entails (i.e., to understand what it is all about), it is important to connect the dots, from beginning to end, even though, at present, there is no ending dot in sight. In the simplest of terms, SWIFT can be described as an innovative, far-thinking initiative that is truly a work in progress. Therefore, and because of the need for the reader to obtain an understanding of this complex operation, this book connects, or more specifically strings dots together one after the other with no exact connection to the end dot. Again, the idea is to connect the dots; therefore, the first dot has to do with history; actually, we are talking about *prehistory*. This is the case because an understanding of southeastern Virginia, the formation of the Chesapeake Bay, and the subsequent discoveries about the Bay's formation are germane to gaining a clear understanding of the purpose, function, and operation of SWIFT; simply, it explains the purpose of SWIFT, what SWIFT is all about.

Note: The following information is provided as part of setting the stage for the formation of the Chesapeake Bay.

A VISIT FROM OUTER SPACE[1]

Timeline 33,000,000 B.C.E

They were all gone—well, at least the non-avian ones: *Tyrannosaurus* (*aka T-Rex*), *Velociraptor, Apatosaurus, Stegosaurus*, and *Triceratops*; only their bones remain. They had been gone for ~32 million years; actually, a bit more than 32,000,000 million years before the present timeframe referred to herein. We would later know they were actually here on the planet because we discovered their preserved skeletal remains, and discoveries continue here and there at various locations on Earth. When the remains are discovered and carefully preserved, they are often displayed to the delight and wonder of on-lookers.

Don't you just love them?

Them?

Yes. We will get to them.

For the moment consider that many experts say (or at least speculate) that it was either a combination of volcanic activity, asteroid impact (creating the Chicxulub crater in the Yucatan; about the size of Staten Island), and/or climate change that effectively ended 76% of life on Earth 66–65 million years ago (based on marine fossils).

Others speculate that it might have been only one of these events that brought the mighty rulers of the Earth, the Terrible Lizards (aka dinosaurs) to their terminal end. Well, you can take your pick of the actual causal factor(s); the fact is that much of life as known during that time period was extinguished in what is known as the K–T event, but recently renamed the K-Pg for the Cretaceous-Paleogene Extinction event—and also known as (and the title most often thought of, if thought about at all): the Fifth Massive Extinction Event on Earth (there are those who feel we are undergoing the Sixth Extinction [aka Holocene—human-caused]). Importantly, meteors have been recorded throughout history.

For the purpose of this book, we will assume, speculate, guess, or believe (take your pick) that it was the meteor that created the Chicxulub Crater, the 125-m hole off the coast of the Yucatan Peninsula, deep beneath the depths of the Gulf of Mexico. The meteor, or meteorite (i.e., a meteor that survives burning up in the atmosphere and reaches land is called a meteorite) is responsible for the largest mass extinction event in history. The impact was so intense that it started wildfires hundreds of miles away from the impact site. So much ash, sulfur, and fragmented debris was flung into the air that the Sun was blotted out. The atmospheric disturbance was so great that Earth sat in perpetual darkness for months, creating a very long and unexpected winter that changed the way of the planet forever. Approximately 75% of life on Earth (including the major dominant species, dinosaurs) died off at this time. Note that many mammals survived this apocalyptic event by being small, needing little food, and warm-blooded. And keep in mind that when an animal's chief predators are no longer around, the chance of maintaining life and procreation is enhanced exponentially. Moreover, the lack of predators following the meteor impact led to the rise of the evolution of all mammals alive today, including humans.

LET'S GET BACK TO 33,000,000 B.C.E

Before getting to the point of why this discussion begins with an era long past, it is important to point out that this book is a geologically based presentation of facts as we know them or as we think we know them. To begin this discussion, we need to begin at the beginning—the first dot to connect in the chain. Thus, Table 13.1 is presented in order to give the reader an understanding of the beginning related to this book. For us, it is the Eocene (*Greek*: meaning dawn of modern fauna) epoch that we are concerned with; it is the beginning of our presentation herein and sets the stage for subsequent material presented and is highlighted in the table.

A BOLT FROM THE BLUE

About 35 million years ago (we have determined this from information derived from drilling and geophysical surveys), a large comet or meteorite (actually it was a bolide—explained later) slammed into the western Atlantic Ocean on a shallow shelf, creating the Chesapeake Bay impact crater. At that catastrophic moment in Earth's history, it was as if Mother Nature had reached into her endless glove (the universe) grabbed hold of that fiery, flaming, scorching mass and wound up her arm, and threw Earth a massive knuckleball. Yes, it was an arching, white-hot knuckle-ball at least

TABLE 13.1
Geologic Time Scale

Erathem or Era	System, Subsystem or Period, Subperiod	Series or Epoch
Cenozoic 65 million years ago to present "age of recent life"	Quaternary 1.8 million years ago to the present	**Holocene** 11,477 years ago (±85 years) to the present—greek "holos" (entire) and "ceno" (new)
		Pleistocene 1.8 million to ~11,477 (±85 years) years ago—the great ice age—greek words "pleistos" (most) and "ceno" (new)
		Pliocene 5.3–1.8 million years ago—greek "pleion" (more) and "ceno" (new)
		Miocene 23.0–5.3 million years ago—greek "meion" (less) and "ceno" (new)
	Tertiary 65.5–1.8 million years ago	**Oligocene** 33.9–23.0 million years ago—greek "oligos" (little, few) and "ceno" (new)
		Eocene 55.8–33.9 million years ago—greek "eos" (dawn) and "ceno" (new)
		Paleocene 65.5–58.8 million years ago—greek "palaois" (old) and "ceno" (new)
	Cretaceous 145.5–65.5 million years ago "the age of dinosaurs"	Late or upper

Early or lower |
| Mesozoic 251.0–65.5 million years ago—greek means "middle life" | Jurassic 199.6–145.5 million years ago | Late or upper
Middle
Early or lower |
| | Triassic 251.0 in 199.6 million years ago | Late or upper
Middle
Early or lower |
| | Permian 299.0–251.0 million years ago | Lopingian
Guadalupian
Cisuralian |

(Continued)

TABLE 13.1 (*Continued*)
Geologic Time Scale

Erathem or Era	System, Subsystem or Period, Subperiod	Series or Epoch
	Pennsylvanian	Late or upper
	318.1–299.0 million years ago	
	The coal age	Middle
		Early or lower
Paleozoic		
542.0–251.0 million	**Mississippian**	Late or upper
Years ago	359.2–318.1 million years ago	
Age of ancient life		Middle
		Early or lower
	Devonian	
	416.0–359.2 million years ago	Late or upper
		Middle
		Early or lower
	Silurian	Pridoli
	443.7–416.0 million years ago	
		Ludlow
		Wenlock
		Llandovery
	Ordovician	Late or upper
	488.3–443.7 million years ago	
		Middle
		Early or lower
	Cambrian	Late or upper
	542.0–488.3 million years ago	
		Middle
		Early or lower
Precambrian		
Approximately 4 billion years ago to 542.0 million years ago.		

2 times brighter than Earth's moon. It was at least 2 miles wide; from the northwest horizon, it crossed paths with Earth at more than 76,000 miles per hour, more than 1,260 miles per minute—roughly, 21 miles per second. It moved too quickly to be heard and its white-hot light would have blinded had it not killed before optic nerves could signal brain matter (Tennant and Hall, 2001). This was the age, the time, and, in some cases, the world of prehistoric sharks, whales, tiny camels, modern ungulates, bats, sea cows, eagles, pelicans, quail, and vultures. Were all these lifeforms present when Mother Nature's fireball impacted Earth?

Probably not.

But will we never really know who was there and who was not there—humans (as we call them) weren't even?

Probably not.

What we do know is those who were present died in a wink [for those who love dinosaurs (remember, don't we all?) not to worry, mighty T-Rex, nasty Utah raptor, and that affable, chicken-sized Velociraptor (who would gobble you and me up in an instant) were nowhere to be seen; they had all perished at least 30 million years earlier during the KT Event—mass extinction. Some might call the Chesapeake Bay bolide impact the ultimate bolt from the blue, the ultimate shock and awe event; however, because the event was instantaneously deadly to all those within impact-affected areas it would be better stated to say that the event was a whole bunch of shock (to minerals, especially quartz) with very little awe. When death is instant, who is around, at least in the immediate area, to be awed?

So, what about the crater?

Good question.

The crater is now covered by Virginia's central to outer Coastal Plain sediments and the lower Chesapeake Bay. The Chesapeake Bay impact crater is a 56-mile wide, complex peak-ring crater with an inner and outer rim, a relatively flat-floored annular trough, and an inner basin that penetrates the basement to a depth of a least 1.2 miles (Powars and Bruce, 2000b).

Let's fast-forward to the bolide's impact result. For millennia, humans had no knowledge that this event had occurred or of the literal impact it would have on Southeastern Virginia, the formation of Chesapeake Bay, and the cause of many present-day local problems with groundwater quality and land subsidence; that it was not until a handful of modern-day Sherlock Holmes,' with Dr. C.W. Poag in the lead, had figured it out. Through intuition, bore-holing operations, application of scientific protocols, and a lot of common sense they determined that at the end of the Eocene Mother Nature's fiery curveball impacted the coastline of what today is known as Virginia's Lower Chesapeake Bay region.

DID YOU KNOW?

Descriptions of the location and geometry of the Chesapeake Bay impact crater are based on the correlation of lithostratigraphic and biostratigraphic data from cores and well cuttings, borehole geophysical logs, and seismic-reflection data (Powars and Bruce, 2000a).

Today we call this event the Chesapeake Bay Meteor incident. Those scientists in the know, however, understand that the term meteor is best replaced with the term bolide. Well, that brings us to the point, and need to explain the term bolide and many other relevant meteorite terms.

So, let us do just that.

Meteorite Terms (USGS, 1998; PWNET, 2016)

Asteroid: A rocky body orbiting the sun, usually >100 m in diameter. Most asteroids orbit between Mars and Jupiter.

Bolide: Lisa Randall (2015) in her bestselling book, *Dark Matter and the Dinosaurs* defines bolide as "an object from space that disintegrates in the atmosphere (p. 127)." USGS (1998) points out that there is no consensus on the definition of a bolide; however, it uses the term to mean an extraterrestrial body in the 1–10 km size range, which impacts the Earth at velocities literally faster than a speeding bullet (Mach 75 = 20–70 mg/s), explodes upon impact, and creates a large crater.

   The bottom line: Bolide is a generic term, used to imply that we do not know the precise nature of the impacting body (and we do not) ... whether it is a rocky or metallic asteroid, or any icy comet for example (see Figure 13.1). Note that in this book we call a bolide an extraterrestrial body of some size and some composition that impacts Earth and creates a crater; for our purposes, it is the descriptor "impact crater" that is the keyword.

Cape Charles, Virginia: Is the location of a huge peaking impact crater whose center is located near this Eastern Shore Virginia Town (see Figure 13.2).

Central peak: A small mountain that forms at the center of a crater in reaction to the force of the impact.

Chesapeake Bay Bolide: This is one of the largest known impact structures found in North America. This event has been dated at about 35–33 million

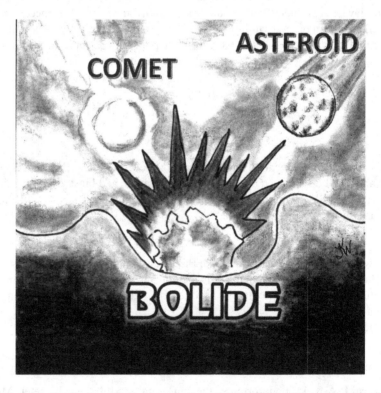

FIGURE 13.1 Bolide impact. (Adaptation from USGS (1998), *Chesapeake Bay Bolide*. Illustration by F.R. Spellman and Kat Welsh (2019).)

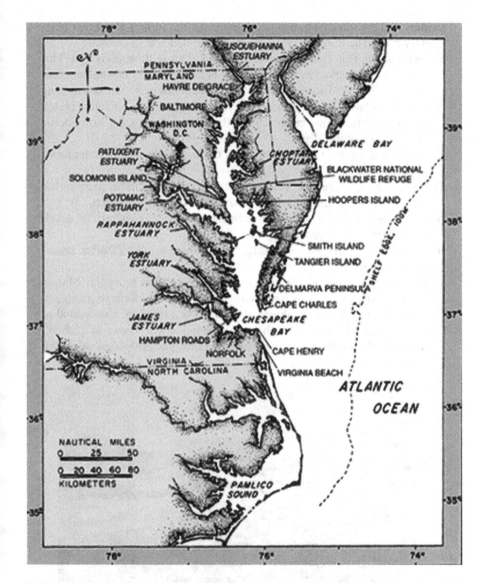

FIGURE 13.2 Chesapeake Bay Area. (Adapted from USGS (1998), *Chesapeake Bay Bolide: Modern Consequences of an Ancient Cataclysm.* Washington, DC: U.S. Geological Survey.)

years ago during the Eocene Epoch of the Cenozoic Era. (*Note*: For the purpose of simplicity and continuity in this text we assume bolide impact occurred 35 million years ago). The nature of the impact substantially affected the geology of the Atlantic continental crust and is suspected to affect the nature and quality of groundwater in southeastern Virginia. Estimated at 90 km (~55.8 miles), the crater may be about 1.3 km (~8.1 miles) deep.

Crater: Is the result of a bolide body impacting the surface of another planetary body. The resulting explosion leaves a round hole or crater.

Eocene epoch: This time period occurred 58–33.8 million years ago. This time period is marked by the emergence of mammals as the dominant land animals. The fossil record reveals many mammals quite unlike anything we have today. However, there were increasing numbers of forest plants, freshwater fish, and insects, much like those today. In fact, the term Eocene means "dawn of the recent."

Ejecta: This is the debris that shoots out of the impact site when a crater forms (Figure 13.3).

Floor: This is the bottom part of an impact crater. It can be flat or rounded and is often lower than the surrounding surface of the planet or moon.

Impact breccia: This is a rubble sediment that contains a mix of debris resulting from an impact event.

Iridium: This is a very hard and brittle metal, atomic number 77, often associated with meteorite impacts.

Mass: This is the measure of an object's inertia, that is, how heavy it is. Mass is different from weight, which measures the gravitational force on an object.

Meteor: This is a bright streak of light in the sky caused by a meteoroid or small icy particle entering Earth's atmosphere. It is also known as a "shooting star." Meteor showers occur when the Earth passes through debris left behind by an orbiting comet.

FIGURE 13.3 Fault fracture/land fissure due to land subsidence. (Adapted from USGS (1998), Illustration by F.R. Spellman and Kat Welsh (2019). *Chesapeake Bay Bolide: Modern Consequences of an Ancient Cataclysm*, U.S. Geological Survey, Washington, DC.)

Meteorite: This consists of small rocky remains of meteoroids that survive a fiery journey through Earth's atmosphere and land on Earth.

Metric units: 1 km = 0.621; 1 m = 3.28 feet; 4,000 km³ = 960 miles³.

Micrometeorite: This is a very small meteorite with a diameter of <1 mm.

Ray bright: These are lines of debris projecting from the edges of craters.

Rim: This is the highest point along the edge of a crater hole.

Rubble bed: Consists of the jumbled sediments and aged dated fossils that are associated with the Exmore beds of the Chesapeake impact structure.

Shocked materials: These are minerals, especially quartz, that show the result of tremendous forces, such as those found in impact events that alter and distort the normal optical qualities of a quartz crystal.

Tektites: These are millimeter to centimeter size glass beads (see Figure 13.4) derived from sediments melted by a bolide impact.

Tsunami: This is a very large ocean wave usually associated with underwater earthquakes or volcanic eruptions. Tsunamis may also be associated with large meteorite impacts in the oceans.

Wave material: This is material left in the trajectory of a meteor after the head of the meteor has passed.

Wall: This refers to the sides of the bowl of a crater.

FIGURE 13.4 Tektites. (From the collection of and photo by F.R. Spellman.)

With regard to the bowl of the Crater and the basin created by the bolide impact event, the inner basin includes a central uplift surrounded by a series of concentric valleys and ridges.

A contour tracing of seismic-reflection data, including basement data down to 6.0 s two-way travel time, shows the seismic "fingerprint" of a bowl-shaped zone of intensely shocked basement rocks down to about 3.5 s two-way travel time (about 33,000–37,0000 ft; 6.2–7 miles). The outer rim of the crater traverses the lower York-James Peninsula (Powars and Bruce, 2000a).

DID YOU KNOW?

In this text, the work of Dr. C. Wylie Poag is referenced often. This is only fitting due to his flagship work on investigating and finding the Chesapeake Bay bolide crater. Therefore, it is only fitting that Dr. Poag, a senior research scientist with the U.S. Geological Survey be highlighted here by sharing a short version of his biography. Dr. Poag's research emphasizes the integration of subsurface geophysical, geological, and paleontological data to reconstruct the stratigraphic framework and depositional history of the Atlantic and Gulf Coast margins of the United States. His 30-year geological career includes experience as a petroleum explorationist, a university professor, and a project coordinator for the National Science Foundation's Deep Sea Drilling Project. D. Poag has published more than 200 abstracts, articles, and books. A recent highlight of Dr. Poag's research has been the identification of the largest impact craters in the United States buried beneath the lower part of Chesapeake Bay and its surrounding peninsulas (USGS, 1997).

THE HORRIBLE GASH

Earth-changing events occur almost daily; they are on-going processes that are beyond human control: we cannot control earthquakes; we cannot prevent or stop iceberg calving; we cannot prevent volcano eruption or collapse—remember the 1980 lateral eruption of Mt. St. Helens when 14 miles of animals and trees were totally destroyed, not forgetting that 57 humans also lost their lives; we cannot, in most cases, prevent landslides, avalanches, megafloods; and we cannot prevent meteor strikes. While it is true that today we can witness or review news and film reports about such events it is also true that millions of these events occurred in ancient times, before humans were even thought of. Because many of these ancient events occurred millions and even a billion or more years ago, today many are not visible because the master architect, Mother Nature, continues to remove, degrade, eliminate or cover up past earth-shaping and changing events. Only through discovery, research, intuition, and, in some cases, common sense are we able to discern earth-changing events of the past (we study the geologic record). This is certainly the case with the discovery of the Chesapeake Bay Crater.

Here is what we know, or think we know now about the creation of the Chesapeake Crater and its subsequent formation of the Lower Chesapeake Bay Region.

It was certainly an out of the blue event. However, most experts just refer to it as a bolide blasting into the Earth and into the shallow sea that covered Virginia from Cape Henry to Richmond; thus, this certainly can be classified as an out of the blue Earth-changing event. It certainly changed things in southwest to upper Virginia. The Chesapeake Bay bolide exploded with more force than the combined nuclear arsenal of today's world powers (Tennant and Hall, 2001). The impact cracked the Earth as deep as 7 miles. The bolide blasted into creation a crater 85 miles wide, creating a flash of evaporating ocean water (millions of gallons evaporated instantly; millions more were hurled 60 miles into the atmosphere) and a volume of ejected bedrock that may have risen in a towering about 30 miles high (National Geographic, 2001). Most of the debris fell back into crater. Some "shocked quartz" and tiny glass beads—"tektites" (see Figure 13.4)—were scattered as far as New Jersey and rained down on Texas and the Caribbean. The bolide acting like a giant drill or jack hammer and unimpeded by ocean, ripped through almost a mile of sediment and sand and penetrated the 600-million-year-old granite bedrock and pounded, hammered, milled, pulverized and minced it. Huge chunks, boulders and grains of solid Earth were propelled upward. Keep in mind that when the bolide was drilling its way into Earth the front end slowed down a bit but the back end was still flying at supersonic speed. Up to 50 miles away the velocity of the intensely hot air blast was theater than 1,500 miles per hours (USGS, 2015). The bedrock, at the front end, and the back end squeezed together like a colossal sponge, and then rebounded with a vengeance. The bedrock splintered and massive faults split open. In the chaos of searing heat and ferocity the bolide vaporized, leaving a crater 55+ miles wide with a network of fractures and fissures spread more than 40 miles beyond its rim.

As the rocks blasted out into the air, they were ignited by friction and subsequently sparked firestorms for hundreds of miles—the surrounding area was literally an inferno of unimaginable proportions. And then the heavy hand of Mother Nature using her gravity grabbed and reclaimed the boulders and water from the sky and returned them to the gaping crater. We all know that when we toss a small stone into a river pool concentric circles ripple outward as the stone drops through to pool bottom. For a brief instant, we are struck by the obvious: the stone sinks to the bottom, following the laws of gravity. Eventually the ripples die away leaving as little mark as the usual human lifespan creates in the waters of the world, and then disappears as if it had never been. Well, this is a scene, happening or experience many of us are familiar in the calmness and normality of our lives. Imagine, however, the bolide and it ejected materials rolling out creating swells forming concentric circles, headed across the ocean to Greenland, Europe, and the East Coast of the United States. Those swells (large tsunamis) raced at extraordinary speeds across the ocean floor and they rose with the land and exploded. Even the Blue Ridge Mountains felt the impact from 1,000 ft high tsunamis (Tennant and Hall, 2001). And then came the run back, adhering to the time-worn axiom applied within the realm of the bounds of Earth and based on scientific fact that what goes up will eventually come back. And come back it did, with a settling. Remember, there are very few substances, if any, that are or can be more destructive of water on the move (give enough time, water destroys anything and everything; again, it is only time that stands in the way). All that tsunami water that ran up the Blue Ridge and licked and then sucked at the Peaks of Otter (and other places) and loosened house sized boulders and other objects and then the water rushed back in a whoosh transporting tons of rock, soil, trees, and wildlife of that time and filled the gigantic empty crater as deep as the Grand Canyon and the rest went off to the sea riding at a gallop the subterranean wave.

And then, oh it was so quiet, so peaceful, the surrounding area steaming, barren; it was over and there was no one around to remember that the impact had happened at all. Unlike the stone thrown into a pond with ripples that form and disappear in rings and leave no evidence that they were ever formed, the Chesapeake Bay bolide was somewhat different; so it can be said again: it was like the common river pool momentarily disturbed by a pebble thrown into its surface in that its bolide-struck Atlantic coastline was still, quiet, at peace—however, below its surface, in the murky depths, it left scars and the effects of its occurrence, but it took time for us to discover this, some of which we are still discovering today and hopefully more tomorrow. Hopefully, discoveries of this sort will be ongoing.

You probably noticed that it was mentioned above that the bolide struck the Atlantic coastline and not the Chesapeake Bay. This was the case, of course, because, at the time of impact, there was no Chesapeake Bay or Eastern shore; it was certainly not a coastline formation as we know it today. The Atlantic Ocean at that time hugged the area near what is now known as Richmond, Virginia, roughly in the path of Interstate-95 today. During the late Eocene and at the time of the bolide impact the area was covered by ocean (sea level was much higher than it is today due to the climate being much warmer than it is at present). The Chesapeake Bay itself did not form until after the Wisconsin Glaciation ice sheet melted 18,000 years ago. As described in a story in the Richmond Times-Dispatch (2005):

> The waves nearly overlapped the Blue Ridge Mountains before washing back into the horrible gash, and then covered the super-heated water beneath a thick blanket of debris, rock and sediment. Over time, as new geologic formation settled, it set the stage for Virginia's baffling groundwater system, with its pockets of salty groundwater. USGS geologist Wylie Poag, another co-discover of the bay's ancient depression, has called it "probably the most dramatic geological event that ever took place in the Atlantic margin of North America."

Over the last 35 million years, erosion has deposited sediments on top of the water, and shifts in the path of the Susquehanna River have formed the peninsula of the Eastern Shore. Today, the impact crater is buried under 1,500 ft of sand, silt, clay, and gravel, with the center of the crater underneath the Delmarva Peninsula (GSA, 2009). Most people know what an impact crater looks like. When we look up to Earth's moon it is quite evident to us (or so we think is the case) on what an impact crater looks like. Also, the image was derived from Meteor Crater in Arizona, some 38 miles east of Flagstaff, the archetypical example of what cratering experts call a *simple* crater. It is a shallow, bowl-shaped excavation, 1 km in diameter, with an upraised subcircular rim, and is extraordinarily well-preserved. Craters wider than 10 km are classified as complex craters, because they exhibit additional features. Complex craters are different than simple bowl-shaped craters because the object that created them hit hard and fast enough to melt the rock and splash it tall in the center like a skyscraper, where it hardened. Like simple craters, the outer margin of complex craters is marked by a raised rim. Inside the rim is a broad, flat, circular plain, called the annual trough. Lager slump blocks fall away from the center's outer wall and slide out over the floor of the annular trough toward the crater center. The inner edge of the annular trough is marked by either a central mountainous peak, a ring of peaks (a peak ring), or both. Inside the peak ring is the deepest part of the crater, called the inner basin. The Chesapeake Bay crater has all the characteristics of a peak-ring crater and is said to look like an upside-down sombrero, with its upturned outer, rim, a trough, and then a high peak in the center (see Figures 13.5 and 13.6).

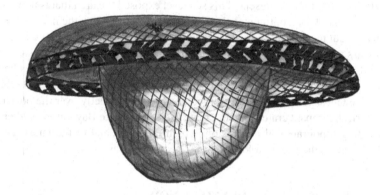

FIGURE 13.5 Illustration of upside-down sombrero. (Illustration by F.R. Spellman and Kat Welsh.)

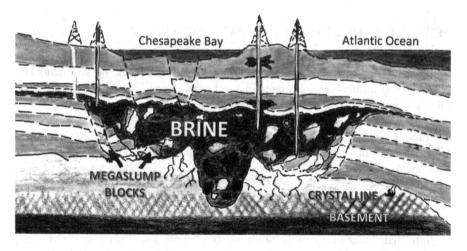

FIGURE 13.6 Adaptation of USGS illustration of the upside-down sombrero-shaped Chesapeake Bay bolide impact crater (1997). (Illustration by F.R. Spellman and Kat Welsh.)

Before we move onto the effects of the Chesapeake Bay Bolide strike, we need to make an important point clear. Namely and specifically, the Chesapeake Bay Bolide did not create the Chesapeake Bay.

No?

Yes, no.

Here is what we know "The Bay" is nowhere near 35 million years old. The fact is as late as 18,000 years ago; the Bay region was dry land.

Dry land?

Yes.

It was the last ice sheet that created what we know of Chesapeake Bay today. This last great ice sheet was at its maximum over North America, and the sea level was about 200 m lower than at present. This sea level exposed the area that is now the bay bottom and part of the continental shelf. With sea level this low the major east coast rivers had to cut narrow valleys across the region all the way to the shelf edge. About 10,000 years ago, however, the ice sheets began to melt rapidly, causing sea levels to rise and flood the shelf and the coastal river valleys—these flooded valleys because of the major modern estuaries, like Delaware Bay and Chesapeake Bay. Note that the rivers of the Chesapeake region converged at a location directly over the buried crater. In short, the impact crater did not create the Chesapeake Bay but instead created a long-lasting topographical depression, which, after the end of the latest Ice Age, helped determine the eventual location of Chesapeake Bay.

DID YOU KNOW?

It is interesting to note that the Chesapeake Bay impact targeted three layers: an upper layer of seawater, a middle layer of sediments; and a lower layer of igneous and metamorphic rocks, called basement rocks.

EFFECTS OF THE CHESAPEAKE BAY BOLIDE IMPACT

One might think that after the bolide hit, water blasted out of the area would be sucked back in along with whatever debris was nearby—including sediments laid down over 35 million years scattered in a jackstraw clutter and crushed debris—and all this would be enough to fill the crater's enormous gorge.

Well, not exactly.

Although it is true that water and debris were sucked back into the chasm it is also true that it was not enough to eliminate the low spot in the floor of what is now the Chesapeake Bay.

DID YOU KNOW?

Note that the impactor structure consists of three parts: (1) a central crater, which, upon impact, briefly existed as hole (transient crater) that was about 16 miles wide and 5 miles deep before it walls collapsed and its floor rebounded (central uplift), partly filling the hole in a few minutes; (2) and annular trough, in which widespread impact deformation primarily was limited to the sediment and uppermost basement layers; and (3) an outer fracture zone, where isolated faults and areas of disruption occur also in the same layers (USGS, 2015).

Keep in mind that we did not know until recently that the crater existed on the floor of Chesapeake Bay and the surrounding area. Dr. Poag's (and others') discovery of the giant crater has completely revised our understanding of the Atlantic Coastal Plain evolution. What many people in the Hampton Roads (Tidewater) Region do not know is that Poag's studies revealed several consequences of the ancient cataclysm that still affect citizens in the Bay Area today. These consequences include the location of the Chesapeake Bay, river diversion, ground instability, disruption of coastal aquifers, and land subsidence (pointed out in Figure 13.7). Although this text focuses primarily on the disruption of local aquifers and land subsidence related to bolide impact and relative sea level rise and the ongoing rebound efforts brought about by Hampton Roads Sanitation District's (HRSD's SWIFT initiative) and United States Geological Survey (USGS), it is also important to point out that the effects of the bolide impact are multiple and these are briefly discussed below.

LOCATION OF CHESAPEAKE BAY

Thirty-five million years ago the Chesapeake Bay did not exist. In fact, as late as 18,000 years ago, the bay region was dry land; the last great ice sheet was at its maximum over North America, and sea level was about 200 m lower than at present. This sea level exposed the area that is now the bay bottom and continental shelf. Because the sea level was so low, the major east coast rivers had to cut narrow valleys across the region all the way to the shelf edge. About 10,000 years ago, however, the ice sheets began to melt rapidly, causing sea levels to rise and flood the shelf and the coastal river valleys. The flooded valleys became the major modern estuaries, like Delaware Bay and Chesapeake Bay. To come to the point, the impact crater created a long-lasting topographic depression, which helped determine the eventual location of Chesapeake Bay (USGS, 2016).

RIVER DIVERSION

The rivers of the Chesapeake region converged at a location directly over the buried crater. Some might think the convergence of these rivers is merely a coincidence.
   Is it?
   The short answer: is no, it is not a coincidence. Notice that in Figure 13.8, the important river channels in the area change course significantly just after they cross the rim of the buried bolide crater. These channels are actually successive buried ice-age channels of the ancient Susquehanna River (formed from 450,000 to

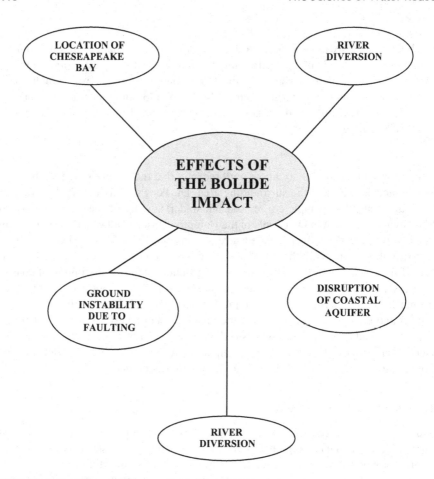

FIGURE 13.7 Effects of Chesapeake bolide impact.

20,000 years ago). Combined with seismic evidence that past impact units sag and thicken over the crater and with the river diversion indicates that the ground surface over the crater remained lower than the areas outside the crater for 35 million years.

The question is, why?

Why does the Rappahannock River flow southeastward to the Atlantic and in contrast, the York and James Rivers make sharp turns to the northeast near the outer rim of the crater?

What is the answer?

Well, the course of the York and James Rivers in the Lower Bay region is the result of the ongoing influence of differential subsidence over the bolide crater.

Differential subsidence?

Yes. Absolutely. Two factors cause subsidence in the region. First, subsidence is the result of loading during the past 35 million years since the impact. Second, subsidence is also due to compaction of the breccia; that is, a rock composed of broken fragments of minerals or gravel cemented together by a fine-grained matrix. The crater breccia is 1.2 km thick and was deposited as water-saturated sandy, rubble-bearing,

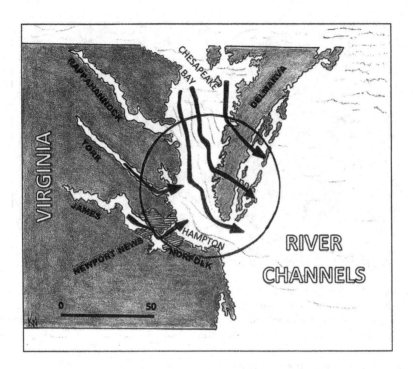

FIGURE 13.8 River channels Chesapeake Bay region. (Adapted from USGS (1997), Illustration by F.R. Spellman and Kat Welsh.)

non-jelled jello-like slurry. The sediment layers surrounding it were already partly consolidated, so the mushy breccia would compact much more rapidly under its subsequent sediment load than the surrounding strata.

You may be asking what all this has to do with the local bay rivers? The combination of the two factors detailed above produced a subsidence differential, causing the land surface over the breccia to remain lower than the land surface outside the crater. Therefore, the river valleys covered the crater and were located in those particular places when rising sea levels flooded them. In short, the impact crater created a long-lasting topographic depression, which helped predetermine the eventual location of Chesapeake Bay (USGS, 1997). Finally, it is important to point out that one of the main focuses of this book and the SWIFT initiative is land subsidence in the Chesapeake Bay area; this will be discussed in detail later.

For now, the important point to know and remember is the continued influence of differential subsidence over the crater.

DID YOU KNOW?

Note that the collapsed central crater is about 23 miles in diameter. If the widths of the annular trough and outer fracture zone are included, the full diameter of the structure is 83–85 miles.

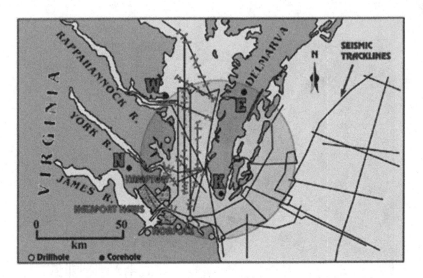

FIGURE 13.9 The location of faults (dashes) where they cross seismic profiles. The large circle shows the extent of the buried crater. The brick pattern shows the three main cities of the lower Chesapeake Bay. Capital letters mark the locations of Newport News, Windmill Point, Exmore, and Kiptopeke coreholes (USGS, 2016).

GROUND INSTABILITY DUE TO FAULTING

Seismic profiles across the crater show many faults that cut the sedimentary beds above the breccia and extend upward toward the bay floor (see Figure 13.9). The current resolution of our seismic profiles allows us to trace the faults to within 10 m of the bay floor. These faults represent another result of the differential compaction and subsidence of the breccia. As the breccia continues to subside under a load of post-impact deposits, it subsides unevenly due to its viable content of sand and huge clasts. This eventually causes the overlying beds to bend and break, and to slide apart along the fault planes. These faults are zones of crustal weakness and have the potential for continued slow movement, or sudden large offsets if reactivated by earthquakes.

Some might ask why it is important to know about the geological faults (planar fractures or discontinuities) and their location. It is important for us to know in detail the location, orientation, and amount of offset of these compaction faults because of the potential for the faulting to separate adjacent sides of the confining unit over the salt-water reservoir. If this occurred, it could allow the salty water to flow upward and contaminate the freshwater supply.

Using the seismic profiles on hand, Dr. Poag has identified and is mapping more than 100 faults or fault clusters around and over the crater, which reaches to or near the bay floor (USGS, 2016).

DISRUPTION OF COASTAL AQUIFERS

The hydrogeological framework thought to be typical of southeastern Virginia, in cross-section, consists of groundwater aquifers alternating with confining beds. The aquifers are mainly sands, which contain water-filled pore spaces between the sand grains. The pore spaces are connected, which allows the water to flow slowly through the aquifers. The confining beds are mainly clay beds, which have only very fine pores, and these are poorly interconnected, which greatly retards or prevents the flow of water. Before we knew about the Chesapeake Bay crater, this framework of alternating aquifers and confining units was applied to models of groundwater flow and water-quality assessments in the lower Chesapeake Bay region.

Based on core samples researchers have determined that in the crater area itself, the orderly stack of aquifers seen outside the crater does not exist; instead, they were truncated and excavated by the bolide impact. In place of those aquifers, there is now a single huge reservoir with a volume of 4,000 km³. That is enough breccia to cover all of Virginia and Maryland with a layer 30 m thick. But the most startling part is that this huge new reservoir does not contain fresh water like the aquifers it replaced; the pore spaces are filled with briny water that is 1.5 times saltier than normal seawater. This water is too salty to drink or to use in industry (USGS, 2016). It is interesting to note that for decades geohydrologists and others in the Hampton Roads region scratched their collective heads and wondered why locations away from the crater water wells yielded good quality freshwater suitable for potable purposes, however, whenever wells were drilled within the crater ring or close to it salty water was all that could be found.

DID YOU KNOW?

The parameters for saline water are:

- **Fresh water**: <1,000 ppm
- **Slightly saline water**: From 1,000 to 3,000 ppm
- **Moderately saline water**: From 3,000 to 10,000 ppm
- **Highly saline water**: From 10,000 to 35,000 ppm
- **Ocean water**: Contains about 35,000 ppm of salt
- **Chesapeake Bay crater water**: Contains about 1.5 times more salt than normal seawater

The presence of this hypersaline aquifer has some practical implications for groundwater management in the Lower Bay region. For example, we need to know how deeply buried the breccia is in order to avoid drilling into it inadvertently and contaminating the overlying freshwater aquifers. Its presence also limits the availability of freshwater. On the Delmarva Peninsula, over the deepest part of the crater, only the aquifers above the breccia are available for freshwater. The crater investigation shows that we need to be especially conservative about groundwater use in the area (USGS, 2016).

LAND SUBSIDENCE

Land subsidence and the potential for land rebound provided by injecting treated wastewater to drinking water quality into the Potomac Aquifer is one of the central points of this book. Later much will be said about this topic and how it is related to the Chesapeake Bay bolide impact crater and its effect, along with the relative sea level rise occurring at present. For now, it is important to point out that there is growing evidence that accelerated land subsidence is reflected in the geology and topography of the modern land surfaces around the bolide crater. The breccia is 1.3 km thick and was deposited as water-saturated, sandy, rubble-bearing slurry (like concrete before it hardens). The sediment layers surrounding the crater, on the other hand, were already partly consolidated, and so the mushy breccia compacted much more rapidly under its subsequent sediment load than the surrounding strata. The compaction differences produce a subsidence differential (i.e., the difference in subsidence between two points on the crater), causing the land surface over the breccia (due to breccia compaction) to remain lower than the land surface over sediments outside the crater.

During Dr. Poag's investigation, he and his team observed that the boundary between older surface rocks and younger surface rocks coincides with the position and orientation of the crater rim on all three peninsulas that cross the rim. The older beds have sagged over the subsiding breccia, and the younger rocks have been deposited in the resulting topographic depression. The topography also reflects the differential subsidence. The Suffolk Scarp and the Ames Ridge are elevated landforms (10–15 m high) located at, and oriented parallel to, the crater rim.

Crater-related ground subsidence also may play a role in the high rate of relative sea-level rise documented for the Chesapeake Bay region. One of the locations of the highest relative sea-level rise is Hampton Roads (the lower part of the James River, located over the crater rim).

THE BOTTOM LINE

This chapter has set the first row of foundation blocks (or the first dot in the chain) for the material to follow. Specifically, this chapter described the late Eocene period in the Virginia Coastal Plain area when the formerly quiescent geological regime was dramatically transformed when a bolide struck in the vicinity of the Delmarva Peninsula. The Chesapeake Bay Impact Crater has greatly influenced the structural, stratigraphic, and hydrogeologic framework of the Lower Chesapeake Bay Region. This consequential event produced the following principal consequences (USGS, 1998):

- The bolide carved a roughly circular crater twice the size of the state of Rhode Island (~6,400 km^2), and nearly as deep as the Grand Canyon (1.3 km deep).
- The excavation truncated all existing groundwater aquifers in the impact area by gouging ~4,300 km^3 of rock from the upper lithosphere, including Proterozoic and Paleozoic crystalline basement rocks and Middle Jurassic to upper Eocene sedimentary rocks.
- A structural and topographic low formed over the crater.
- The impact crater may have predetermined the present-day location of Chesapeake Bay.

- A porous breccia lens, 600–1,200 m thick, replaced local aquifers, resulting in groundwater ~1.5 times saltier than normal seawater.
- Long-term differential compaction and subsidence of the breccia lens spawned extensive fault systems in the area, which are potential hazards for local population centers in the Chesapeake Bay area.

NOTE

1 The information in this part is adapted from Spellman, F.R. (2020). *Sustainable Water Initiative for Tomorrow*. Lanham, MD: Bernan Press.

REFERENCES

GSA. (2009). Chesapeake Bay impact structure: Postimpact sediments. In G.S. Gohn (ed.), *ICDP-USGS Deep Drilling Project in the Chesapeake Bay Impact Structure: Results from the Eyreville Core Holes.* Boulder, CO: The Geological Society of America.

National Geographic. (2001). Chesapeake Bay Crater Offers Clues to Ancient Cataclysm. November 13. Accessed at https://newsnationalgeographic.com/news/2001/1113_chesapeakecrater.html.

Powars, D.S., and Bruce, T.S. (2000). *The Effects of the Chesapeake Bay Impact Crater on the Geological Framework and Correlation of Hydrogeologic Units of the Lower York-James Peninsula, Virginia.* Washington, DC: USGS. Accessed 09/29/19 @ https://pubs.usgs.gov/pp/p1612/powars.html.

Powars, D.S. and Bruce, T.S. (2000b). *Chesapeake Bay Impact Crater.* Washington, DC: USGS.

PWNET. (2016). The impact crater. Prince William NET accessed at https://meteor.pwnet.org/impact_event/impact_crater.htm.

Randall, L. (2015). *Dark Matter and the Dinosaurs.* New York: Harper Collins.

Richmond Times Dispatch. (2005). Drill Explores Blast: Research Seeks Insight into Explosion That Carved Huge Crater under the Chesapeake. September 8. No longer online.

Tennant, D., and Hall, M. (2001). The Chesapeake Bay meteor: A mystery, meteors and one man's quest for the truth. *The Virginian-Pilot*, June 24.

USGS. (1997). Location of Chesapeake Bay. Accessed at https://woodshole.er.usgs.gov/epubs/bolide/location_of_bay.html.

USGS. (1998). *The Chesapeake Bay Bolide.* Washington, DC: US Geological Survey.

USGS. (2015). The Chesapeake Bay Impact Structure. Accessed 2/4/24 @ https://pubs.usgs.gov/fs/2015/fs20153071.pdf.

USGS. (2016). The Chesapeake Bay Bolide Impact: A New View of Coastal Plain Evolution: Fact Sheet 049-98. Accessed https://pubs.usgs.gov/fs/fs/49-98/.

14 Groundwater Hydraulics
$Q = kiA$

Note to reader: Groundwater hydraulics (aka groundwater fluid mechanics), the movement of water from the surface to underground aquifers is Mother Nature's natural process of stocking and restocking, in many areas, the ocean of fresh water beneath our feet. This chapter about groundwater hydraulics is important to readers in allowing the gaining of understanding about one of the main purposes of the SWIFT initiative; that is, the human process of restocking an aquifer—in this particular instance, it is the Potomac Aquifer being restocked by HRSD's SWIFT and discussed in detail later. Figure 14.1 shows three SWIFT process discharge pipes conveying water for sampling and test purposes of the SWIFT water discharged into the Potomac Aquifer. The site to the right side is the testing point for the Upper Potomac Aquifer, the middle site is for testing water within the Middle Potomac

FIGURE 14.1 HRSD's SWIFT sampling and testing sites at the Nansemond treatment plant in Suffolk, Virginia. The right side test site tests water in the Upper Potomac Aquifer; the middle test site tests water in the Middle Potomac Aquifer; the left test site is for testing water in the Lower Potomac Aquifer (Photo by F.R. Spellman).

DOI: 10.1201/9781003498049-17

Aquifer, and the site to the left side is the testing the water in the Lower Potomac Aquifer. Note that the SWIFT imitative project has a state-of-the-art laboratory and appropriate equipment for testing SWIFT effluent as it enters the Potomac Aquifer. The laboratory personnel consist mostly of graduate students working on their master's Degrees and PhDs.

OCEAN BELOW OUR FEET

Unbeknownst to most of us, our Earth possesses an unseen ocean; it constitutes 99% of usable freshwater on Earth. This ocean, unlike the surface oceans that cover most of the globe, is freshwater: it is the groundwater that lies contained in aquifers beneath the Earth's crust. Whether we call groundwater an aquifer, porewater, subsurface water, subterranean water, underground water, well water, or phreatic water (Greek for spring or well water), this gigantic water source forms a reservoir that feeds all the natural fountains and springs of Earth. But how does water travel into the aquifers that lie under the Earth's surface?

Groundwater sources are replenished from a percentage of the average ~3 ft of water that falls to Earth each year on every square foot of land. Water falling to Earth as precipitation follows three courses. Some runoff directly to rivers and streams (roughly 6 inches of the 3 ft), eventually working back to the sea. Evaporation and transpiration through vegetation take up about 2 ft. The remaining 6 inches seeps into the ground, entering and filling every interstice, each hollow and cavity. Gravity pulls water toward the center of the Earth. That means that water on the surface will try to seep into the ground below it. Although groundwater comprises only one-sixth of the total, (1,680,000 miles of water), if we could spread out this water over the land, it would blanket it to a depth of 1,000 ft.

The science of groundwater hydraulics (aka fluid mechanics) is concerned with evaluating the occurrence, availability, and quality of groundwater. In particular, groundwater hydraulics is concerned with the natural or induced movement of water through permeable rock formations. To understand groundwater hydraulics, it is important to understand the operation of the natural plumbing system within it. Moreover, Earth's natural plumbing system can only be understood if the geologic framework—flow through permeable rock formations—is understood.

To get even close to understanding basic groundwater hydraulics, it is necessary to have a basic comprehension of the fundamental properties of unconfined and confined aquifers.

Part of the precipitation that falls on land infiltrates the land surface, percolates downward through the soil under the force of gravity, and becomes groundwater. Groundwater, like surface water, is extremely important to the hydrologic cycle and to our water supplies. Almost half of the people in the U.S. drink public water from groundwater supplies. Overall, more water exists as groundwater than surface water in the U.S., including the water in the Great Lakes. But sometimes, pumping it to the surface is not economical, and in recent years, pollution of groundwater supplies from improper disposal has become a significant problem.

DID YOU KNOW?

There might be as many as 200 times the total annual flow in all rivers stored in freshwater aquifers and there are saline and briny aquifers present at depth below many freshwater aquifers. This certainly is the case of the aquifers below the Chesapeake Bolide Crater. It was this finding that made scientists wonder why—how did the saline and briny water get inside the aquifers? This question led to the discovery of the Bolide Crater.

We find groundwater in saturated layers called aquifers under the Earth's surface. Hydrologists call groundwater aquifers hydrologic shock absorbers that lessen the impact of rainfall events; they level out the flow in rivers—keep in mind that the base flow of rivers is due to gradual groundwater discharge. Three types of aquifers exist: unconfined, confined, and springs. Aquifers are made up of a combination of solid materials such as rock and gravel and open spaces called pores. Regardless of the type of aquifer, the groundwater in the aquifer is in a constant state of motion. This motion is caused by gravity or by pumping.

The actual amount of water in an aquifer depends upon the amount of space available between the various grains of material that make up the aquifer. The amount of space available is called porosity. The ease of movement through an aquifer is dependent upon how well the pores are connected. For example, clay can hold a lot of water and has high porosity, but the pores are not connected, so water moves through the clay with difficulty. The ability of an aquifer to allow water to infiltrate is called permeability.

UNCONFINED AQUIFERS

The aquifer that lies just under the Earth's surface is called the zone of saturation, an unconfined aquifer (water table) (see Figure 14.2). This type of aquifer is composed of granular materials, such as mixtures of clay, silt, sand, and gravel. The top of the zone of saturation is the water table. An unconfined aquifer is only contained on the bottom and is dependent on local precipitation for recharge. Unconfined aquifers are a primary source of shallow well water (see Figure 14.2); these shallow wells are not desirable as a public drinking water source. When a well is sunk a few feet into an unconfined aquifer, the water level remains, for a time, at the same altitude at which it was first reached in drilling, but of course, this level may fluctuate later in response to many factors. They are subject to local contamination from hazardous and toxic materials—fuel and oil, and septic tanks and agricultural runoff providing increased levels of nitrates and microorganisms. These wells may be classified as groundwater under the direct influence of surface water (GUDISW), and therefore require treatment for control of microorganisms. The water level in wells sunk to greater depths in unconfined aquifers may stand at, above, or below the water table, depending upon whether the well is in the discharge or recharge area of the aquifer.

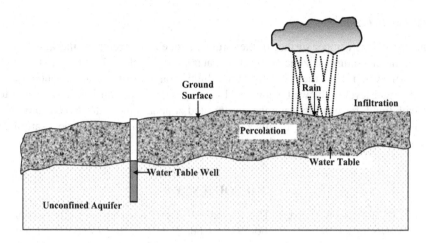

FIGURE 14.2 Unconfined aquifer.

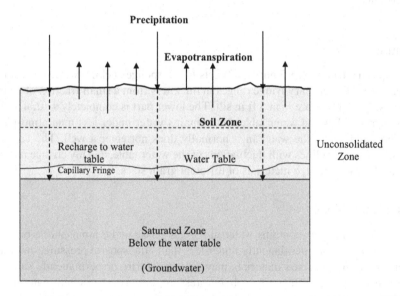

FIGURE 14.3 The unsaturated zone, capillary fringe, water table, and saturated zone.

This type of aquifer is often called a water table aquifer. The unconfined groundwater below the water table is under pressure greater than atmospheric. Unconfined aquifers consist of an unsaturated and a saturated zone (see Figure 14.3). In the unsaturated zone, the spaces between particle grains and the cracks in rocks contain both air and water. Although a considerable amount of water can be present in the unsaturated zone, this water cannot be pumped by wells because capillary forces hold it too tightly. In contrast to the unsaturated zone, the voids in the saturated zone are completely filled with water.

SATURATED ZONE

The approximate upper surface of the saturated zone is referred to as the water table. Water in the saturated zone below the water table is referred to as groundwater. Below the water table, the water pressure is high enough to allow water to enter a well as the water level in the well is lowered by pumping, thus permitting groundwater to be withdrawn for use. Between the unsaturated zone and the water table is a transition zone, capillary fringe. In this zone, the voids are saturated or almost saturated with water that is held in place by capillary forces (USGS, 2016a).

DID YOU KNOW?

It's all about slow seepage. The movement of groundwater normally occurs as slow seepage through the pore spaces between particles of unconsolidated Earth materials or through networks of fractures and solution openings in consolidated rocks.

CAPILLARY FRINGE

The capillary fringe (see Figure 14.3) acts like a sponge sucking water up from the water table. The capillary fringe ranges in thickness from a small fraction of an inch in coarse gravel to more than 5 ft in silt. The lower part is completely saturated, like the material below the water table, but contains water under less than atmospheric pressure, and hence the water in it normally does not enter a well. The capillary fringe rises and declines with fluctuations of the water table, and my change in thickness as it moves through materials of different grain sizes.

UNSATURATED ZONE

The *unsaturated zone* contains water in the gas phase under atmospheric pressure, water temporarily or permanently under less than atmospheric pressure, and air or other gases. Fine-grained materials may be temporarily or permanently saturated with waste under less than atmospheric pressure, but coarse-grained materials are unsaturated and generally contain liquid water only in sites surrounding the contacts between grains.

CAPILLARITY

Have you ever observed a paper towel dunked into a see-through container of water? The water magically climbs up the towel, seemingly defying gravity (and it is). This is capillary action in action.

Well, this capillary action is present in soil too. The rise of water in the interstices in rocks or soil may be considered to be caused by (1) the molecular attraction (adhesion) between the solid material and the fluid, and (2) the surface tension of the fluid, an expression of the attraction (cohesion) between the molecules of the fluid.

Have you ever noticed that water will wet and adhere to a clean floor, whereas it will remain in drops without wetting a dust-covered floor? This, along with the molecular attraction between the solid material and the fluid, is important because it also points to the tendency of the fluid's attraction to the cleanliness of the material. In addition, the height of the capillary rise is governed by the size of the opening.

DID YOU KNOW?

Water molecules are sticky; they like to stay close (cohesion) and adhere to other substances. When it encounters a porous substance, it can climb upward against gravity.

The surface of water resists considerable tension without losing its continuity. Thus, a carefully placed greased needle floats on water, as do certain insects having greasy pads on the feet or water-resistant hair on the underbodies. A good insect example is the Water Strider ("Jesus bugs") (Order: Hemiptera). These ride the top of the water, with only their feet making dimples in the surface film. Like all insects, the water striders have a three-part body (head, thorax, and abdomen), six jointed legs, and two antennae. It has a long, dark, narrow body (see Figure 14.4).

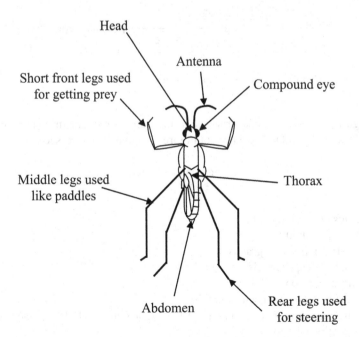

FIGURE 14.4   Water strider ("Jesus bugs") (Order: Hemiptera).

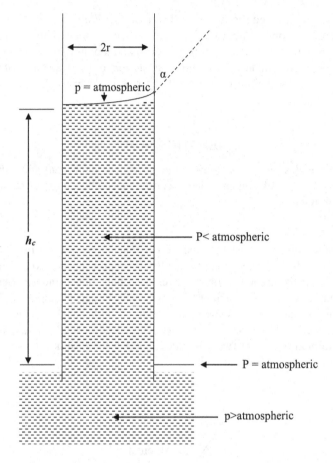

FIGURE 14.5 Capillary rise of water in a tube (exaggerated).

In Figure 14.4, the water has risen to h_c in a tube of radius r immersed in a container of water. The relations shown in Figure 14.5 may be expressed

$$\pi r^2 p g h_c = 2 \pi r T \cos \alpha \qquad (14.1)$$

where
r = radius of capillary tube
p = density of fluid
g = acceleration due to gravity
h_c = height of capillary rise
T = surface tension of fluid
α = angle between meniscus and tube

Note that, according to Equation 14.1, weight equals lift by surface tension. Solving Equation 14.1 for h_c,

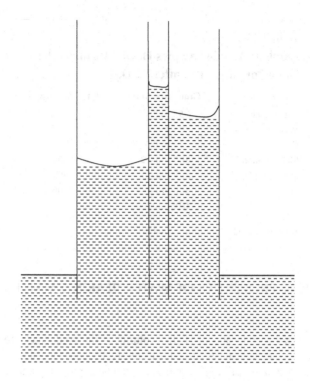

FIGURE 14.6   Rise in capillary tubes of different diameters (exaggerated).

$$h_c = \frac{2T}{rpg} \cos \alpha \tag{14.2}$$

For pure water in a clean glass, $\alpha = 0$, and $\cos \alpha = 1$. At 20°C, $T = 72.8$ dyne/cm, which may be taken as 1 g/cm, and $g = 980.665$ cm/s², whence

$$h_c = \frac{0.15}{r} \tag{14.3}$$

Surface tension is sometimes given in grams per centimeter and for pure water in contact with air, at 20°C; its value is 0.074 g/cm. In order to express it in grams per centimeter, we must divide 72.8 by g, the standard acceleration of gravity; thus 72.8 dyne/cm/980.665 cm/s² = 0.074 g/cm.

From Equation 14.3, it is seen that the height of capillary rise in tubes is inversely proportional to the radius of the tube. The rise of water in interstices of various sizes in the capillary fringe may be likened to the rise of water in a bundle of capillary tubes of various diameters, as shown in Figure 14.6. In Table 14.1, note that the capillary rise is nearly inversely proportional to the grain size.

TABLE 14.1

Capillary Rise in Samples Having Virtually the Same Porosity, 41%, after 72 Days

Material	Grain Size (mm)	Capillary Rise (cm)
Fine gravel	5 to 2	2.5
Very coarse sand	2 to 1	6.5
Coarse sand	1 to 0.5	13.5
Medium sand	0.5 to 0.2	24.6
Fine sand	0.2 to 0.1	42.8
Silt	0.1 to 0.05	105.5
Silt	0.05 to 0.02	200

Source: Based on Atterberg, cited in Terzaghi (1942).

HYDROLOGIC PROPERTIES OF WATER-BEARING MATERIALS

PorosiTY

Porosity is defined as the ratio of (1) the volume of the void spaces to (2) the total volume of the rock or soil mass. Stated differently, the porosity of a rock or solid is simply its property of containing interstices. It can be expressed quantitatively as the ratio of the volume of the interstices to the total volume and may be expressed as a decimal fraction or as a percentage. Thus,

$$\theta = \frac{v_i}{V} = \frac{v_w}{V} = \frac{V - v_m}{V} = 1 - \frac{v_m}{V} [\text{dimensionless}] \qquad (14.4)$$

where
θ = porosity, as a decimal fraction
v_i = volume of interstices
V = total volume
v_w = volume of water (in a saturated sample)
v_m = volume of mineral particles
Porosity may be expressed also as

$$\theta = \frac{p_m - p_d}{p_m} = 1 - \frac{p_d}{p_m} [\text{dimensionless}] \qquad (14.5)$$

where
p_m = mean density of mineral particles (grain density)
p_d = density of dry sample (bulk density)
Multiplying the right-hand aides of Equations 14.4 and 14.5 by 100 gives the porosity as a percentage.

Primary Porosity

Primary porosity is porosity associated with the original depositional texture of the sediment. That is, in soil and sedimentary rocks the primary interstices are the spaces between grains or pebbles. In intrusive igneous rocks, the few primary interstices result from cooling and crystallization. Extrusive igneous rocks may have large openings and high porosity resulting from the expansion of gas, but the openings may or may not be connected. With time, the metamorphism of igneous or sedimentary rocks generally reduces the primary porosity and may virtually obliterate it.

Secondary Porosity

Secondary porosity is porosity that develops after the deposition and burial of the sediment in the sedimentary basin. Fractures such as joints, faults, and openings along planes of bedding or schistosity in consolidated rocks having low primary porosity and permeability may afford appreciable secondary porosity.

Controlling Porosity of Granular Materials

In describing the conditions that control the porosity of granular materials, it is convenient to use the time-proven approach originally used by Slichter (1899). He explained that if a hypothetical granular material were composed of spherical particles of equal size, the porosity would be independent of particle size (whether the particles were the size of silt or the size of the Earth) but would vary with the packing arrangement of the particles. Slichter explained that the lowest porosity of 25.95 (about 26) percent would result from the most compact rhombohedral arrangement (Figure 14.7a) and the highest porosity of 47.64 (about 48) percent would result from the least compact cubical arrangement (Figure 14.7c). The porosity of the other arrangements, such as that shown in Figure 14.7b, would be between these limits.

In addition to the arrangement of grains (as shown in Figure 14.7) having an impact on controlling the porosity of granular materials the shape of the grains (i.e., their angularity) and their degree of assortment (i.e., range in particle size) also have an impact on porosity. The angularity of particles causes wide variations in porosity and may increase or decrease it, according to whether the particles tend to bridge openings or pack together like pieces of mosaic. The greater the range in particle size the lower the porosity, as the small particles occupy the voids between the larger ones.

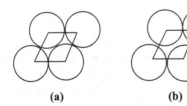 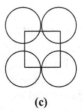

(a)   **(b)**   **(c)**

FIGURE 14.7 Sections of four contiguous spheres of equal size. A most compact arrangement, lowest porosity; (b) less compact arrangement, higher porosity; (c) least compact arrangement, highest porosity. (Modified form Slichter (1899, p. 1).)

Void Ratio

The void ratio of a rock or soil is the ratio of the volume of its interstices to the volume of its mineral particles. It may be expressed as:

$$\text{Void ratio} = \frac{v_i}{v_m} = \frac{v_w}{v_m} = \frac{\theta}{1-\theta} \, [\text{dimensionless}] \tag{14.6}$$

where
θ = porosity, as a decimal fraction
v_i = volume of interstices
v_w = volume of water (in a saturated sample)
v_m = volume of mineral particles

Permeability

The permeability of a rock or soil is a measure of how easily water can travel (i.e., how easily it transmits fluid) such as water, under a hydropotential gradient. Soil and loose sediments, such as sand and gravel, are porous and permeable. They can hold a lot of water, and it flows easily through them. The permeability is approximately proportional to the square of the mean grain diameter,

$$K \approx Cd^2 \tag{14.7}$$

where
k = intrinsic permeability
C = a dimensionless constant depending upon porosity, range, and distribution of particle size, shape of grains, and other factors
d = the mean grain diameter of some workers and the effective grain diameter of others

---

DID YOU KNOW?

Although clay and shale are porous and can hold a lot of water, the pores in these fine-grained materials are so small that water flows very slowly through them. Clay has a low permeability.

Intrinsic Permeability

The term intrinsic permeability, adopted by the U.S. Geological Survey, states that the permeability in question is an intensive property (not a spatial average of a heterogeneous block of material), that it is a function of the material structure only (and not of the fluid), and explicitly distinguishes the value from that of relative permeability. Intrinsic permeability may be expressed

$$k = -\frac{qv}{g(dh/dl)} = -\frac{qv}{(dp/dl)} \tag{14.8}$$

where
   k = intrinsic permeability
   q = rate of flow per unit area = Q/A
   v = kinematic viscosity
   g = acceleration of gravity
   dh/dl = gradient, or unit change in head per unit length of flow
   dp/dl = potential gradient, or unit change in potential per unit length of flow

From Equation 14.8, it may be stated that a porous medium has an intrinsic permeability of one unit of length squared if it will transmit in unit time a unit volume of fluid of unit kinematic viscosity through a cross-section of the unit area measured at right angles to the flow direction under a unit potential gradient.

HYDRAULIC CONDUCTIVITY

Hydraulic conductivity, represented as K, is a property of soils and rocks that describes the ease with which water can move through pore spaces or fractures. It is said that a medium has a hydraulic conductivity of unit length per unit time if it will transmit in unit time a unit volume of groundwater at the prevailing viscosity through a cross-section of unit area, measured at right angles to the direction of flow, under a hydraulic gradient of unit change in head through unit length of flow. The suggested units are:

$$K = \frac{q\left(\text{rate of flow per unit area} = Q/A\right)}{dh/dl\left(\text{gradient}\right)} \tag{14.9}$$

$$K = \frac{\text{ft}^3}{\text{ft}^2\,\text{day}\left(-\text{ft/ft}\right)} = \text{ft/day} \tag{14.10}$$

or

$$K = \frac{\text{m}^3}{\text{m}^2\,\text{day}\left(-\text{m/m}\right)} = \text{m/day} \tag{14.11}$$

Note that the minus signs in Equations 14.10 and 14.11 result from the fact that the water moves in the direction of decreasing head.

TRANSMISSIVITY (T)

The *Transmissivity* (T) is the rate at which water of the prevailing kinematic viscosity flows horizontally through an aquifer, such as to a pumping well. Transmissivity is a preferred parameter because aquifers are not uniform; they have variable hydraulic conductivity (HC), gradients, and cross-sectional area. The standard equation for transmissivity is shown as follows.

$$T = Q / W * I \tag{14.12}$$

where
   T = transmissivity
   W = the width of the aquifer
   Q = is flow
   I = hydraulic gradient

STORAGE (S)—WATER YIELDING AND RETAINING CAPACITY

Storage capacity refers to the volume of water that an aquifer can hold. It can include both the specific yield and the specific retention.

Specific Yield

Specific yield (S_y) (the portion of water that drains freely under gravity) is defined as the ratio of (1) the volume of water that a saturated rock or soil will yield by gravity to (2) the total volume of the rock or soil. This may be expressed

$$S_y = \frac{v_g}{V}\,[\text{dimensionless}] \qquad (14.13)$$

where
   S_y = specific yield, as a decimal fraction
   v_g = volume of water drained by gravity
   V = total volume

DID YOU KNOW?

Specific yield is usually expressed as a percentage. The value is not definitive, because the quantity of water that will drain by gravity depends on variables such as duration of drainage, temperature, mineral composition of the water, and various physical characteristics of the rock or soil under consideration. Values of specific yield, nevertheless, offer a convenient means by which hydrologists can estimate the water-yielding capacities of Earth materials and, as such, are very useful in hydrologic studies.

Specific Retention

The *specific retention* of a rock or soil with respect to water has been defined by Meinzer (1923) as the ratio that will be retained against gravity drainage from a saturated rock to the volume of the rock. It may be expressed

$$S_r = \frac{v_r}{V} = \theta - S_y \qquad (14.14)$$

where
   S_r = specific retention, as a decimal fraction
   v_r = volume of water retained against gravity, mostly by molecular attraction
   Note: Fine-grained materials contribute significantly to storage.

Grain Size

With *grain size*, the texture of geologic materials varies widely.

CONFINED AQUIFERS

A confined aquifer is an aquifer below the land surface that is saturated with water. A confined aquifer is sandwiched between two impermeable layers that block the flow of water.

The water in a confined aquifer is under hydrostatic pressure. It does not have a free water table (see Figure 14.8). Confined aquifers are called artesian aquifers. Wells drilled into artesian aquifers are called artesian wells and commonly yield large quantities of high-quality water. An artesian well is any well where the water in the well casing would rise above the saturated strata. Wells in confined aquifers are normally referred to as deep wells and are not generally affected by local hydrological events.

A confined aquifer is recharged by rain or snow in the mountains where the aquifer lies close to the surface of the Earth. Because the recharge area is some distance from areas of possible contamination, the possibility of contamination is usually very low. However, once contaminated, confined aquifers may take centuries to recover.

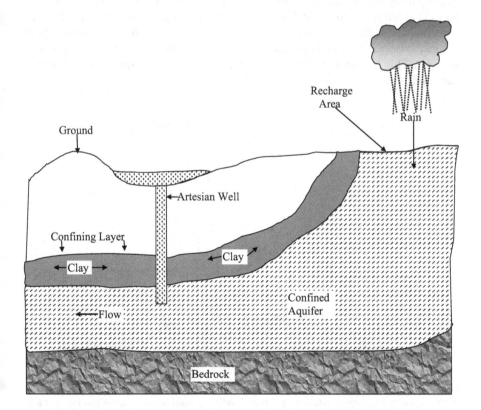

FIGURE 14.8 Confined aquifer.

Groundwater naturally exits the Earth's crust in areas called springs. The water in a spring can originate from a water table aquifer or a confined aquifer. Only water from a confined spring is considered desirable for a public water system.

STEADY FLOW OF GROUNDWATER

In a steady flow of groundwater through permeable material, there is no change in the head with time. Mathematically, this statement is symbolized by $dh/dt = 0$, which says that the change in head, dh, with respect to the change in time, dt, equals zero. Note that steady flow generally does not occur in nature, but it is a very useful concept in that steady flow can be closely approached in nature and aquifer tests, and this condition may be symbolized by $dh/dt = 0$.

Figure 14.9 shows a hypothetical example of true steady radial flow. Here steady radial flow will be reached and maintained when all the recoverable groundwater in the cone of depression has been drained by gravity into the well discharging at a constant rate Q.

Darcy's Law

Hagen (1839) and Poiseuille (1846) found that the rate of flow through capillary tubes is proportional to the hydraulic gradient; however, it was Darcy in 1856 who experimented with the flow of water through sand that determined that the rate of laminar (viscous) flow of water through sand also is proportional to the hydraulic gradient. This is known as *Darcy's law* and it is generally expressed as shown in Equation 14.15,

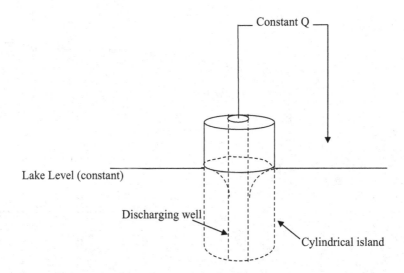

FIGURE 14.9 Hypothetical example of steady flow (well discharging at constant rate Q from a cylindrical island in a lake of constant level). (*Ground Water Hydraulics*, Department of Interior Geological Survey Paper 708 (1975).)

$$q = \frac{Q}{A} = -\frac{Kdh}{dl} \qquad (14.15)$$

It will be noted that K, the constant of proportionality in Darcy's law, is the hydraulic conductivity. Darcy's Law is also often shown in Equation 14.16.

$$Q = KIA \qquad (14.16)$$

where
Q = rate of flow
K = hydraulic conductivity
I = slope
A = cross-sectional area

Units for Darcy's law (Equation 14.16): If Q = m³/s, then K must be in m/s, A in m² (I is a unitless quantity: a 1% slope would be 0.01).

To illustrate the use of Equation 14.15 (Darcy's law), assume that we wish to compute the total rate of groundwater movement in a valley where A, the cross-sectional area, in 100 ft deep times 1 mile wide, where K = 500 ft/day, and dh/dl = 5 ft/mile. Then

$$Q = -(100\,\text{ft})(5,280\,\text{ft})(500\,\text{ft/day})\left(\frac{-5\,\text{ft}}{5,280\,\text{ft}}\right)$$

$$= 250,000\,\text{ft}^3/\text{day}$$

VELOCITY

Because the hydraulic conductivity, K, has the dimensions of velocity some might mistake this for the particle velocity of the water, whereas K is actually a measure of the volume rate of flow through a unit cross-sectional area. For the average particle velocity, v, we must also know the porosity of the material. Thus,

$$Q = \bar{v}A\theta = -KA\frac{dh}{dl}$$

where
$\bar{v}$ = average velocity, in feet per day
θ = porosity, as a decimal fraction

GROUNDWATER FLOW AND EFFECTS OF PUMPING

Water pumped from the groundwater system not only causes the water table to lower but alters the direction of groundwater movement leads to land subsidence but also can have a debilitating impact on the quality of the water within an aquifer. Generally, groundwater possesses high physical quality. When pumped from an aquifer composed of a mixture of sand and gravel, if not directly influenced by surface water, groundwater is often used without filtration. It can also be used without disinfection if it has a low coliform count (depending on location). However, groundwater can become contaminated. When septic systems fail, saltwater intrudes, improper disposal of wastes occurs, improperly stockpiled chemicals leach, underground storage tanks leak, hazardous materials spill, fertilizers, and pesticides are misplaced, and when mines are improperly abandoned, groundwater can become contaminated.

Note: Word to the wise: if you are planning on digging a well you need to tap productive aquifers; locate near users; locate near power; and drill as shallow as possible.

To understand how an underground aquifer becomes contaminated, you must understand what occurs when pumping is taking place within the well. When groundwater is removed from its underground source (i.e., from the water-bearing stratum) via a well, water flows toward the center of the well. In a water table aquifer, this movement causes the water table to sag toward the well. This sag is called the cone of depression. The shape and size of the cone depend on the relationship between the pumping rate and the rate at which water can move toward the well. If the rate is high, the cone is shallow, and its growth stabilizes. The area that is included in the cone of depression is called the cone of influence (see Figure 14.9), and any contamination in this zone will be drawn into the well.

THE 411 ON WELLS

Water supply wells may be characterized as shallow or deep. In addition, wells are classified as follows:

1. **Class I**: cased and grouted to 100 ft
2. **Class II A**: cased to a minimum of 100 ft and grouted to 20 ft
3. **Class II B**: cased and grouted to 50 ft

Note: During the well development process, mud/silt forced into the aquifer during the drilling process is removed, allowing the well to produce the best-quality water at the highest rate from the aquifer.

Issues for Any Well

- Type of well
- Location
- Design
- Installation
- Operation
- Abandonment

Shallow Wells

Shallow wells are those that are <100 ft deep. Such wells are not particularly desirable for municipal supplies since the aquifers they tap are likely to fluctuate considerably in depth, making the yield somewhat uncertain. Municipal wells in such aquifers cause a reduction in the water table (or phreatic surface) that affects nearby private wells, which are more likely to utilize shallow strata. Such interference with private wells may result in damage suits against the community. Shallow wells may be dug, bored, or driven.

1. **Dug wells**: Dug wells are the oldest type of well and date back many centuries; they are dug by hand or by a variety of unspecialized equipment. They range in size from ~4 to 15 ft in diameter and are usually about 20–40 ft deep. Such wells are usually lined or cased with concrete or brick. Dug wells are prone to failure from drought or heavy pumpage. They are vulnerable to contamination and are not acceptable as a public water supply in many locations.

2. **Driven wells**: Driven wells consist of a pipe casing terminating in a point slightly greater in diameter than the casing. The pointed well screen and the lengths of pipe attached to it are pounded down or driven in the same manner as a pile, usually with a drop hammer, to the water-bearing strata. Driven wells are usually 2–3 inches in diameter and are used only in unconsolidated materials. This type of shallow well is not acceptable as a public water supply.

3. **Bored wells**: Bored wells range from 1 to 36 inches in diameter and are constructed in unconsolidated materials. The boring is accomplished with augers (either hand or machine-driven) that fill with soil and then are drawn to the surface to be emptied. The casing may be placed after the well is completed (in relatively cohesive materials) but must advance with the well in noncohesive strata. Bored wells are not acceptable as a public water supply.

Deep Wells

Deep wells are the usual source of groundwater for municipalities. Deep wells tap thick and extensive aquifers that are not subject to rapid fluctuations in water (piezometric surface—the height to which water will rise in a tube penetrating a confined aquifer) level and that provide a large and uniform yield. Deep wells typically yield water of more constant quality than shallow wells, although the quality is not necessarily better. Deep wells are constructed by a variety of techniques; we discuss two of these techniques (jetting and drilling) as follows.

1. **Jetted wells**: Jetted well construction commonly employs a jetting pipe with a cutting tool. This type of well cannot be constructed in clay, hardpan, or where boulders are present. Jetted wells are not acceptable as a public water supply.

2. **Drilled wells**: Drilled wells are usually the only type of well allowed for use in most public water supply systems. Several different methods of drilling are available; all of which are capable of drilling wells of extreme depth and diameter. Drilled wells are constructed using a drilling rig that creates a hole into which the casing is placed. Screens are installed at one or more levels when water-bearing formations are encountered.

Well Casing

A well is a hole in the ground called the borehole. The hole is protected from collapse by placing a casing inside the borehole. The well casing prevents the walls of the hole from collapsing and prevents contaminants (either surface or subsurface) from entering the water source. The casing also provides a column of stored water and housing for the pump mechanisms and pipes. Well casings constructed of steel or plastic material are acceptable. The well casing must extend a minimum of 12 inches above grade.

Grout

To protect the aquifer from contamination, the casing is sealed to the borehole near the surface and near the bottom where it passes into the impermeable layer with grout. This sealing process keeps the well from being polluted by surface water and seals out water from water-bearing strata that have undesirable water quality. Sealing also protects the casing from external corrosion and restrains unstable soil and rock formations. Grout consists of near cement that is pumped into the annular space (it is completed within 48 hours of well construction); it is pumped under continuous pressure starting at the bottom and progressing upward in one continuous operation.

Well Pad

The well pad provides a ground seal around the casing. The pad is constructed of reinforced concrete 6 ft × 6 ft (6-inch thick) with the well head located in the middle. The well pad prevents contaminants from collecting around the well and seeping down into the ground along the casing.

Sanitary Seal

To prevent contamination of the well, a sanitary seal is placed at the top of the casing. The type of seal varies depending upon the type of pump used. The sanitary seal contains openings for power and control wires, pump support cables, a drawdown gauge, discharge piping, a pump shaft, and an air vent while providing a tight seal around them.

Well Screen

The well screen is the most important single factor affecting the efficiency of a well, and because of its importance, it is often called the heart of the well. Screens can be installed at the intake point(s) on the end of a well casing or the end of the inner casing on gravel packed well. These screens perform two functions: (1) supporting the borehole and (2) reducing the amount of sand that enters the casing and the pump. They are sized to allow the maximum amount of water while preventing the passage of sand/sediment/gravel.

DID YOU KNOW?

With regard to screen length, for confined aquifers, 80%–90% of the thickness of the water-bearing zone should be screened. The best results are obtained by centering the screen section in the aquifer. For unconfined aquifers, maximum specific capacity is obtained by use of the longest screen possible but more available drawdown results from using the shortest screen possible. These factors are optimized by screening the bottom 30%–50% of the aquifer (Driscoll, 1986).

Casing Vent

The well casing must have a vent to allow air into the casing as the water level drops. The vent terminates 18 inches above the floor with a return bend pointing downward. The opening of the vent must be screened with #24 mesh stainless steel to prevent the entry of vermin and dust.

Drop Pipe

The drop pipe or riser is the line leading from the pump to the wellhead. It assures adequate support so that an aboveground pump does not move and so that a submersible pump is not lost down the well. This pipe is either steel or PVC. Steel is the most desirable.

WELL HYDRAULICS

When the source of water for a water distribution system is from a groundwater supply, knowledge of well hydraulics is important to the operator. Basic well hydraulics terms are presented and defined, and they are related pictorially (see Figure 14.10).

- **Static water level**: the water level in a well when no water is being taken from the groundwater source (i.e., the water level when the pump is off; see Figure 14.10). Static water level is normally measured as the distance from the ground surface to the water surface. This is an important parameter because it is used to measure changes in the water table.
- **Pumping water level**: the water level when the pump is off. When water is pumped out of a well, the water level usually drops below the level in the surrounding aquifer and eventually stabilizes at a lower level; this is the pumping level (see Figure 14.10).
- **Drawdown**: the difference, or the drop, between the Static Water Level and the Pumping Water Level, measured in feet. Simply, it is the distance the water level drops once pumping begins (see Figure 14.10).
- **Cone of depression**: in unconfined aquifers, there is a flow of water in the aquifer from all directions toward the well during pumping. The free water surface in the aquifer then takes the shape of an inverted cone or curved

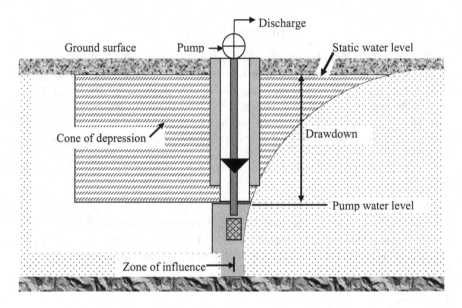

FIGURE 14.10 Hydraulic characteristics of a well.

funnel line. The curve of the line extends from the Pumping Water Level
to the Static Water Level at the outside edge of the Zone (or Radius) of
Influence (see Figure 14.10).

Note: The shape and size of the cone of depression are dependent on the relationship
between the pumping rate and the rate at which water can move toward the well. If
the rate is high, the cone will be shallow and its growth will stabilize. If the rate is
low, the cone will be sharp and continue to grow in size.

- **Zone (or radius) of influence**: the distance between the pump shaft and
  the outermost area affected by drawdown (see Figure 14.10). The distance depends on the porosity of the soil and other factors. This parameter
  becomes important in well fields with many pumps. If wells are set too
  close together, the zones of influence will overlap, increasing the drawdown
  in all wells. Obviously, pumps should be spaced apart to prevent this from
  happening.

Two important parameters not shown in Figure 14.10 are well yield and specific
capacity.

1. *Well yield* is the rate of water withdrawal that a well can supply over a long
   period, or, alternatively, the maximum pumping rate that can be achieved
   without increasing the drawdown. The yield of small wells is usually measured
   in gallons per minute (liters per minute) or gallons per hour (liters per hour).

For large wells, it may be measured in cubic feet per second (cubic meters per second).

2. *Specific capacity* is the pumping rate per foot of Drawdown (gpm/ft), or

$$\text{Specific capacity} = \text{Well yield} \div \text{drawdown} \tag{14.17}$$

Example 14.1

Problem: If the well yield is 300 gpm and the drawdown is measured to be 20 ft, what is the specific capacity?

Solution:

$$\text{Specific capacity} = 300 \div 20$$

$$\text{Specific capacity} = 15 \, \text{gpm per ft of drawdown}$$

Specific capacity is one of the most important concepts in well operation and testing. The calculation should be made frequently in the monitoring of well operation. A sudden drop in specific capacity indicates problems such as pump malfunction, screen plugging, or other problems that can be serious. Such problems should be identified and corrected as soon as possible.

Well Drawdown Calculations

As mentioned and shown in Figure 14.10, *drawdown* is the drop in the level of water in a well when water is being pumped. Drawdown is usually measured in feet or meters. One of the most important reasons for measuring drawdown is to make sure that the source water is adequate and not being depleted. The data that is collected to calculate drawdown can indicate if the water supply is slowly declining. Early detection can give the system time to explore alternative sources, establish conservation measures, or obtain any special funding that may be needed to get a new water source. Well drawdown is the difference between the pumping water level and the static water level.

$$\text{Drawdown(ft)} = \text{Pumping water level(ft)} - \text{Static water level(ft)} \tag{14.18}$$

Example 14.2

Problem: The static water level for a well is 70 ft. If the pumping water level is 90 ft, what is the drawdown?

Solution:

$$\text{Drawdown(ft)} = \text{Pumping water level(ft)} - \text{Static water level(ft)}$$

$$= 90 \, \text{ft} - 70 \, \text{ft}$$

$$= 20 \, \text{ft}$$

Example 14.3

Problem: The static water level of a well is 122 ft. The pumping water level is determined using the sounding line. The air pressure applied to the sounding line is 4.0 psi and the length of the sounding line is 180 ft. What is the drawdown?

Solution:

First, calculate the water depth in the sounding line and the pumping water level:

1. Water depth in sounding line $= (4.0\,\text{psi})(2.31\,\text{ft/psi})$

$$= 9.2\,\text{ft}$$

2. Pumping water level $= 180\,\text{ft} - 9.2\,\text{ft} = 170.8\,\text{ft}$

Then calculate the drawdown as usual:

Drawdown(ft) = Pumping water level(ft) − Static water level(ft)

$$= 170.8\,\text{ft} - 122\,\text{ft}$$

$$= 48.8\,\text{ft}$$

Well Yield Calculations

Well yield is the volume of water per unit of time that is produced from the well pumping. Usually, well yield is measured in terms of gallons per minute (gpm) or gallons per hour (gph). Sometimes, large flows are measured in cubic feet per second (cfs). Well yield is determined by using the following equation.

$$\text{Well yield(gpm)} = \frac{\text{Gallons produced}}{\text{Duration of test(minutes)}} \qquad (14.19)$$

Example 14.4

Problem: Once the drawdown level of a well stabilized, it was determined that the well-produced 400 gallons during a 5-min test.

Solution:

$$\text{Well yield(gpm)} = \frac{\text{Gallons produced}}{\text{Duration of test(minutes)}}$$

$$= \frac{400\,\text{gallons}}{5\,\text{minutes}}$$

$$= 80\,\text{gpm}$$

Example 14.5

Problem: During a 5-min test for well yield, a total of 780 gallons are removed from the well. What is the well yield in gpm? In gph?

Solution:

$$\text{Well yield(gpm)} = \frac{\text{Gallons removed}}{\text{Duration of test(minutes)}}$$

$$= \frac{780 \, \text{gallons}}{5 \, \text{minutes}}$$

$$= 156 \, \text{gpm}$$

Then convert gpm flow to gph flow : $(156 \, \text{gal/minutes})(60/\text{hours}) = 9{,}360 \, \text{gph}$

Specific Yield Calculations

Specific yield is the discharge capacity of the well per foot of drawdown. The specific yield may range from 1 gpm/ft drawdown to more than 100-gpm/ft drawdown for a properly developed well. Specific yield is calculated using the Equation 14.20.

$$\text{Specific yield(gpm/ft)} = \frac{\text{Well yield(gpm)}}{\text{Drawdown(ft)}} \qquad (14.20)$$

Example 14.6

Problem: A well produces 260 gpm. If the drawdown for the well is 22 ft, what is the specific yield in gpm/ft, what is the specific yield in gpm/ft of drawdown?

Solution:

$$\text{Specific yield(gpm/ft)} = \frac{\text{Well yield(gpm)}}{\text{Drawdown(ft)}}$$

$$= \frac{260 \, \text{gpm}}{22 \, \text{ft}}$$

$$= 11.8 \, \text{gpm/ft}$$

Example 14.7

Problem: The yield for a particular well is 310 gpm. If the drawdown for this well is 30 ft, what is the specific yield in gpm/ft of drawdown?

Solution:

$$\text{Specific yield(gpm/ft)} = \frac{\text{Well yield(gpm)}}{\text{Drawdown(ft)}}$$

$$= \frac{310 \, \text{gpm}}{30 \, \text{ft}}$$

$$= 10.3 \, \text{gpm/ft}$$

DEPLETING THE GROUNDWATER BANK ACCOUNT

Normally, as shown in Figure 14.11, water is recharged to the groundwater system by percolation of water from precipitation and then flows to a stream or other water body through the groundwater system. Where surface water, such as lakes and rivers, is scarce or inaccessible, groundwater supplies many of the hydrologic needs of people everywhere. In the United States, it is the source of drinking water for about half the total population and nearly all of the rural population, and it provides over 50 billion gallons per day for agricultural needs. Water stored in the ground for use can be compared to money in a bank account. However, water pumped from the groundwater system causes the water table to lower and alters the direction of groundwater movement. Some water that flowed to a stream or other water body no longer does so and some water may be drawn from the steam into the groundwater system, thereby reducing the amount of streamflow. Groundwater depletion, the term used to define long-term water level, is a key issue associated with groundwater use. Many areas of the United States are experiencing groundwater depletion (USGS, 2016a).

Let's get to the point here: If you withdraw money at a faster rate than you deposit new money in your bank account you will eventually start having account-supply problems. Likewise, if you pump water out of the ground faster than it is replenished over the long-term similar problems will develop. The volume of groundwater in storage is decreasing in many areas of the United States in response to pumping. Groundwater depletion is primarily caused by sustained groundwater pumping. Some of the negative effects of groundwater depletion and the focus of this book are listed below and depicted in Figure 14.12

- drying up of wells
- reduction of water in streams and lakes

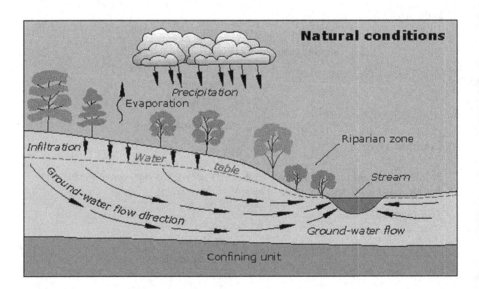

FIGURE 14.11   Normal groundwater conditions. (USGS, 2016a.)

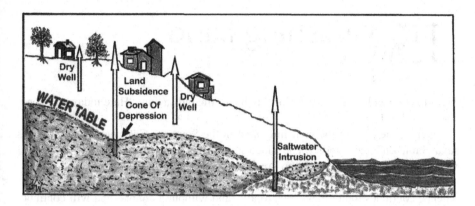

FIGURE 14.12 Impacts of over-pumping of groundwater and groundwater depletion. (Adapted from USGS (2016b). Illustration by F.R. Spellman and Kat Welsh).

- deterioration of water quality
- increased pumping costs
- land subsidence

Although the negative effects of groundwater depletion listed above and shown in Figure 14.12 are of major concern to those affected by the depletion of groundwater bank accounts, in this book, our major concern related to groundwater depletion is the lowering of the water table and land subsidence. In the next chapter, we focus on land subsidence and its implications for those affected.

THE BOTTOM LINE

The absolute value and need, the ever-increasing need, for quality groundwater cannot be overstated. The bottom line on groundwater is simple: Without groundwater planet Earth and all its inhabitants could not survive.

REFERENCES

Driscoll, F. (1986). *Groundwater and Wells*. St. Paul: Johnson Division.
Meinzer, O.E. (1923). Outline of groundwater hydrology, with definitions: U.S. *Geological Survey Water-Supply Paper* 494: 71 p.
Slichter, C.S. (1899). Theoretical investigations of the motion of ground waters. U.S. Geological Survey 19th Annual Report, pt. II-C, pp. 295–384.
Terzaghi, K. (1942). Soil moisture and capillary phenomena in soils: Chapter 10-A. In O.E. Meinzer (ed.), *Hydrology*. New York: McGraw Hill Book Co., pp. 331–363.
USGS. (2016a). Groundwater Flow and Effects of Pumping. Accessed from https://water.usgs.gov/edu/earthgwdeclien.html.
USGS. (2016b). Groundwater Depletion. Accessed @ https://water.usgs.gov/edu/gwdepletion.html.

15 Vanishing Land

Note to the reader: You have probably heard the old saying, "Here today and gone tomorrow."

Well, in the case of the vanishing land in the Southern Chesapeake Bay Region (aka Hampton Roads and/or The Tidewater Region), this is not precisely the case; this old saying does not precisely apply. In this region, the vanishing land area is occurring, but slowly—at a rate of 1.1–4.8 millimeters per year (mm/yr), and subsidence continues today inevitably, totally, and without a doubt—and will continue if some sort of mitigation is not employed. This is where HRSD's SWIFT initiative applies. SWIFT, as explained earlier, has multiple intended functions or purposes. One of these functions or purposes is to attempt to retard land subsidence which is explained below but first standard conversion factors and datum (i.e., an item of data) is supplied.

CONVERSION FACTORS AND DATUM

Multiply	By	To Obtain
	Length	
millimeter (mm)	0.03937	inch (in.)
meter (m)	3.281	foot (ft)
kilometer (km)	0.6214	mile (mi)
	Area	
square kilometer (km²)	0.004047	acre
	Flow Rate	
Millimeter per year (mm/yr)	25.4	inch per year (in/yr)

LAND SUBSIDENCE[1]

From the Lower Chesapeake Bay Region (Hampton Roads or Tidewater) of Virginia to San Francisco Bay/Delta to the Florida Everglades and from upstate New York to Houston, people are dealing with a common problem in these diverse locations—vanishing land as a result of land subsidence due to the withdrawal of groundwater. Vanishing land due to subsidence is not an isolated problem: an area of more than 17,000 square miles in 45 States, an area roughly the size of New Hampshire and Vermont combined, has been directly affected by land subsidence (USGS, 2013). More than 80% of the identified subsidence in the Nation is a consequence of our exploitation of underground water. Moreover, it seems certain that the increasing development of land and water resources threatens to exacerbate existing land-subsidence

150

DOI: 10.1201/9781003498049-18

problems and initiate new ones. This chapter focuses on three principal processes causing land subsidence: the compaction of aquifer systems because of groundwater withdrawal, the oxidation of organic soils, and the collapse of cavities in carbonate and evaporite rocks. Additionally, in this chapter, we point out the value of applying science and engineering innovations used in effectively mitigating or limiting damages from land subsidence. One thing is certain: scientific understanding is critical to the formulation of balanced decisions about the management of land and water resources. When scientific information is presented in plain English, and a conversational approach and when obstacles are not present, understanding can flow like unimpeded groundwater.

By the way, did you know that land subsidence helps explain why the region has the highest rates of sea-level rise on the Atlantic Coast of the United States? An important fact to keep in mind is that land subsidence and rising water levels combine to cause what is known as *relative sea-level rise*. In contrast, absolute sea level change refers to how the height of the ocean surface is above the center of the Earth, without regard to whatever nearby land is raising or falling.

DID YOU KNOW?

Water budgets provide a means for evaluating the availability and sustainability of a water supply. A water budget simply states that the rate of change in water stored in an area, such as a watershed, is balanced by the rate at which water flows into and out of the area. Keep in mind that water flows into aquifers via infiltration slowly. If you ever visit and hike the Narrows in Zion National Park in Utah (highly recommended) there is a water spout at the Lodge that when opened pours water that took an estimated 3,000 years to flow from the top of the Canyon to the tap. Having personally consumed gallons of this water, I can confirm that it is the tastiest water I have ever consumed.

GROUNDWATER WITHDRAWAL

As mentioned, permanent subsidence can occur when water stored beneath the Earth's surface is removed by pumping. The reduction of fluid pressure in the pores and cracks of aquifer systems, especially in unconsolidated rocks, is inevitably accompanied by some deformation of the aquifer system. Because the granular structure—the so-called "skeleton"—of the aquifer system is not rigid, but more or less complaint, a shift in the balance of support for the overlying material causes the skeleton to deform slightly. Both the aquifers and aquitards that constitute the aquifer system undergo deformation but to different degrees. During the typically slow process of aquitard drainage (when the irreversible compression or consolidation of aquitards occurs) is when almost all the permanent subsidence takes place (Tolman and Poland, 1940). This concept, known as the aquitard-drainage model, has formed the theoretical basis of many successful subsidence investigations.

DID YOU KNOW?

Studies of subsidence in the Santa Clara Valley in California established the theoretical and field application of the laboratory-derived principle of effective stress and the theory of hydrodynamic consolidation to the drainage and compaction of aquitards (Tolman and Poland, 1940; Poland and Green, 1962; Green 1964; Poland and Ireland, 1988).

EFFECTIVE STRESS

The principle of effective stress was first proposed by Terzaghi (1925). For our purpose in this book "effective" means the calculated stress that was effective in moving soil and/or causing displacements. According to this principle, when the support provided by fluid pressure is reduced, such as when groundwater levels are lowered, support previously provided by the pore-fluid pressure is transferred to the skeleton of the aquifer system, which compresses to a degree. On the other hand, when the pore-fluid pressure is increased, such as when groundwater recharges the aquifer system, support previously provided by the skeleton is transferred to the fluid and the skeleton expands. In this way, the skeleton alternatively undergoes compression and expansion as the pore-fluid pressure fluctuates with aquifer-system discharge and recharge. When the load on the skeleton remains less than any previous maximum load, the fluctuations create only a small elastic deformation of the aquifer system and a small displacement of the land surface. This fully recoverable deformation occurs in all aquifer systems, commonly resulting in seasonal, reversible displacements in land surface of up to 1 inch or more in response to the seasonal changes in groundwater pumpage (USGS, 2013).

PRECONSOLIDATION STRESS

Simply stated, preconsolidation stress provides insights into a soil's history and behavior impacting construction and stability. Stated differently, the maximum level of past stressing of a skeletal element is termed the preconsolidation stress. And again, stated differently, preconsolidation stress is the maximum effective vertical overburden stress that a particular soil has sustained in the past. When the load on the aquitard skeleton exceeds the preconsolidation stress, the aquitard skeleton may undergo significant, permanent rearrangement, resulting in irreversible compaction. Because the skeleton defines the pore structure of the aquitards, this results in a permanent reduction of pore volume as the pore fluid is "squeezed" out of the aquitards into the aquifers. In confined aquifer systems subject to large-scale overdraft, the volume of water derived from irreversible aquitard compaction is essentially equal to the volume of subsidence and can typically range from 10% to 30% of the total volume of water pumped. This represents a one-time mining of stored groundwater and a small permanent reduction in the storage capacity of the aquifer system. Alternative names for preconsolidation stress are preconsolidation pressure, pre-compression stress, pre-compaction stress, and preload stress (Dawidowski and Koolen, 1994).

AQUITARDS ROLE IN COMPACTION

In recent decades increasing recognition has been given to the critical role of aquitards in the intermediate and long-term response of alluvial systems to groundwater pumpage. Aquitard systems play an important role in compaction. In many such systems interbedded layers of silt sand clays, once dismissed as non-water yielding, comprise the bulk of the groundwater storage capacity of the confined aquifer system. This is the case based on their substantially greater porosity and compressibility and, in many cases, their greater aggregate thickness compared to the more transmissive, coarser-grained sand and gravel layers (USGS, 2013).

Aquitards are less permeable than aquifers. Thus, the vertical drainage of aquitards into adjacent pumped aquifers may proceed very slowly, and thus lag far behind the changing water levels in adjacent aquifers. The lagged response within the inner portions of a thick aquitard may be largely isolated from the higher frequency seasonal fluctuations and more influenced by lower frequency, longer-term trends in groundwater levels. Because the migration of increased internal stress into the aquitard accompanies its drainage, as more fluid is squeezed from the interior of the aquitard, larger and larger intern stresses from the interior of the aquitard, larger and larger internal stresses propagate farther into the aquitard.

DID YOU KNOW?

Responses to changing water levels following several decades of groundwater development suggest that stresses directly driving much of the compaction are somewhat insulated from the changing stresses caused by short-term water-level variations in the aquifers.

When the preconsolidation stress is exceeded by the internal stresses, the compressibility increases dramatically, typically by a factor of 20–100 times, and the resulting compaction is largely nonrecoverable. At stresses greater than the preconsolidation stress, the lag in aquitard drainage increases by comparable factors, and concomitant compaction may require decades or centuries to approach completion. The theory of hydrodynamic consolidation (Terzaghi, 1925)—an essential element of the *aquitard drainage model*—describes the delay involved in draining aquitards when heads are lowered in adjacent aquifers, as well as the residual compaction that may continue long after drawdowns in the aquifers have essentially stabilized. Numerical modeling based on Terzaghi's theory has successfully simulated complex histories of compaction observed in response to measuring water-level fluctuations (Helm, 1975).

DID YOU KNOW?

Hydrocompaction—compaction due to wetting—is a near-surface phenomenon that produces land-surface subsidence through a mechanism entirely different from the compaction of deep, over-pumped aquifer systems.

THE VANISHING OF HAMPTON ROADS

For our purpose in this text when we refer to Hampton Roads we are referring to both the water body and land region as one because land subsidence and relative sea level rise in the area pertains to both. With regard to the total area it is composed of 527 square miles (1,364 km²) and is made up of nine major cities: Norfolk, Virginia Beach, Chesapeake, Newport News, Hampton, Portsmouth, Suffolk, Poquoson, and Williamsburg; as a combined statistical area it also includes Kitty Hawk, Elizabeth City, North Carolina. The entire area has a population of over 1.7 million. With regard to the body of water known as Hampton Roads, is one of the world's largest natural harbors. It incorporates the mouths of the Elizabeth River, Nansemond River, and James River with several smaller rivers and empties into the Chesapeake Bay—a treasured estuary—near its mouth leading to the Atlantic Ocean.

   Note to readers: I have noted that the Chesapeake Bay region is a treasured estuary as mentioned above but also it is a bastion of early American History. Moreover, it is not a bad place to live; having resided in the area for more than 50 years, I think, I am qualified to state this as fact. Having spent most of those 50 years studying the Bay area and specifically water pollution problems in the area it has become a lifetime project for me to continue the study but my focus has shifted dramatically from the pollution of the Bay and the existence of algal dead zones to a more pressing problem; namely, the literal vanishing of the land area.

HAMPTON ROADS: SEA LEVEL RISE

Of all the potential impacts of natural (cyclical) or human-induced climate change, a global rise in sea level appears to be the most certain and the most dramatic. As shown in Figure 15.1, for the last 5,000 years the rate of sea level rise was only 3 feet per 1,000 years. In the Chesapeake Bay region, the relative rise in sea level has been about 1 foot during the last 100 years (Figure 15.1). While scientists view this rapid rate as possibly a temporary acceleration, many of these same scientists believe that it signals a new trend in response to global warming. The point is if the rate of rise accelerates in the near future as projected, it could have serious repercussions for Chesapeake Bay.

   Because water levels are measured relative to the land, and as stated earlier, relative sea-level rise in the Chesapeake Bay region has two components: global water level increase and land subsidence. Worldwide or eustatic seal-level rise is caused by

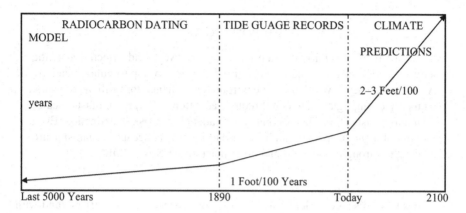

FIGURE 15.1 Sea-level rise in Hampton Roads region past and present (USGS, 1995).

water released from melting glaciers and the thermal expansion of seawater. Both are related to global warming and have amounted to a 6-inch rise in the last century. Let's take a closer look at the global climate change and warming problem.

GLOBAL CLIMATE CHANGE AND WARMING

At present, many feel that humanity is conducting an unintended, uncontrolled, globally pervasive experiment whose ultimate consequences could be second only to nuclear war. The Earth's atmosphere is being changed at an unprecedented rate by pollutants resulting from human activities, inefficient and wasteful fossil fuel use and the effects of rapid population growth in many regions. These changes are already having harmful consequences in many parts of the globe.

—Toronto Conference Statement (June 1988)

The preceding quotation clearly states the issue. But what is global warming? Simply put, global warming is a long-term rise in the average temperature of Earth. This appears to presently be the case, even though the geological record shows abrupt climate changes occur from time to time (Crowley and North, 1988). Here's a second question, is it true that many people used to question the validity of the concept of global warming as an environmental hazard? Is global warming actually occurring? The answer to this accompanying question is of enormous importance to all life on Earth—and is the subject of intense debate throughout the globe. Again, all the debate for the occurrence of global warming can't dispute the historical record that points out that measurements made in central England, Geneva, and Paris from about 1,700 until the present indicate a general downward trend in surface temperature (Thompson, 1995).

DID YOU KNOW?

In regard to biodiversity, climate change can have broad effects on it (the number and variety of plant and animal species in a particular location). Although species have adapted to environmental change for millions of years, a quickly changing climate could require adaption on larger and faster scales than in the past. Those species that cannot adapt are at risk of extinction. Even the loss of a single species can have cascading effects because organisms are connected through food webs and other inactions (USEPA, 2010).

For the sake of discussion, let's assume that global warming is occurring (and with rising sea levels, changing weather patterns, and global ice melt-off this is hard to argue against). With this assumption in place, we must ask other questions, ones that deal with why, how, and what. (1) Why is global warming occurring? (2) How can we be sure it is occurring? (3) What will be the ultimate effects? (4) What can and are we going to do about it? These questions are difficult to answer. The real danger is that we may not be able to definitively answer these questions before it is too late—when we've reached the point that the process has progressed beyond the power of humans to effect prevention or mitigation. This situation raises a red flag—a huge red flag—and additional questions. Are we to stand by and do nothing? Are we to simply ignore the potential impact of this problem? Are we to take the consequences of global warming lightly? Are we not to take precautionary actions now instead of later—much later, when it is too late? Indeed, a red flag has been raised (a cause-and-effect relationship to the greenhouse effect), but there is still time before it begins to wave in the inevitable climate change—when mitigation becomes harder, more expensive, and impossible to effect.

DID YOU KNOW?

In regard to oceans, they and the atmosphere are constantly interacting—exchanging heat, water, gases, and particles. As the atmosphere warms, the ocean absorbs some of this heat. The amount of heat stored by the ocean affects the temperature of the ocean both at the surface and at great depths. Warming of the Earth's oceans can affect and change the habitat and food supplies for many kinds of marine life—from plankton to polar bears. The oceans also absorb carbon dioxide from the atmosphere. Once it dissolves in the ocean, carbon dioxide reacts with seawater to form carbonic acid. As people put more carbon dioxide into the atmosphere, the oceans absorb some of this extra carbon dioxide, which leads to more carbonic acid. An increasingly acidic ocean can have negative effects on animal life, such as coral reefs (USEPA, 2010).

Factors Involved with Global Warming/Cooling

Right now, scientists are able to point to six factors that could be involved in long-term global warming and cooling.

1. Long-term global warming and cooling could result if changes in the Earth's position relative to the sun occur (the Earth's orbit around the sun), with higher temperatures when the two are closer together and lower when further apart.
2. Long-term global warming and cooling could result if major catastrophes (meteor impacts or massive volcanic eruptions) throw pollutants into the atmosphere that can block out solar radiation occur.
3. Long-term global warming and cooling could result if changes in albedo (reflectivity of Earth's surface) occur. If the Earth's surface were more reflective, for example, the amount of solar radiation radiated back toward space instead of absorbed would increase, lowering temperatures on Earth.
4. Long-term global warming and cooling could result if the amount of radiation emitted by the sun changes.
5. Long-term global warming and cooling could result if the shape and relationship of the Land and oceans change.
6. Long-term global warming and cooling could result if the composition of the atmosphere changes.

Note: "If the composition of the atmosphere changes"—this possibility, of course, relates directly to our present concern: Have human activities had a cumulative impact large enough to affect the total temperature and climate of Earth? We are not certain, right now. We are somewhat concerned and alert to the problem, but we are not certain. So the question is: What are we doing about global warming as it affects sea level rise? We answer this pertinent question in the next section.

Global Warming and Sea Level Rise

In the past few decades, human activities (burning fossil fuels, leveling forests, and producing synthetic chemicals such as CFCs) have released into the atmosphere huge quantities of carbon dioxide and other greenhouse gases. These gases are warming the Earth at an unprecedented rate. If current trends continue, they are expected to raise Earth's average surface temperature by at least 1.5°C–4.5°C (or more), in the next century—with warming at the poles perhaps two to three times as high as warming at the middle latitudes (Wigley et al., 1986).

If we assume global warming is inevitable and/or is already underway, what, then, must we do? Obviously, we cannot jump off the planet and head for greener pastures. We live on Earth and are stuck here (we have no effective method or technology to allow us to leave, or a convenient place to go if we tried). If this is the case, understanding the dynamics of change that are evolving around us and taking whatever prudent actions we can to mitigate the situation makes good sense.

We must also take this attitude and approach to the effect global warming is having on the rise in sea level. This rise is already underway, and with it will come increased storm damage, pollution, and subsidence of coastal lands.

"Rise in sea level is already underway?" Absolutely. Consider the following information taken from USEPA's 1995 report, *The Probability of Sea Level Rise.*

1. Global warming is most likely to raise sea levels 15 cm by the year 2050 and 34 cm by the year 2100. There is also a 10% chance that climate change will contribute 30 cm by 2050 and 65 cm by 2100. These estimates do not include sea level rise caused by factors other than greenhouse warming.

2. There is a 1% chance that global warming will raise sea level 1 m in the next 100 years and 4 m in the next 200 years. By the year 2200, there is also a 10% chance of a 2-m contribution. Such a large rise in sea level could occur either if Antarctic ocean temperature warms 5°C or Antarctic ice streams respond more rapidly than most glaciologists expect, or if Greenland temperatures warm by more than 10°C. Neither of these scenarios is likely.

3. By the year 2100, climate change is likely to increase the rate of sea level rise by 4.1 mm/yr. There is also a 1-in-10 chance that the contribution will be >10 mm/yr, as well as a 1-in-10 chance that it will be <1 mm/yr.

4. Stabilizing global emissions in the year 2050 would be likely to reduce the rate of sea level rise by 28% by the year 2100, compared with what it would be otherwise. These calculations assume that we are uncertain about the future trajectory of greenhouse gas emissions.

5. Stabilizing emissions by the year 2025 could cut the rate of sea level rise in half. If a high global rate of emissions growth occurs in the next century, the sea level is likely to rise 6.2 mm/yr by 2100; freezing emissions in 2025 would prevent the rate from exceeding 3.2 mm/yr. If less emissions growth were expected, freezing emissions in 2025 would cut the eventual rate of sea level rise by one-third.

6. Along most coasts, factors other than anthropogenic climate change will cause the sea to rise more than the rise resulting from climate change alone. These factors include compaction and subsidence of land, groundwater depletion, and natural climate variations. If these factors do not change, the global sea level is likely to rise 45 cm by the year 2100, with a 1% chance of a 112 cm rise. Along the coast of New York, which typifies the United States, sea level is likely to rise 26 cm by 2050 and 55 cm by 2100. There is also a 1% chance of a 55 cm rise by 2050 and a 120 cm rise by 2100.

Along with the EPA's findings reported above, additional lines of evidence corroborate that the global mean sea level has been rising for at least the last 100 years. According to Broecker (1987), this evidence is apparent in tide gauge records; erosion of 70% of the world's sandy coasts and 90% of America's sandy beaches; and the melting and retreat of mountain glaciers. Edgerton (1991) points out that the correspondence between the two curves of rising global temperatures and rising sea levels during the last century appears to be more than coincidental.

Major uncertainties are present in estimates of future sea level rise. The problem is further complicated by our lack of understanding of the mechanisms contributing to relatively recent rises in sea level. In addition, different outlooks for climatic warming dramatically affect estimates. In all this uncertainty, one thing is sure. Estimates of sea level rise will undergo continual revision and refinement as time passes and more data is collected.

MAJOR PHYSICAL EFFECTS OF SEA LEVEL RISE

With increased global temperatures, global sea level rise will occur at a rate unprecedented in human history (Edgerton, 1991). Changes in temperature and sea level will be accompanied by changes in salinity levels. For example, a coastal freshwater aquifer is influenced by two factors: pumping and mean sea level. In pumping, if withdrawals exceed recharge, the water table is drawn down and saltwater penetrates inland. With the mean sea level, the problem occurs if the sea level rises and the coastline moves inland, reducing the aquifer area. Additional problems brought about by changes in temperature and sea level are seen in tidal flooding, oceanic currents, biological processes of marine creatures, runoff and landmass erosion patterns, and saltwater intrusion.

The most important direct physical effect of sea level rise is on a coastal beach system. At current rates of sea level rise of 1–2 mm/yr, significant coastal erosion is already produced. Two major factors contribute to beach erosion. First, deeper coastal waters enhance wave generation, thus increasing their potential for overtopping barrier islands. Second, shorelines and beaches will attempt to establish new equilibrium positions according to what is known as the *Bruun rule*; these adjustments will include a recession of shoreline and a decrease in shore slope (Bruun, 1962, 1986).

MAJOR DIRECT HUMAN EFFECTS OF SEA LEVEL RISE

Along with the physical effects of sea level rise, in one way or another, directly or indirectly, accompanying effects have a direct human side, especially concerning human settlements and the infrastructure that accompanies them: highways, airports, waterways, water supply, and wastewater treatment facilities, landfills, hazardous waste storage areas, bridges, and associated maintenance systems. Sea level rise could also cause intrusion of saltwater into groundwater suppliers (Edgerton, 1991).

To point out that this infrastructure will be placed under tremendous strain by a rising sea level coupled with other climatic changes is to understate the possible consequences. Indeed, the impact on infrastructure is only part of the direct human impact. For example, there is widespread agreement among scientists that any significant change in world climate resulting from warming or cooling will (1) disrupt world food production for many years; (2) lead to a sharp increase in food prices; and (3) cause considerable economic damage.

Just how much of a rise in sea level are we talking about? According to USEPA (1995),

> If the experts on whom we relied fairly represent the breadth of scientific opinion, the odds are fifty-fifty that greenhouse gases will raise sea level at least 15 cm by the year 2050, 25 cm by 2100, and 80 cm by 2200. (p. 123).

SIDE BAR 15.1: THE 411 ON GLOBAL CLIMATE CHANGE

a. Is global warming a hoax? Is Earth's climate changing? Are warmer times or colder times on the way? Is the greenhouse effect going to affect our climate, and if so, do we need to worry about it? Will the tides rise and flood New York? Does the ozone hole portend disaster right around the corner?

These and many other questions related to climate change have come to the attention of us all. We are inundated by a constant barrage of newspaper headlines, magazine articles and television news reports on these topics. Recently, we've seen report after report on El Niño and its devastation of the West Coast of the U.S. (and Peru and Ecuador)—and its reduction of the number, magnitude and devastation of hurricanes that annually blast the East Coast of the U.S.

To illustrate just how constant the barrage of newspaper headlines has been, we have selected and listed below a sample of the climate change news, and global warming headlines published in many locations throughout the globe in April 2008 (Carbonify.com 2009 and one of the latest headlines January 2017 Carbonify.com).

April 1, 2008—GLOBAL WARMING AWARENESS AND APATHY
April 2, 2008—OCEANS UNDER STRESS FROM GLOBAL WARMING
April 7, 2008—AUSTRALIAN DROUGHT AFFECTED AREAS GROW
April 13, 2008—FOSSIL FUEL CARBON EMISSIONS OVER 8 GIGATONS
April 22, 2008—UK MIGRATING BIRDS NUMBERS DROP
April 28, 2008—MARCH WARMEST ON RECORD GLOBALLY
January 2017—2016 WAS THE HOTTEST YEAR ON RECORD

Scientists have been warning us of the catastrophic harm that can be done to the world by atmospheric warming. One view states that the effect could bring record droughts, record heat waves, record smog levels, and an increasing number of forest fires.

Another caution put forward warns that the increasing atmospheric heat could melt the world's icecaps and glaciers, causing ocean levels to rise to the point where some low-lying island countries would disappear, while the coastlines of other nations would be drastically altered for ages—or perhaps for all time.

What's going on? We hear plenty of theories put forward by doomsayers, but are they correct? If they are correct, what does it all mean? Does anyone really know the answers? Should we be concerned? Should we invest in waterfront property in Antarctica? Should we panic?

No. While no one really knows the answers—"we don't know what we don't know syndrome"—and while we should be concerned, no real cause for panic exists.

Should we take some type of decisive action—should we come up with quick answers and put together a plan to fix these problems? What really needs to be done? What can we do? Is there anything we can do?

The key question to answer here is "What really needs to be done?" We can study the facts, the issues, and the possible consequences—but the key to successfully combating these issues is to stop and seriously evaluate the problems. We need to let scientific fact, common sense, and cool-headedness prevail. Shooting from the hip is not called for, makes little sense—and could have Titanic consequences for us all.

The other question that has merit here is, "Will we take the correct actions before it is too late?" The key words here are: "correct actions." Eventually, we may have to take some action (beyond hiding in a cave somewhere). But we do not yet know what those actions could be or should be.

From our perspective, one thing is certain; in our college-level environmental health courses, we address, sooner or later, global warming and/or global climate change. Through time and experience we have learned (yes, teachers learn, too) that whether we call it global warming, global climate change (humankind-induced global warming, under a broader label), or an inconvenient truth, the topic is a conundrum (riddle, the answer of which is a pun). As such, before diving into the many emotionally charged, heated class discussions about this "hot" topic (pun intended).

Consider this: any damage we do to our atmosphere affects the other three environmental mediums: water and soil—and biota (us—all living things). Thus, the endangered atmosphere (if it is endangered) is a major concern (a life and death concern) to all of us.

THE PAST

Before we begin our discussion of the past, we need to define the era we refer to when we say, "the past." Tables 15.1 and 15.2 are provided to assist us in making this definition. Table 15.1 gives the entire expanse of time from Earth's beginning to the present. Table 15.2 provides the sequence of geological epochs over the past 65 million years, as dated by modern methods. The Paleocene through Pliocene

TABLE 15.1
Geologic Eras and Periods

Era	Period	Millions of Years Before Present
Cenozoic	Quaternary	2.5-present
	Tertiary	65 to 2.5
Mesozoic	Cretaceous	135 to 65
	Jurassic	190 to 135
	Triassic	225 to 190
Paleozoic	Permian	280 to 225
	Pennsylvanian	320 to 280
	Mississippian	345 to 320
	Devonian	400 to 345
	Silurian	440 to 400
	Ordovician	500 to 440
	Cambrian	570 to 500
Precambrian		4,600 to 570

TABLE 15.2
Geological Epochs.

Epoch	Million Years Ago
Holocene	01 to 0
Pleistocene	1.6 to 0.01
Pliocene	5 to 1.6
Miocene	24 to 5
Oligocene	35 to 24
Eocene	58 to 35
Paleocene	65 to 58

together make up the Tertiary Period; the Pleistocene and the Holocene compose the Quaternary Period.

When we think about climatic conditions in the prehistoric past, two things generally come to mind—Ice Ages and dinosaurs. Of course, in the immense span of time that pre-history covers, those two eras represent only a brief moment in time. So let's look at what we know or what we think we know about the past, and about Earth's climate and conditions. One thing to consider—geological history shows us that the normal climate of the Earth was so warm that subtropical weather reached 60°N and S, and polar ice was entirely absent.

Only during less than about 1% of the Earth's history did glaciers advance and reach as far south as what is now the temperate zone of the northern hemisphere. The latest such advance, which began about 1,000,000 years ago, was marked by geological upheaval and (perhaps) the advent of human life on Earth. During this time, vast ice sheets advanced and retreated, grinding their way over the continents.

OBVIOUS QUESTIONS

    i. Is global warming or global climate change real?
    ii. Are humans responsible for global climate change?
    iii. Is global climate change cyclical or human-driven?

The Bottom Line: At this point does it really matter whether global warming is cyclical or human driven? The evident and undeniable point is that sea levels are rising globally and in Hampton Roads. But there is another problem, a problem that must be factored in to add to the sea level problem. The other problem? Land subsidence in Hampton Roads.

LAND SUBSIDENCE IN HAMPTON ROADS[2]

This section concentrates specifically on the Hampton Roads area and in particular the lower Chesapeake Bay Region. Within the lower Chesapeake Bay Region the continuing appearance of dead zone problems, nutrient pollution, sediment contamination, and sea life decline are without a doubt serious problems that persist to garner the attention of officials responsible for monitoring and managing the health of Chesapeake Bay. As serious as these problems are, it is the on-ongoing rise in the relative sea-level problem that is beginning to impact Hampton Roads and offers a future that can best be described as foreboding and quite wet unless certain mitigation practices are put into place, not tomorrow but at the present and continuing into the future. Part of the relative sea-level rise problem is due to land subsidence, and it is this problem that is addressed in the section.

So, let's get to it.

Local or isostatic factors contribute to *relative* sea-level rise in Hampton Roads, Virginia through subsidence or sinking of the land. In the Chesapeake Bay area, subsidence of land is due to both geologic factors and excessive withdrawal of groundwater which has amounted to 6 inches in the last 100 years at rates of 0.039–0.189 in/yr (1.1–4.8 mm/yr). Consequently, there has been a relative increase in sea level in the Chesapeake Bay area of 1 ft in the last century. More specifically, land subsidence in the region is the result of the flexing of the Earth's crust from glacial isostatic adjustment in response to glacier formation and melting. In addition, more than half of the observed subsidence is the result of the aquifer system in Hampton Roads that has been compacted by extensive groundwater extraction at rates of 1.5- to 3.7-mm/ yr. This helps explain why the southern Chesapeake Bay region has the highest rate of sea-level rise on the Atlantic Coast of the United States (Zervas, 2009). Because the communities in the region must grapple with flooding problems that lead to the disappearance of existing land by a combination of rising sea levels and subsiding land, all of which will continue to worsen in the future, it is important to understand and potentially manage land subsidence.

Data indicate that land subsidence has been responsible for more than half the relative sea-level rise measured in the Hampton Roads area; it also suggests that the problem will be ongoing in the future. This is bad news for those residing in the area. Land subsidence is a serious issue because the increased flooding has and will continue to have important economic, environmental, and human health consequences for the heavily populated and ecologically important southern Chesapeake Bay region.

As mentioned earlier, land subsidence in the region is the result of the flexing of the Earth's crust from glacial isostatic adjustment in response to glacier formation and melting. In addition, more than half of the observed subsidence is the result of the aquifer system in Hampton Roads that has been compacted by extensive groundwater extraction at rates of 1.5- to 3.7-mm/yr. This helps explain why the southern Chesapeake Bay region has the highest rate of sea-level rise on the Atlantic Coast of the United States (Zervas, 2009). Because the communities in the region must grapple with flooding problems that leads to the disappearance of existing land by a combination of rising sea levels and subsiding land, all of which will continue to worsen in the future, it is important to understand and potentially manage land subsidence.

Land subsidence in the region is the result of the flexing of the Earth's crust from glacial isostatic adjustment in response to glacier formation and melting. In addition, more than half of the observed subsidence is the result of the aquifer system in Hampton Roads that has been compacted by extensive groundwater extraction at rates of 1.5- to 3.7-mm/yr. This helps explain why the southern Chesapeake Bay region has the highest rate of sea-level rise on the Atlantic Coast of the United States (Zervas, 2009). Because the communities in the region must grapple with flooding problems that leads to the disappearance of existing land by a combination of rising sea levels and subsiding land, all of which will continue to worsen in the future, it is important to understand and potentially manage land subsidence.

As explained earlier, land subsidence is the sinking or lowering of the land surface. In the United States, most land subsidence is caused by human activities (Galloway et al., 1999). Earlier we described land subsidence problems in the western region of the country. And after recognition of the problems associated with groundwater drawdown and resulting land subsidence, the areas affected set up monitoring networks and ultimately adopted new water-management practices to prevent or arrest land subsidence.

Experience has shown that rates and locations of land subsidence change over time so accurate measurements and predictive tools are needed to improve understanding of land subsidence. Although rates of land subsidence are not as high on the Atlantic Coast as they have been in the Houston-Galveston area or the Santa Clara Valley, land subsidence is important because of the low-lying topography and susceptibility to sea-level rise in the southern Chesapeake Bay region.

In the Lower Chesapeake Bay region, increased flooding, wetland and coastal ecosystem alteration, and damage to infrastructure and historical sites are all the result of land subsidence. This problem is not a new one; it is well known to regional planners who have gained an understanding of what land subsidence is; that is, the why, where, and how fast it is occurring, now and in the future.

LAND SUBSIDENCE CONTRIBUTES TO RELATIVE SEA-LEVEL RISE

As shown in Figure 15.2, land subsidence contributes to the relative sea-level rise that has been measured in the Chesapeake Bay. Tidal-station measurements of sea levels, however, do not distinguish between water that is rising and land that is sinking—the combined elevation changes are termed *relative sea-level rise*. Global sea-level rise and land subsidence increase the risk of coastal flooding and contribute to shoreline retreat.

As relative sea levels rise, shorelines retreat, and the magnitude and frequency of near-shore coastal flooding increase. This is particularly a problem in Norfolk, Virginia (downtown area and Ocean View District) where during a coastal storm event and corresponding high tide, downtown Norfolk streets flood; at times, it can flood to several feet. Although land subsidence can be slow, its effects accumulate over time; this has been an expensive problem in Norfolk and other parts of the southern Chesapeake Bay region. Analysis by McFarlane (2012) found that between 59,000 and 176,000 residents living near the shores of the southern Chesapeake Bay could be either permanently inundated or regularly flooded by 2100. This estimate was based on the 2010 census data, using the spring high tide as a reference elevation and assuming a 1-m relative sea-level rise. Damage to personal property was estimated to be $9 billion to $26 billion, and 120,000 acres of economically valuable land could be inundated or regularly flooded, under these same assumptions. Historic and cultural resources are also vulnerable to increased flooding from relative sea-level

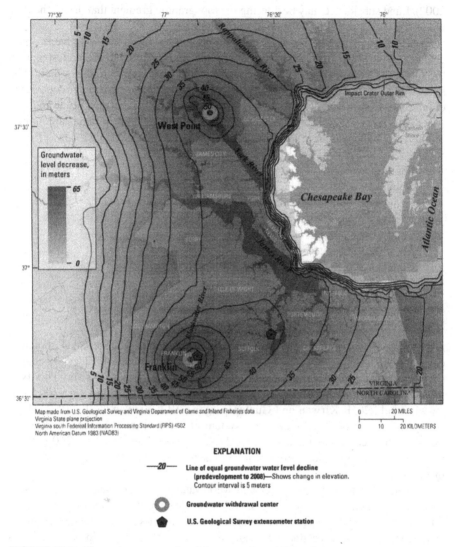

FIGURE 15.2 Groundwater water level decreases from 1990 to 2008 (USGS, 2013; Modified from Heywood and Pope (2009)).

rise in the southern Chesapeake Bay, particularly at shoreline sites near tidal water, such as the 17th-century historic Jamestown site.

It should be pointed out that the shoreline area in southern Hampton Roads is not the only area prone to flooding. Land subsidence can also increase flooding in areas away from the coast in low-lying areas such as Franklin, Virginia. The city of Franklin is about 60 road miles west of Hampton Roads. The Blackwater River Basin, which encompasses Franklin and other local areas can be subject to increased flooding as the land sinks. In fact, Franklin and the counties of Isle of Wight and Southampton have experienced large floods in years (Federal Emergency Management Agency,

2002). Land subsidence may be altering the topographic gradient that drives the flow of the river and possibly contributing to the flooding.

Wetland and marsh ecosystems in low-lying coastal areas are sensitive to small changes in elevation (Cahoon et al., 2009). Salt marshes, which are widespread in the southern Chesapeake Bay region, are dependent on tidal dynamics for their existence. Small changes in either land or sea elevations can alter sediment deposition, organic production and plant growth, and the balance between fresh water and seawater (Morris et al., 2002). The effects of sea-level rise on tidal wetlands are numerous and already apparent in local wetlands. Some of these effects include:

- Shoreline erosion
- Habitat loss
- Changes in tidal amplitude
- Landward migration of tidal waters
- Landward migration of habitats
- More frequent inundation
- Changes in plant and animal species composition
- Changes in tidal flow patterns
- Migration of estuarine salinity gradients
- Changes in sediment transport

Although sea-level rise has one of the most direct effects on tidal wetlands, shoreline environments also are affected by land subsidence. When land subsides, it subjects shorelines to increased waver action, increasing erosion and wash-over. This type of damage is happening in the Chesapeake Bay because of relative sea-level rise (Erwin et al., 2011; Kirwan and Guntenspergen, 2012; Kirwan et al., 2012). Major changes in the coastal and marine ecosystem of the southern Chesapeake Bay are expected to be caused by relative sea-level rise (Spellman and Whiting, 2006); these changes will likely be more severe if land subsidence continues including damage to infrastructure.

Buildings, bridges, canals, water and wastewater treatment plants, electrical substations, communication towers, pipes, and other components that make up a region's infrastructure can be damaged from relative groundwater rise or differential settling in areas with high subsidence gradients (Galloway et al., 1999). As land sinks and sea levels continue to rise, groundwater levels rise towards the land surface in coastal areas, which can cause problems for subterranean structures, septic fields, buried pipes and tanks and cables, and infrastructure not designed for elevated groundwater levels. Storm and wastewater interceptor lines in urban areas are vulnerable because land subsidence can alter the topographic gradient driving the flow through the sewers, causing increased flooding and more frequent sewage discharge from combined sewer overflows.

AQUIFER COMPACTION

When groundwater is pumped from the Potomac aquifer system in southern Chesapeake Bay, pressure decreases. The pressure change is reflected by water levels in wells, with water levels decreasing as aquifer-system pressure decreases.

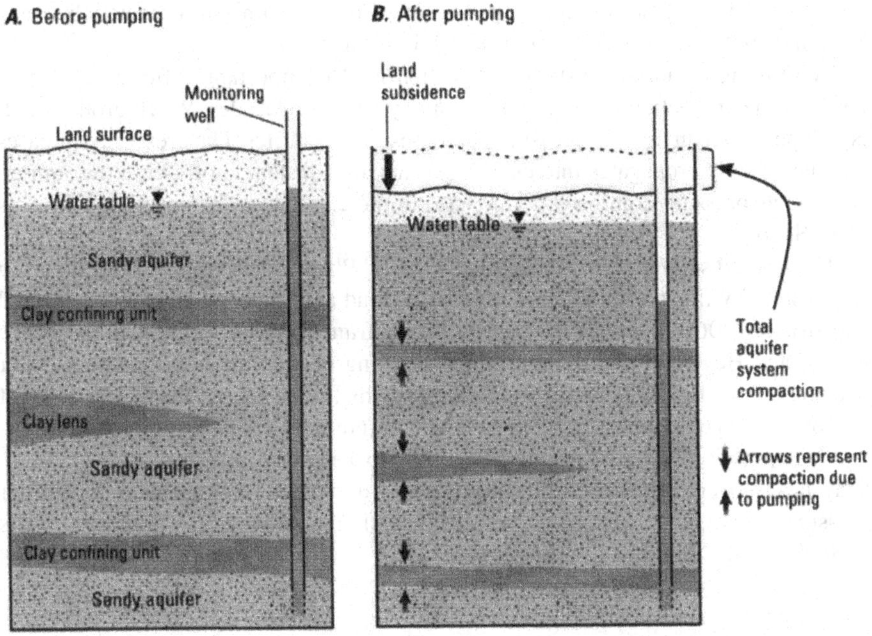

A. Before pumping **B.** After pumping

FIGURE 15.3 Aquifer-system compaction caused by groundwater withdrawals A, before and B, after pumping (USGS, 2013; Modified From Galloway et al. (1999)).

This is happening over most of the southern Chesapeake Bay region, with the greatest water-level decrease seen near the pumping centers of Franklin and West Point, Virginia (see Figure 15.2). As water levels decrease, the aquifer system compacts causing the land surface to subside (see Figure 15.3). Water levels have decreased over the entire Virginia Coastal Plain and in the Potomac aquifer, which is the deepest and thickest aquifer in the southern Chesapeake Bay region and supplies about 75% of groundwater withdrawn from the Virginia Coastal Plain aquifer system (Heywood and Pope, 2009).

Three factors determine the amount of aquifer-system compaction: water-level decline, sediment compressibility, and sediment thickness. If any of these three factors increase in magnitude, then the amount of aquifer-system compaction and land subsidence increases. Because all three of these factors vary spatially across the southern Chesapeake Bay region, rates of load subsidence caused by aquifer-system compaction also vary spatially across the region.

The Virginia Coastal Plain aquifer system consists of many stacked layers of sand and clay. Although groundwater is withdrawn primarily from the aquifers (sandy layers), most compaction occurs in confining units and clay lenses, the relatively impermeable layers sandwiched between and within the aquifers (Pope and Burbey, 2004). The compression of the clay layers is mostly nonrecoverable, meaning that, if groundwater levels later recover and increase, then the aquifer system does not expand to its previous volume and the land surface does not rise to its previous elevations (Pope and Burbey, 2004). Konikow and Neuzil (2007) estimated that 95% of the water removed from storage in the Virginia Coastal Plain aquifer system between 1891 and 1980 was derived from the confining layers.

The timing of aquifer-system compaction is also important. After groundwater levels drop compaction can continue for many years or decades. When groundwater is pumped from an aquifer, pressure decreases in the aquifer. The pressure decreases and then slowly propagates into clay layers that are adjacent to or within the aquifer. As long as pressure continues to decrease in the clay layers, compaction continues (USGS, 2013).

The layered sediments of the Virginia Coastal Plain aquifer system range in grain size from very fine (silts and clays) to coarse (sand and shell fragments) (McFarland and Bruce, 2006). Based on the hydrogeologic framework of McFarland and Bruce (2006) and Heywood and Pope (2009), confining outside the bolide impact crater occupies about 16% of the total aquifer-system thickness, an average of 100m out of the total average thickness of 619m. These continuing layers have high specific storage (compressibility) estimated to be 0.00015 per meter (Pope and Burbey, 2004). Clay layers overlying and within the Potomac aquifer are compressing as aquifer pressure decreases migrating vertically and laterally from pumping wells (USGS, 2013).

DID YOU KNOW?

NOAA's National Geodetic Survey continuously monitors the movements of the Earth through its nationwide network of nearly 2,000 permanent GPS stations; this important system is called Continuously Operating References Stations or CORS.

GLACIAL ISOSTATIC ADJUSTMENT

Earth is always active, on the move, continually, if bit by bit, ever-changing. Temperatures rise and fall in cycles over millions of years. The last ice age occurred about 16,000 years ago when great sheets of ice covered much of Earth's Northern Hemisphere. Though the ice melted long ago, the land once under and around the ice is still rising and falling in reaction to its ice burden. This ongoing movement of land is called glacial isostatic adjustment or postglacial rebound (see Figure 15.4). Here's how it works: Imagine that you are lying down on a soft mattress and then get up from the same spot. You will likely see an indentation in the mattress where your body had been and a puffed-up area around the indentation where the mattress rose. Once you get up, the mattress takes a little time before it releases back to its original shape. Earth acts similar to that mattress and is constantly on the move. We can't always see the movement unless we witness an earthquake or its aftermath or a volcanic lateral eruption like Mount St. Helens' event in 1980 or land fissures and sinkholes opening in the Earth around us (NOAA, 2015). Truth be told even the strongest materials like Earth's crust budge, shift, move, buckle, warp, and deform when enough pressure is applied.

FIGURE 15.4 Glacier on retreat in Alaska allowing for isostatic land adjustment (Photo by F.R. Spellman while canoeing the area in 2015).

The Virginia Coastal Plain aquifer consists of layered sediments overlaying crystalline bedrock. Bedrock is not solid and unyielding but actually flexes and moves in response to stress. Bedrock in the mid-Atlantic region is moving slowly downward in response to the melting of the Laurentide ice sheet that covered Canada and the northern United States during the last ice age (Sella et al., 2007; Boon et al., 2010). When the ice sheet still existed, the weight of the ice pushed the underlying Earth's crust downward and, in response, areas away from the ice sheet were forced upward (called glacial fore bulge—a flexural bulge in front as a result of the load behind it; see Figure 15.4). The southern Chesapeake Bay region is in the glacial fore bulge area and was forced upward by the Laurentide ice sheet. The ice sheet started melting about 18,000 years ago and took many thousands of years to disappear entirely. As the ice melted and its weight was removed, glacial fore bulge areas, which previously had been forced upward, began sinking and continue to sink today. Again, this movement of the Earth's crust in response to ice loading or melting is called glacial isostatic adjustment. Data from GPS and carbon dating of marsh sediments indicate that regional land subsidence in response to glacial isostatic adjustment in the southern Chesapeake Bay region may have a current rate of about 1 mm/yr (Engelhart, 2010; Engelhart et al., 2009). This downward velocity rate is uncertain and probably not uniform across the region.

THE BOTTOM LINE

According to DeJong et al. (2015), currently, relative sea-level rise (3.4 mm/yr) is faster in the Lower Chesapeake Bay Region than any other location on the Atlantic coast of North America, and twice the global average eustatic rate (1.7 mm/yr).

NOTES

1 Based on information from USFWS (1995). Vanishing Lands. Accessed 12/20/22 @ http://www.fws.gov/s/amer/Vanishinglands.Sealevel.
2 Based on information from Eggleston and Pope. USGS (2013) Circular 2013-1992.

REFERENCES

Boon, J.D., Brubaker, J.M., and Forest, D.M. (2010). Chesapeake Bay land subsidence and sea-level change: An evaluation of past and present trends and future outlook. Virginia Institute of Marine Science Special Report 425 in Applied Marine Science and Ocean Engineering, 41 p. plus appendixes, accessed at https://web.vims.edu/GreyLit/VIMS/stramsoe425.pdf.
Broecker, W. (1987). Unpleasant surprises in the greenhouse? *Nature* 328: 123–126.
Bruun, P. (1962). Sea level rise as a cause of shore erosion. *Proceedings of the American Society of Engineers and Journal Waterways Harbors Division* 88: 117–130.
Bruun, P. (1986). Worldwide impacts of sea level rise on shorelines. In: J.G. Titus (ed.), *Effects of Changes in Stratospheric Ozone and Global Climate*, vol. 4. New York: UNEP/EPA, pp. 99–128.
Cahoon, D.R., Reed, D.J., Kolker, A.S., Brinson, M.M., Stevenson, J.C., Riggs, S., Christian, R., Reyes, E., Voss, C., and Kunz, D. (2009). Coastal wetland sustainability, Chapter 4. In: J.G. Titus, K.E. Anderson, D.R. Cahoon, D.B. Gesch, S.K. Gill, B.T. Gutierrez, E.R. Thieler, and S.J. Williams (eds.), *Coastal Sensitivity to Sea-Level Rise: A Focus on the Mid-Atlantic Region*, A Report by the U.S. Climate Change Science Program and the Subcommittee on Global Change Research. Washington, DC: U.S. Environmental Protection Agency U.S. Climate Change Science Program Synthesis and Assessment Product 4.1, pp. 57–72.
Carbonify.com. (2009/2017). Global Warming: A Hoax? Accessed 11/07/19 @ https://www.carbonify.Com/articles/global-warming-hoax.htm.
Crowley, T.J., and North, G.R. (1988). Abrupt climate change and extinction events in earth's history. *Science* 240: 241–245.
Dawidowski, J.B., and Koolen, J.J. (1994). Computerized determination of the preconsolidation stress in compaction texting of field core samples. *Soil and Tillage Research* 31(2): 277–282.
DeJong, B.D., Bierman, P.R., Newell, W.L., Rittenour, R.M., Mahan, S.A., Balco, G., and Rood, D.H. (2015). Pleistocene relative sea levels in the Chesapeake Bay region and their implications for the next century. *GSA Today* 25(8): 4–10.
Edgerton, L. (1991). *The Rising Tide: Global Warming and World Sea Levels*. Washington, DC: Island Press.
Engelhart, S.E. (2010). Sea-level changes along the U.S. Atlantic Coast: Implications for glacial isostatic adjustment models and current rates of sea-level change. Publicly Accessible Penn Dissertations, 407. https://repositioryupenn.edu/edisserations/407.
Engelhart, S.E., Horton, B.P., Douglas, B.C., Peltier, W.R., and Tornqvist, T.E. (2009). Spatial variability of late Holocene and 20th century sea-level rise along the Atlantic coast of the United States. *Biology* 37: 1115–1118.

Erwin, R.M., Brinker, D.F., Watts, B.D., Costanzo, G.R.., and Morton, D.D. (2011). Islands at bay-Rising seas, eroding islands, and waterbird habitat loss in Chesapeake Bay, USA. *Journal of Coastal Conservation* 15: 51–60.

Federal Emergency Management Agency. (2002). Flood Insurance Study of Franklin, Virginia, Community 510060 (revised September 4, 2002). Federal Emergency Management Agency, 16 p.

Galloway, D.I., Jones, D.R., and Ingebritsen, S.E. (eds.) (1999). Land subsidence in the United States. U.S. Geological Survey Circular 1182, 177 p. accessed at https://pubs.usgs.gov/circ/circ1182/.

Green, J.H. (1964). Compaction of the aquifer system and land subsidence in the Santa Clara Valley, California, U.S. Geological Survey Water-Supply Paper 1779-T, 11 p.

Helm, D.C. (1975). Once-dimensional simulation of aquifer system compaction ear Pixley, Calif., part 1. *Constant Parameters: Water Resource Research* 11: 465–478.

Heywood, C.E., and Pope, J.P. (2009). Simulation of groundwater flow in the coastal plain aquifer system of Virginia. U.S. Geological Survey Scientific Investigations Report 2009-5039, 115 p. accessed at https://pubs.usgs.gov/sir/2009/5039/.

Kirwan, M.L., and Guntenspergen, G.R. (2012). Feedbacks between intimidation, root production, and shoot growth in a rapidly submerging brackish marsh. *Journal of Ecology* 100(3): 760–770.

Kirwan, M.L., Langley, J.A., Guntenspergen, G.R., and Megonigal, J.P. (2012). The impact of sea-level rise on organic matter decay rates in Chesapeake Bay brackish tidal marshes. *Biogeosciences Discussions* 9(10): 14689–14708.

Konikow, L.F., and Neuzil, C.E. (2007). A method to estimate groundwater depletion from confining layers. *Water Resources Research* 43(7): W07417, 15 p.

McFarlane, B.J. (2012). Climate change in Hampton roads-phase III-sea level rise in Hampton roads, Virginia. Chesapeake, Virginia, Hampton Roads Planning District Commission report PET13-06, July, 102 p. accessed at https://www.hrpdeva.gov/uploads/docs/HRPDC_ClimateChangeReport2012_Rull_Reduced.pdf.

McFarland, E.R., and Bruce, T.S. (2006). The Virginia coastal plain hydrogeologic framework. U.S. Geological Survey Professional Paper 1731, 118 p. 25 pls., accessed at https:pubs.water.usgs.gov/pp1731/.

Morris, J.T., Sundareshwar, P.V., Nietch, C.T., Kjerfve, Bjorm, and Cahoon, D.R. (2002). Responses of coastal wetlands to rising sea level. *Ecology* 83(10): 2869–2877.

NOAA. (2015). What is glacial isostatic adjustment? Accessed from National Oceanic and Atmospheric Administration: https://oceanserive.noass.gov/facts/galicial-adjustment.html.

Poland, J.F., and Green, J.H. (1962). Subsidence in the Santa Clara Valley, California: A progress report. U.S. Geological Survey Water-Supply Paper 1619-C, 16 p.

Poland, J.F., and Ireland, R.L. (1988). Land subsidence in the Santa Clara Valley, California, as of 1982. U.S. Geological Survey Professional Paper 497-F, 61 p.

Pope, J.P., and Burbey, T.J. (2004). Multiple-aquifer characteristics from single borehole extensometer records. *Ground Water* 42(1): 45–58.

Sella, G.F., Stein, S., Dixon, T.H, Craymer, M., James, T.S., Mazzotti, St, and Dokka, R.L. (2007). Observation of glacial isostatic adjustment in "stable" North America with GPS. *Geophysical Research Letters* 34(2): 1.02306, 6 p.

Spellman, F.R., and Whiting, N. (2006). *Environmental Science and Technology: Concepts and Applications*. Boca Raton, FL: CRC Press.

Terzaghi, K. (1925). Principles of soil mechanics, IV-settlement and consolidation of clay. *Engineering New-Record* 95(3): 874–878.

Thompson, D.J. (1995). The seasons, global temperature, and precession. *Science* 268: 59.

Tolman, C.F., and Poland, J.F. (1940). Ground-water infiltration, and ground-surface recession in Santa Clara Valley, Santa Clara County, California. Transactions American Geophysical Union, vol. 21, pp. 23–24.

USEPA. (1995). *The Probability of Sea Level Rise.* Washington, DC: Environmental Protection Agency.

USEPA. (2010). Climate Change and Ecosystems. Accessed 06/02/11 @ www.epa.gov/climate change.

USGS (1995). *Sea Level Rise in Hampton Roads Region.* Washington, DC: USGS

USGS. (2013). Land subsidence in the United States. Circular 1182. U.S. Department of the Interior, U.S. Geological Survey. Washington, D.C.

Wigley, T.M., Jones, P.D., and Kelly, P.M. (1986). Empirical climate studies: Warm world scenarios and the detection of climatic change induced by radioactively active gases. In B. Bolin et al. (eds.), *The Greenhouse Effect, Climatic Change, and Ecosystems.* New York: Wiley, pp. 201–205.

Zervas, C. (2009). Sea level variations of the United States, 1854-2006. National Oceanic and Atmospheric Administration Technical Report NOS CO-OPS 053, 76 p. plus appendixes, accessed at https://www.co-ops.nos.noas.gov/publicatins/Tech_r[r_53.pdf.

16 Gauging and Observing Land Subsidence

IN A NUTSHELL

Measurements of aquifer system compaction, land-surface elevations, and water levels are used to improve understanding of the processes responsible for changes in the elevation of the land's surface. We are fortunate today in our practice of gauging and observing land subsidence due to technical logical advances and practices such as using interferometric synthetic aperture radar (InSAR), continuous GPS (CGPS) measurements, campaign global positioning system (GPS) surveying, and spirit-leveling surveying. You will find in this chapter that aquifer compaction is measured by using extensometers. The practice of using extensometers is good engineering practice because the resulting measurements have the added benefit of being depth-specific due to the instruments being anchored at specific depths of interest. With regard to preciseness, using spirit-leveling surveys and extensometers gives the best measurements.

OUTSIDE THE SHELL—GAUGING SUBSIDENCE[1]

Land subsidence can be effectively and accurately measured units several reliable and proven techniques. Multiple measuring or monitoring techniques are often used together to understand different aspects of land subsidence (Table 16.1 and Figure 16.1). Because rates and locations of land subsidence change over time, repeat measurements at multiple locations are often needed to improve understanding of the complex phenomenon and guide computer models that forecast future subsidence. Extensometers measure changes in aquifer-system thickness, whereas other methods measure land surface elevation, from which subsidence is calculated by subtracting measurements over time.

BOREHOLE EXTENSOMETERS

An *extensometer* is a device that is used to measure changes in the length of an object. Used in applications related to land subsidence measurement, a borehole extensometer measures the compaction or expansion of an aquifer system independently of other vertical movements, such as crustal and tectonic Motions (Galloway et al., 1999). An extensometer measures changes in aquifer system thickness by recording changes in the distance between two points in a well (see Figure 16.2).

DOI: 10.1201/9781003498049-19

TABLE 16.1

Land Subsidence Monitoring Methods

Method	Type of Data	Measures Aquifer-System Compaction Independently	Spatial Coverage	Temporal Detail
Borehole extensometer	Aquifer-system thickness at one location, continuous record	Yes	Low	High
Tidal station	Sea elevations at one location, continuous record	No	Low	High
Geodetic surveying	Land elevations at one or several locations, multiple times or continuous record	No	Low to moderate	Low to high
Remote sensing (InSAR)	Land elevations over a times	No wide area, at multiple	High	Moderate

Source: USGS (2013).

GPS, Global Positioning System; InSAR, Interferometric Synthetic Aperture Radar.

Usually, the two measurement points are established at the top and bottom of a well to measure the total aquifer-system compaction between the land surface and the bottom of the aquifer system. Alternatively, specific intervals within a well can be measured, for example, to measure the compaction of just one aquifer within a layered aquifer system. Extensometer measurements are often combined with surface monitoring techniques to determine the portion of total land subsidence attributable to aquifer-system compaction (Poland, 1984).

HRSD's extensometer used as part of its SWIFT project, installed at Nansemond Treatment Plant in Suffolk, Virginia, and owned, operated, recorded, and maintained by United States Geological Survey (USGS) is shown in Figures 16.3 and 16.4.

TIDAL STATIONS

A tidal station measures sea elevation at one location. To determine long-term trends, sea-level measures are averaged over time to remove the effect of waves, tides, and other short-term fluctuations. In the southern Chesapeake Bay, tidal stations have been in operation for many decades (Table 16.2). Tidal stations are significant and valuable because they indicate relative sea-level rise.

GEODETIC SURVEY

Geodetic surveying is the measurement of land surface coordinates. Geodetic surveying is most commonly performed either with Global Positioning System (GPS) technology that reads signals from satellites to obtain very detailed location and time information or with traditional optical leveling equipment. Because historical geodetic survey records are available for the southern Chesapeake Bay region,

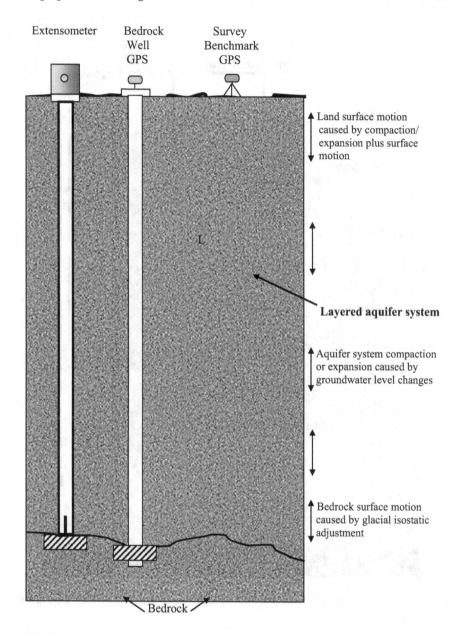

FIGURE 16.1 Subsidence monitoring methods. Survey Benchmark GPS to measure land surface motion. Bedrock Well GPS to measure bedrock surface motion and Extensometer to measure aquifer system compaction or expansion (USGS, 2016).

geodetic surveying can be used to determine cumulative land subsidence over many decades. Benchmark stations are established and, for as long as they remain undisturbed, can be surveyed multiple times to determine elevation changes between surveys (USGS, 2013).

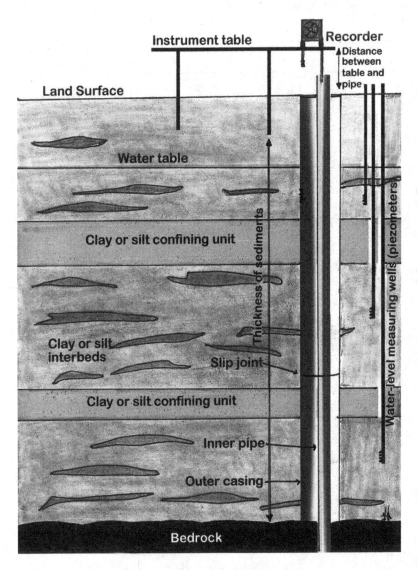

FIGURE 16.2 A borehole extensometer with recording equipment (USGS, 2013; Illustration by F.R. Spellman and Kat Welsh).

The Continuously Operating Reference Station (CORS) network, a network of long-term GPS stations throughout the United States that includes stations in the Southern Chesapeake Bay region is managed by the National Geodetic Survey. Each CORS station continuously records three-dimensional position data (north–south, east–west, and up-down), allowing rates of change to be calculated over time (Snay and Soler, 2008). A CORS station records ground position at one site and is designed to operate for many years.

FIGURE 16.3   USGS HRSD Nansemond pipe extensometer with a total depth of 1,960 ft (Photo by D.L. Nelms Hydrologist, USGS, 2018).

Besides stationary GPS sites such as the CORS stations, portable GPS receivers can be used to expand spatial coverage. In the Houston-Galveston area, GPS receivers mounted on trailers have been used to collect data at up to four different sites each month (Galloway et al., 1999; Bawden et al., 2012). The portable GPS approach had acceptable subcentimeter accuracy, gave greater spatial coverage than stationary GPS would have, and had a lower cost than InSAR technology.

FIGURE 16.4 The housing of the Nansemond Treatment Plant extensometer showing the triangular table and instrument bridge above the extensometer. The piers that support the table extend down 65 ft. The movement of the table relative to the extensometer is how land-surface movement is measured (Photo by D.L. Nelms, USGS).

TABLE 16.2

Relative Sea-Level Rise at Selected National Oceanic and Atmospheric Administration Tidal Stations in the Southern Chesapeake Bay Region

| | | | Rate of Relative Sea-Level Rise | |
ID	Site Name	Period	Measured (mm/yr)	95% CI
8632200	Kiptopeake, Virginia	1951–2006	3.5	±0.42
8637624	Gloucester Point, Virginia	1950–2006	3.8	±0.47
8638610	Sewells Point, Virginia	1927–2006	4.4	±0.27
8638660	Portsmouth, Virginia	1935–2006	3.8	±0.45
	Average		**3.9**	**±0.40**

Source: Data are from Zervas (2009) and USGS (2013).
CI, confidence interval; mm, millimeters; mm/yr, millimeters per year; %, percent.

InSAR

InSAR is a radar technique used in geodesy and remote sensing. InSAR has been used to investigate surface deformation resulting from land subsidence (Galloway and Hoffmann, 2007). With InSAR, as little as 5 mm of elevation change can be measured over hundreds or thousands of square kilometers with a horizontal spatial resolution down to 20 m (Pritchard, 2006). Interferograms (maps) show land-surface elevation changes that are produced by combining two synthetic aperture rate (SAR) images acquired by multiple satellite or airborne passes over the same area at different times. InSAR analysis has the advantage of measuring subsidence over a large area, whereas traditional geodetic leveling and GPS surveying are performed at only one or a handful of locations, during a survey (Sneed et al., 2002; Stork and Sneed, 2002).

Using InSAR in the Chesapeake Bay region has potential limitations. Subsidence rates determined by InSAR might have errors that are larger than the subsidence rates observed in the region (1.1–4.8 mm/yr). The region's high humidity and dense vegetation would create spurious radar signals, require the use of persistent scatter techniques, and result in lower measurement resolution than is found in more arid regions (Rancoules et al., 2009). Also, available satellite data cover only a relatively short time span. The best available synthetic aperture radar (SAR) satellite data for the southern Chesapeake Bay region covers from 1992 to 2000, so the time of accumulated subsidence determined from these data would be no more than 8 years. Despite these limitations, InSAR could be used to identify hotspot areas of subsidence. Such mapping could be useful for identifying unexpected areas of subsidence, focusing attention on important areas, and picking locations for other ground-based subsidence monitoring techniques (USGS, 2013).

IMPORTANCE OF LAND SUBSIDENCE MONITORING

Land subsidence, or sinking of the land surface, has occurred and is still occurring in many locations throughout the United States. Earlier, we pointed out a few of the most impacted areas in the United States but the focus of this book is on Hampton Roads, Virginia (the area of southern Chesapeake Bay, the largest estuary in the United States) where land subsidence is an ongoing and serious issue, but also where innovation and technology and far thinking might be able to arrest or correct the problem. In Hampton Roads, the boundary between land and water is low-lying. Land subsidence is important in the region because it causes increased flooding, alters wetland and coastal ecosystems, and damages infrastructure and historical sites such as Jamestown, Virginia.

As population densities continue to increase in the Hampton Roads region, the flood hazard unfortunately is increasing as well due to locally high rates of land subsidence with global sea level rise. The combination—global sea level rise and coastal subsidence—is doubly significant even in locations where the occurrence of major hurricanes is relatively rare compared to other regions of the country. In the Hampton Roads region in particular, extratropical cyclones or "nor'easters" that have not caused significant flooding in the past will begin to do so—and with greater frequency—as sea level continues to rise *relative* to the land. Flood hazard mitigation compounded by land subsidence and relative sea level rise at a minimum requires an understanding of the land and by-ocean processes that contribute to it in complex and often unpredictable ways.

The fact is that rates and locations of land subsidence are not well known throughout the Hampton Roads area because monitoring has been insufficient in recent decades. Monitoring data is needed to better understand rates and locations of land subsidence and to plan for preventing or mitigating its potentially damaging effects.

Again, recurrent flooding problems have prompted concern about land subsidence in Hampton Roads (Sweet et al., 2014). In addition, these concerns are compounded by evidence that groundwater pumping and associated aquifer depressurization have caused past land subsidence (Pope and Burbey, 2004; Holdahl and Morrison, 1974) and measurements showing that relative sea-level rise is faster in Hampton Roads than elsewhere on the Atlantic Coast (Sallenger et al., 2012). As mentioned, rates and locations of land subsidence are not well known in the Hampton Roads area because monitoring has been insufficient. Therefore, risks commonly associated with coastal land subsidence—increased flooding, alteration of wetland and coastal ecosystem, and damage to infrastructure and historical sites—cannot be accurately assessed. More frequent monitoring at multiple locations using multiple complementary methods is needed to build an understanding of subsidence and to plan how to avoid or mitigate the effects of subsidence.

Before land subsidence can be understood it must be monitored. Monitoring data provides the foundation for understanding why, where, and how fast land subsidence is occurring, both now and in the future. Because rates of land subsidence change over time and vary from one location to another, monitoring should be done at multiple locations for multiple years. Monitoring data are used (USGS, 2016)

- To avoid or mitigate problems caused by land subsidence—Urban planners, resource managers, and politicians use monitoring data to guide their decisions.
- To answer questions, such as—Why is subsidence occurring?
- To predict future land subsidence—Predictive models that can test mitigation strategies require monitoring data for accuracy and reliability.
- To make maps—Maps showing critical areas for mitigating land subsidence are based on monitoring data.

Land subsidence monitoring measures

- land surface motion
- bedrock surface motion
- changes in aquifer system thickness

MONITORING METHODS

Land subsidence is detected by measuring land surface positions over time and calculating rates of change by subtraction. As pointed out earlier there are several reliable and accurate techniques for measuring land subsidence in Hampton Roads: Borehole extensometers, tidal stations, geodetic surveying, and remote sensing (InSAR).

BOREHOLE EXTENSOMETERS–ONGOING MONITORING (2016)

As shown in Figures 16.1–16.4, borehole extensometers are wells designed for measuring compact or expansion of an aquifer system (Galloway et al., 1999). Extensometers typically are paired with monitoring wells so that the correlation between groundwater-level changes and aquifer compaction can be determined.

In Hampton Roads from 1996 to 2016, no borehole extensometers were active. However, historic extensometer data are available, covering the period 1979–1995 for an extensometer located at Franklin Virginia, and 1982–1995 for an extensometer located at Suffolk, Virginia (Pope and Burbey, 2004).

These older existing extensometers at Franklin and Suffolk have recently been equipped by the USGS with digital potentiometers, dial gages, and satellite telemetry to provide aquifer compaction measurements with sub-millimeter (0.01 mm) accuracy. Data are being collected to test if the extensometer stations can be reactivated to detect aquifer compaction and expansion. The extensometers will be monitored for several months and, if monitoring results are successful, the extensometers may be reactivated on a long-term basis. The possibility of installing GPS antennas on the extensometers, to determine contributions to subsidence from glacial isostatic rebound, will also be investigated (USGS, 2016).

Michelle Sneed, a USGS expert on subsidence and extensometers was brought in to consult on land subsidence monitoring options in Hampton Roads. She described how, in California, extensometers provide the basis for understanding how land subsidence is related to groundwater withdrawals, for calibrating InSAR estimates of land subsidence, and for calibrating predictive models of land subsidence. Extensometers there provide data used for water-resource planning and subsidence-mitigation planning (USGS, 2016).

GEODETIC SURVEYING—ONGOING MONITORING 2016

Geodetic surveying is the measurement of land surface position. Global positioning system (GPS) technology is now widely used to perform geodetic surveying. Permanent GPS stations, such as the network of Continuously Operating Reference Stations (CORS) operated by the National Geodetic Survey (NGS), provide continuous information about land surface motion at single locations. CORS stations typically achieve centimeter-scale accuracy for absolute vertical position measurement and millimeter-scale accuracy for differential vertical position measurement. Permanent geodetic stations, such as CORS, also provide valuable information for calibrating remote sensing measurements of subsidence (USGS, 2016).

Survey networks, consisting of multiple high-integrity monuments (benchmarks) that are installed on land periodically occupied with GPS antennas to measure land surface position, can also provide valuable regional estimates of land subsidence. Dr. Philippe Hensel (NGS) has offered his expertise to help design and implement such a survey network for Hamptons Roads.

A separate type of geodetic surveying that would be valuable for understanding land subsidence in Hampton Roads is using GPS antennas on bedrock wells to measure bedrock surface motion (Figure 16.1). This can be done at any new extensometer that is being constructed. Existing bedrock wells, such as those at Franklin and Suffolk, may also be available as platforms for this type of monitoring.

The NGS, the lead US federal agency for surveying and geodetic science, operates the CORS network of benchmark stations that continuously record land surface positions in fine detail in 3-dimensions. The CORS network includes five benchmark sites in Hampton Roads.

Various other organizations have established continuous monitoring GPS antennas at benchmark stations in Hampton Roads that are not part of the CORS network. For example, the NASSA Langley Research Center in Hampton, Virginia, established four benchmark sites with GPS antennas in 2015. In some cases, data from these non-CORS stations are available and, if a site has been constructed and operated following NGS guidelines (NGS, 2013; Floyd, 1978), the resulting data can be of high quality and useful for subsidence calculation.

The NGS is currently (2017) analyzing historic surveys of first-order benchmark sites on the Atlantic Coast, including in Hampton Roads, to determine rates of subsidence over the past century. This study will produce maps of subsidence rates over multiple time periods.

TIDAL STATIONS—ONGOING MONITORING 2016

For many decades, the National Oceanic & Atmospheric Administration (NOAA) has operated tidal stations to provide continuous water-level data at four sites in Hampton Roads. Data are publically available at no cost from NOAA's website.

REMOTE SENSING—ONGOING MONITORING 2016

InSAR is a remote sensing technique used to measure land surface elevation changes over wide areas, for example, over the entire Hampton Roads area. InSAR can be used to determine the map critical areas of land subsidence, select locations for detailed geodetic surveying, and plant strategies for preventing and mitigating land subsidence (Bawden et al., 2003). Accuracy of InSAR subsidence estimates will be important in Hampton Roads, because subsidence rates in the area have been measured at 1.1–4.8 mm, as compared to the typical error for InSAR of 5–10 mm. The high atmospheric humidity and dense vegetation found in Hampton Roads can reduce InSAR accuracy. Problems with error can be overcome by analyzing a large number of satellite scenes, applying persistent scatter analysis techniques, using INSAR data collected over multiple years, and using L-band or X-band rather than C-band InSAR data (USGS, 2016).

Probably the most valuable aspect of InSAR remote sensing is its capacity to input valuable data for detailed mapping of regional subsidence over time. The type of remote sensing data used to map subsidence, InSAR, has been collected for Hampton Roads by various satellites since 1992 and is currently collected by several international satellites. IN 2017, a new United States satellite, NISAR, will collect InSAR data over Hampton Roads (USGS, 2016).

THE BOTTOM LINE

The U.S. Geological Survey is cooperating with Federal, State, and local government agencies to study and better understand the problem of land subsidence in the southern Chesapeake Bay region. In order to make informed decisions, local resource managers, planners, politicians, and regulators need in-depth knowledge of the particulars involved with relative sea-level rise and land subsidence in the Hampton Roads area. This knowledge is necessary in the planning of increased flood risks and preventing land subsidence.

THE REAL BOTTOM LINE

The intended purpose of this presentation and this book is to satisfy the elements described above—to explain what relative sea-level and land subsidence are and to describe and provide nuts-and-bolts explanations (in the chapters to follow) for planners, managers, and others a methodology and technology that can potentially prevent, provide rebound from, or mitigate the impact of land subsidence in the Hampton Roads region.

NOTE

1 Material in this chapter is based on USGS (2013). *Land Subsidence in the United States*. Circular 1182. U.S. Department of the Interior, U.S. Geological Survey. Washington, D.C.

REFERENCES

Bawden, G.W., Johnson, M.R., Kasmarek, M.C., Brandt, J., and Middleton, C.S. (2012). Investigation of land subsidence in the Houston-Galveston Region of Texas by using the global positioning system and interferometric synthetic aperture radar, 1993-2000. U.S. Geological Survey Scientific Investigations Report 2012-5211, 88 p. accessed at https://pubs.ugsu.gov/sir/2012/5211.

Bawden, G.W., Sneed, M., Stork, S.V., and Galloway, D.L. (2003). Measuring human-induced land subsidence from space. U.S Geological Survey Fact Sheet 069-03, 4 p., https://pubs.usgs.gov/fs/fs/fs069003/.

Floyd, R.P. (1978). Geodetic bench marks, NOAA manual NOS NGS 1, U.S. Dept of Commerce, National Oceanic and Atmospheric Administration, Rockville, MD, September 1978, 52 pp. https://www.ngs.noaa.gov/PUBS_LIB/GeodeticBMs.pdf.

Galloway, D.L., and Hoffmann, J. (2007). The application of satellite differential SAR interferometry-derived ground displacements in hydrogeology. *Hydrogeology Journal* 15(1): 133–154.

Galloway, D.L., Jones, D.R., and Ingebritsen, S.E. (eds.) (1999). Land Subsidence in the United States. U.S. Geological Survey Circular 1182, 177 p. accessed at https://pubs.usgs.gov/circ/circ1182/.

Holdahl, S.R., and Morrison, N. (1974). Regional investigations of vertical crustal move-ments in the U.S., using precise relevelings and mareograph data. *Tectonophysics* 23(4): 373–390.

National Geodetic Survey. (2013). Guidelines for new and existing Continuously Operating Reference Stations (CORS) National Geodetic Survey National Ocean Survey, NOAS Silver Spring, Maryland, January 2013. https://ngs.noaa.gov?PUBS_LIB/CORS_grfuid-lines.pdf.

Poland, J. F. (ed.) (1984). *Guidebook to Studies of Land Subsidence Due to Ground-Water Withdrawal.* United Nations Educational, Scientific and Cultural Organization, 305 p. plug appendices. Accessed at https://wwwrcamni.wr.usgs.go/rgws/Unesco/PDF-Cahpters/Guidebook.pdf.

Pope, J.P., and Burbey, T.J. (2004). Multiple-aquifer characterization from single borehole extensometer records. *Ground Water* 42(1): 45–58.

Pritchard, M.E. (2006). InSAR, a tool for measuring Earth's surface deformation. *Physics Today* 59(7): 68–69.

Rancoules, D., Bourgine, B., de Michele, M., Le Cozannet, G., Closset, L., Bremmer, C., Veldkamp, H., Tragheim, D., Bateson, I., Crosetto, M., Agudo, M., and Engdahl, M. (2009). Validation and intecomparison of persistent scatterrers interferometry-PSIC4 project results. *Journal of Applied Geophysics* 68(3): 335–347.

Sallenger, A.H., Doran, K.S., and Howd, P.A. (2012). Hotspot of accelerated sea-level rise on the Atlantic Coast of North America. *Nature Climate Change* 2(12): 884–888.

Snay, R.A., and Soler, T. (2008). Continuously operating reference station (CORS): History, applications, and future enhancements. *Journal of Surveying Engineering* 134(4): 95–104.

Sneed, M., Stork, S.V., and Ikehara, M.E. (2002). Detection and measurement of land subsid-ence using global position system and interferometric synthetic aperture radar, Coachella Valley, California, 1998-2000. U.S. Geological Survey Water-Resources Investigations Report 02-4239, 29 p.

Stork, S.V., and Sneed, M. (2002). Houston-Galveston Bay Area, Texas, from space: A new tool for mapping land subsidence. U.S. Geological Survey Fact Sheet 2002-110, 6 p. https://pubs.usgs.gov/fs/fs-110-02/.

Sweet, W., Park, J., Marra, J., Zervas, C., and Gill, S. (2014). Sea level rise and nuisance flood frequency changes around the United States, NOAS Tech. Report NOS CO-OPS 073, 58 pp. Https://tidesandcurrents.noaa.gov/publications/NOAA_Technical_Report_NOS_COOP_073.pdf.

USGS. (2013). Land subsidence in the United States. Circular 1182. U.S. Department of the Interior, U.S. Geological Survey. Washington, DC.

USGS. (2016). Land subsidence monitoring in Hampton roads: Progress report. Jack Eggleston, USGS Virginia Water Science Center. Accessed @ www.hrpdc.gov/uploads/docs/04A_attachment_USGS_HRPDC_Subsi...

USGS (2018). United States (Lower 48) Seismic Hazards Long-Term Model. Accessed 12.11/23 @ https://www.usgs.gov/programs/earthquake-hazards/science/2018.

Zervas, C. (2009). Sea level variations of the United States, 1854-2006. National Oceanic and Atmospheric Administration Technical Report NOS CO-OPS 053, 76 p. plus appen-dixes, accessed at https://www.co-ops.nos.noas.gov/publicatins/Tech_/r_53.pdf.

17 HRSD and the Potomac Aquifer

When we use up one resource and then another resource and then another, there is one critical resource mankind never runs out of. What is that one item that sustains us? Well, in a word it is innovation, innovation, innovation—innovators focus on solutions and find solutions rather than being bogged down with problems.

—F.R. Spellman (2021)

HAMPTON ROADS SANITATION DISTRICT

Its genesis was driven by oysters. No, not the genesis of Chesapeake Bay. Again, its genesis was driven by a heavy, unstoppable, all-knowing hand; a sculptor without equal. Hampton Roads Sanitation District (HRSD), arguably the premier wastewater treatment district on the globe, became a viable governor-appointed state commission-monitored entity because of a significant decline in the oyster population in Chesapeake Bay. As a case in point, consider that in the Hampton Roads region of Chesapeake Bay in 1607 when Captain John Smith and his team settled in Jamestown, oysters up to 13 inches in size were plentiful—more than could ever be harvested and consumed by the handful of early settlers. And this population of oysters and other aquatic lifeforms remained plentiful until the human population gradually increased in the Bay region.

Over-harvesting of oysters by the increased numbers of humans living in the Chesapeake Bay region was (and might still be) a major issue with the decline of the oyster population. However, the real culprit causing the decline in the oyster population is pollution. Before the Bay became polluted from sewage, sediment, and garbage disposal, oysters could handle natural pollution from stormwater run-off and other sources. Ninety years ago, when there was a much larger oyster population than today, it is estimated that the large oyster population could filter pollutants from the Bay and clean it in as little as 4 days—each oyster filtering about 50 gallons of water each day. By the 1930s however, the declining oyster population was overwhelmed by the increasing pollution levels.

For years and in many written accounts, the author has stated that pollution is a judgment call. That is, pollution as viewed by one person may not be pollution observed by another. You might shake your head and ask a couple of questions, "Pollution is a judgment call? Why is pollution a judgment call?" A judgment is based on an opinion; it is an opinion because people differ in what they consider to be a pollutant based on their assessment of the accompanying benefits and/or risks to their health and economic well-being posed by the pollutant. For example, visible

DOI: 10.1201/9781003498049-20

and invisible chemicals spewed into the air or water by an industrial facility might be harmful to people and other forms of life living nearby. However, if the facility is required to install expensive pollution controls, forcing the industrial facility to shut down or move away, workers who would lose their jobs and merchants who would lose their livelihoods might feel that the risks from polluted air and water are minor weighed against the benefits of profitable employment and business opportunity. The same level of pollution can also affect two people quite differently. Some forms of air pollution, for example, might cause slight irritation to a healthy person but cause life-threatening problems to someone with chronic obstructive pulmonary disease (COPD) like emphysema. Differing priorities lead to differing perceptions of pollution (concern at the level of pesticides in foodstuffs generating the need for wholesale banning of insecticides is unlikely to help the starving). No one wants to hear that cleaning up the environment is going to have a negative impact on them. The fact is public perception lags behind reality because the reality is sometimes unbearable. This perception lag is clearly demonstrated in Case Study 17.1.

CASE STUDY 17.1: CEDAR CREEK COMPOSTING

Cedar Creek Composting (CCC) Facility was built in 1970. A 44-acre site designed to receive and process to compost wastewater biosolids from six local wastewater treatment plants, CCC composted biosolids at the rate of 17.5 dry tons per day. CCC used the aerated static pile (ASP) method to produce pathogen-free, humus-like material that could be beneficially used as an organic soil amendment. The final compost product was successfully marketed under a registered trademark name.

Today, the CCC Facility is no longer in operation. The site was shutdown in early 1997. From an economic point of view, CCC was highly successful. When a fresh pile of compost had completed the entire composting process (including curing), dump truck after dump truck would line the street outside the main gate, waiting in the hope of buying a load of the popular product. Economics was not the problem. In fact, CCC could not produce enough compost fast enough to satisfy the demand. No, economics was not the problem.

What was the problem, then? The answer to this is actually twofold: social and then eventually legal considerations. Social limitations were imposed by the community in which the compost site was located. In 1970, the 44 acres CCC occupied were located in an out-of-town, rural area. CCC's only neighbor was a regional, small airport on its eastern border. CCC was completely surrounded by woods on the other three sides. The nearest town was 2 miles away. But, by the mid-1970s, things started to change. Population growth and its accompanying urban sprawl quickly turned forested lands into housing complexes and shopping centers. CCC's western border soon became the site of a two-lane road that was upgraded to four and then to six lanes. CCC's northern fence separated it from a mega-shopping mall. On the southern end of the facility, acres of houses, playgrounds, swimming pools, tennis courts, and a golf course were built. CCC became an island surrounded by urban growth. Further complicating the situation was the airport; it expanded to the point that by 1985, three major airlines used the facility.

CCC's ASP composting process was not a problem before the neighbors moved in. We all know dust and odor control problems are not problems until the neighbors complain—and complain they did. CCC was attacked from all four sides. The first complaints came from the airport. The airport complained that dust from the static piles of compost was interfering with air traffic control.

The new, expanded highway brought several thousand new commuters right up alongside CCC's western fence line. Commuters started complaining anytime the composting process was in operation; they complained primarily about the odor—that thick, earthy smell permeated everything.

After the enormous housing project was completed, and people took up residence there, complaints were raised on a daily basis. The new homeowners complained about the earthy odor and the dust that blew from the compost piles onto their properties anytime they were downwind from the site. The shoppers at the mall also complained about the odor.

City Hall received several thousand complaints over the first few months before they took any action. The city environmental engineer was told to approach CCC's management and see if some resolution of the problem could be effected. CCC management listened to the engineer's concerns but stated that there was not a whole lot that the site could do to rectify the problem.

As you might imagine, this was not the answer the city fathers were hoping to get. Feeling the increasing pressure from local inhabitants, commuters, shoppers, and airport management people, the city brought the local state representatives into the situation. The two state representatives for the area immediately began a campaign to close down the CCC facility.

CCC was not powerless in this struggle—after all, CCC was there first, right? The developers and those people in those new houses did not have to buy land right next to the facility—right? Besides, CCC had the USEPA on its side. CCC was taking a waste product no one wanted, one that traditionally ended up in the local landfill (taking up valuable space), and turning it into a beneficial reuse product. CCC was helping to conserve and protect the local environment, a noble endeavor.

The city politicians did not really care about noble endeavors but they did care about the concerns of their constituents, the voters. They continued their assault through the press, electronic media, legislatively, and by any other means they could bring to bear.

CCC management understood the problem and felt the pressure. They had to do something, and they did. Their environmental engineering division was assigned the task of coming up with a plan to mitigate not only CCC's odor problem but also its dust problem. After several months of research and a pilot study, CCC's environmental engineering staff came up with a solution. The solution included enclosing the entire facility within a self-contained structure. The structure would be equipped with a state-of-the-art ventilation system and two-stage odor scrubbers. The engineers estimated that the odor problem could be reduced by 90% and the dust problem reduced by 98.99%. CCC management thought they had a viable solution to the problem and was willing to spend the 5.2 million dollars to retrofit the plant.

After CCC presented its mitigation plan to the city council, the council members made no comment but said that they needed time to study the plan. Three weeks later, CCC received a letter from the mayor stating that CCC's efforts to come up with a plan to mitigate the odor and dust problems at CCC were commendable and to be applauded but were unacceptable.

From the mayor's letter, CCC could see that the focus of the attack had now changed from a social to a legal issue. The mayor pointed out that he and the city fathers had a legal responsibility to ensure the good health and well-being of local inhabitants and that certain legal limitations would be imposed and placed on the CCC facility to protect their health and welfare.

Compounding the problem was the airport. Airport officials also rejected CCC's plan to retrofit the compost facility. Their complaint (written on FAA official paper) stated that the dust generated at the compost facility was hazarding flight operations, and even though the problem would be reduced substantially by engineering controls, the chance of control failure was always possible, and then an aircraft could be endangered. From the airport's point of view, this was unacceptable.

Several years went by, with local officials and CCC management contesting each other on the plight of the compost facility. In the end, CCC management decided that they had to shutdown its operation and move to another location, so they closed the facility.

After the shutdown, CCC management staff immediately started looking for another site to build a new wastewater biosolids-to-compost facility. They are still looking. To date, their search has located several pieces of property relatively close to the city (but far enough away to preclude any dust and odor problems) but they have had problems finalizing any deal. Buying the land is not the problem—getting the required permits from various county agencies to operate the facility is. CCC officials were turned down in each and every case. The standard excuse? Not in my backyard. Have you heard this phrase before? It is so common now, it is usually abbreviated—NIMBY. Whether back in the day or at present, NIMBY is alive and well. To this very year (2017), CCC officials are still looking for a location for their compost facility; they are not all that optimistic about their chances of success in this matter.

So, now you, the reader is asking "What does composting biosolids have to do with oyster depletion, pollution of Chesapeake Bay, and HRSD? Good question. Well, it does not take a leap of faith or an understanding of hieroglyphics to recognize and appreciate a couple of key points being made here. First, along with the over-harvesting of oysters (including crabs and other species of Bay life) pollution has degraded the waters of Chesapeake Bay and directly contributed to the decrease in the oyster population therein. Second, pollution did not become an issue, like it did with the compost facility, until the "neighbors" complained and they did complain. Not only in the 1930s were they complaining about the reduced population of sea life in southern Chesapeake Bay (and other regions of the Bay) but the accumulation of floating sewage, the amplification of nasty odors emanating from the biodegradation of the sewage got passersby, would-be swimmers, and boaters and fisherpeople's attention. Third, again, pollution is a judgment call and in 1940 the judgment was made by voters to authorize the governor to appoint a representative commission to oversee pollution mitigation of the southern Chesapeake Bay region. Thus, HRSD came into being. HRSD is a political subdivision of the Commonwealth of Virginia with a service area that includes 17 counties and cities encompassing 2,800 square miles in its southeastern Virginia service area (see Figure 17.1). HRSD's collection

HRSD Service Area
A Political Subdivision of the Commonwealth of Virginia

Facilities include the following:

1. Atlantic, Virginia Beach
2. Chesapeake-Elizabeth, VA. Beach
3. Army Base, Norfolk
4. Virginia Initiative, Norfolk
5. Nansemond, Suffolk
6. Lawnes Point, Smithfield
7. County of Surry
8. Town of Surry

09. Boat Harbor, Newport News
10. James River, Newport News
11. Williamsburg, James City County
12. York River, York County
13. West Point, King William County
14. King William, KingWilliam County
15. Central Middlesex, Middlesex County
16. Urbanna, Middlesex County

Serving the Cities of
Chesapeake, Hampton,
Newport News, Norfolk,
Poquoson, Portsmouth, Suffolk,
Virginia Beach, Williamsburg and the
Counties of Gloucester,
Isle of Wight, James City,
King and Queen, King William,
Mathews, Middlesex, Surry* and York
*Excluding the Town of Claremont

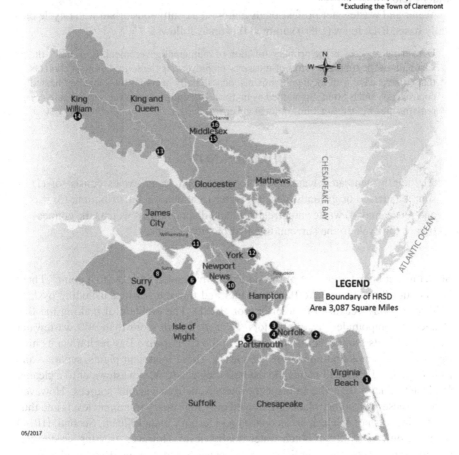

FIGURE 17.1   HRSD service area.

system consists of more than 500 miles of piping, 6–66 inches in diameter. HRDS possesses more than 100 active pumping operations that pump raw wastewater to nine major treatment plants in Hampton Roads and four smaller plants in the Middle Peninsula region. The combined capacity of HRSD facilities is 249 million gallons per day (MGD).

Probably another question rumbling through the reader's brain matter at this point is "Has HRSD solved the pollution problem in the Chesapeake Bay?" This question leads us to another question, "Have the oysters rebounded in quantity?" The answer to both these questions is yes; that is, to a point. On an ongoing, 24 to 7 basis, HRSD treats wastewater to quality better than is contained in the James River, Elizabeth River, York River, and other river systems in the region. Those who have no knowledge of wastewater treatment, HRSD, and/or the conditions of the rivers in this region might have second thoughts about this statement. However, it is true. It is all about the human-made water cycle. In this case, we are talking about the urban water cycle.

Water and wastewater professionals maintain a continuous urban water cycle on a daily basis. B.D. Jones (1980) summed this up as follows:

> Delivering services is the primary function of municipal government. It occupies the vast bulk of the time and effort of most city employees, is the source of most contacts that citizens have with local governments, occasionally becomes the subject of heated controversy, and is often surrounded by myth and misinformation. Yet, service delivery remains the "hidden function" of local government.

DID YOU KNOW?

Artificially generated water cycles or the urban water cycles consist of (1) source (surface or groundwater), (2) water treatment and distribution, (3) use and reuse, and (4) wastewater treatment and disposition, as well as the connection of the cycle to the surrounding hydrological basins.

So, let us get back to the part of the statement that ended with "—to a point." There is no doubt that Chesapeake Bay is cleaner and that the sea life, including oysters, is happier today because of the efforts of HRSD. The problem of making the Bay cleaner is compounded by two factors. First, there are more than 300 wastewater treatment plants that outfall treated water to the Chesapeake Bay region via its nine major river systems and numerous tributaries. These treatment plants, separate and isolated from HRSD's 13 plants, do the best they can to treat wastewater to a cleaner product (effluent) than the influent they received from various sources. However, some of these 300+ other plants are treated only to primary treatment levels and thus their effluent is not as clean as secondary and tertiary plant effluent. Second, HRSD treats wastewater to a top-notch water quality level. However, treating wastewater to remove nutrients is a complicated and expensive undertaking. Biological nutrient removal and other nutrient removal technologies are available and in use in many locales, but the technology is expensive—expensive to the point where the treatment technology needed and used might overtax ratepayers.

Earlier, it was pointed out that the Chesapeake Bay occasionally suffers from dead zones due to algal blooms. The *algae bloom* is a phenomenon whereby excessive nutrients within the Bay cause an explosion of plant life that results in the depletion of the oxygen in the water needed by fish and other aquatic life. Algae bloom is usually the result of urban runoff (of lawn fertilizers, etc.). The potential tragedy is that of a "fish kill," where the bay life dies in one mass execution.

Algal bloom and dead zones and the resulting fish kill events are a major issue, of course. However, when you add this problem to relative sea-level rise and land subsidence, it can be readily seen that the issue and problems with maintaining the health of the Chesapeake Bay and its inhabitants are multifaceted.

THE SOLUTION TO POLLUTION IN CHESAPEAKE BAY[1]

Every Problem has a Solution. And that has proven to be true in most cases with many more worldwide problems that need solutions to be found, hopefully. With regard to the problems with the Chesapeake Bay, nutrient pollution, saltwater intrusion, land subsidence, and relative sea-level rise in the Hampton Roads region, HRSD has developed the innovative Sustainable Water Initiative For Tomorrow (SWIFT) program (a work in progress; a decadal project). Do not confuse the acronym swift with the adjectives for fast, speedy, rapid, hurried, immediate, or quick. SWIFT is a deliberate, well-thought-out, long-term project that is being developed on a timeline that is set for the installation of the technical equipment and operational procedures with a completion date of 2030.

Okay, so what is HRSD's SWIFT?

SWIFT is a program to inject treated wastewater into the subsurface; specifically, it is designed to inject treated wastewater to drinking water quality into the Potomac Aquifer. Injection of water into the subsurface is expected to raise groundwater pressures, thereby potentially expanding the aquifer system, raising the land surface, and counteracting land subsidence occurring in the Virginia Coastal Plain. In 2016, a pilot project site was under construction at the HRSD Nansemond Wastewater Treatment Plant in Suffolk, Virginia, to test injection into the aquifer system. HRSD has asked USGS to prepare a proposal for the installation of an extensometer monitoring station at the test site to monitor groundwater levels and aquifer compaction and expansion.

THE PROBLEM

SWIFT is designed to counter land subsidence at various locations in the Hampton Roads area of southern Chesapeake Bay where land subsidence rates of 1.1–4.8 millimeters per year have been observed (Eggleston and Pope, 2013; Holdahl and Morrison, 1974).

Injection of treated wastewater (treated to drinking water quality) is expected to counteract land subsidence or raise land surface elevations in the region. Careful monitoring of aquifer system compaction and groundwater levels can be used to optimize the injection process and to improve a fundamental understanding of the relation between groundwater pressures and aquifer-system compaction and expansion.

There is more to HRSD's treated wastewater injection project, SWIFT, than just arresting or mitigating land subsidence and relative sea-level rise in the Hampton Roads region. One of the additional goals of the project is to stop the discharge of treated wastewater from seven of its plants which would mean 18 million pounds a year less of nitrogen, phosphorus, and sediment out-falling into the bay. Assuming SWIFT works as designed this is a huge benefit to Chesapeake Bay in that it may help to prevent or reduce the formation of algal bloom dead zones. Not only would success as a result of treated wastewater injection benefit the bay, but it would also be a huge benefit for the ratepayers at HRSD. To meet regulatory guidelines to remove nutrients from discharged treated wastewater would cost hundreds and millions of dollars and almost non-stop retrofitting at the treatment plants to keep up with advances in treatment technology and regulatory requirements. Another goal of HRSD's SWIFT project is to restore or restock potable groundwater supplies in the local aquifers. The drawdown of water from the groundwater supply has not only contributed to land subsidence but to a reduction of water available for potable use.

HRSD's planned restocking of Hampton Roads groundwater supply with injected wastewater treated to potable water quality is not without its critics. The critics state that HRSD's wastewater injection project would contaminate potable water aquifers. For the critics and others, as discussed earlier, this is where the so-called "yuck factor" comes into play. This is the common view of many of the critics who feel HRSD's SWIFT project is nothing more than a direct reuse of wastewater; that is, a pipe-to-pipe connection of toilet water to their home water taps.

What the critics and others do not realize is that we are already using and drinking treated and recycled toilet water. As far as HRSD's SWIFT project contaminating existing aquifers with toilet water, it is important to point out that this water is to be treated (and already is at Nansemond Treatment Plant in Suffolk, Virginia) to drinking water quality—to drinking water quality is the key phase here. This sophisticated and extensive train of unit drinking water quality treatment processes—treated wastewater that a former HRSD General Manager and several others drank right out of the process recently, and, by the way, they are doing just fine today, thank you very much—is discussed in detail later in the text. The bottom line: Statements about the yuck factor involved in drinking treated toilet water are grossly over-stated.

The SWIFT project includes the construction of an extensometer monitoring station with the ability to accurately measure land-surface elevations, bedrock-surface elevations, and changes in aquifer-system thickness. Monitoring of groundwater levels and aquifer-system elastic response will benefit the operation of the wastewater injection system at the Nansemond Treatment Plant in Suffolk, Virginia, and provide guidance on future wastewater injection facilities for the SWIFT project.

OBJECTIVES

The objectives of the SWIFT project are to:

- Design and construct an extensometer station for collection of aquifer-system thickness data with sub-millimeter accuracy for long-term (decadal) operation.
- Operate and maintain the station for at least 3 years to collect data describing groundwater levels, land-surface and bedrock vertical motion, and changes in aquifer-system thickness.

Nansemond Treatment Plant Extensometer Plan

The HRSD extensometer station at the Nansemond Treatment Plant will be designed and constructed to produce reliable and accurate data describing aquifer thickness, groundwater levels, land-surface elevation, and bedrock for at least 20 years and likely for 40 years or more.

The USGS, from its National Research Drilling Program (NDRP), has obtained unique and extensive experience in designing and installing extensometers in many locations around the United States, including California, Texas, and Louisiana. Based on this past experience USGS recommended including the following features in the extensometer instrument:

- Casing slip joints to accommodate the vertical stress and strain that accompany aquifer compaction/expansion and that can cause casing failure.
- Casing centralizers and a counterweighted fulcrum at the surface to reduce friction and striking between the extensometer pipe and casing.
- Combined mechanical strain gages and digital potentiometers to achieve both accuracy and long-term record continuity.
- Deep surface mount support piers for the reference table to exclude surface soil compaction and heaving from the aquifer compaction/expansion signals.
- Heavy steel beam frames to support surface equipment.

Under the proposed plan and as discussed earlier, the USGS has constructed an extensometer station at HRSD's Nansemond wastewater treatment facility in Suffolk, Virginia. USGS Water Science Center personnel oversaw all aspects of the construction and operation of the extensometer system, including the drilling and down-hole work. USGS also deployed its own drill rig, support equipment, and drill crew. A staff geologist with the USGS and a geologist for the Virginia Department of Environmental Quality (DEQ) shared on-site geologist duties during the approximate 1 month of drilling (see Figure 17.2). Installation of surface reference-frame structures and monitoring equipment was performed or overseen by USGS staff. The work lasted from July 2016 through August 2019. Drilling and construction occurred in Year 1 and USGS operated the station in Years 1–3. Station operation can continue beyond Year 3 by mutual agreement.

The extensometer station was constructed within the fenced perimeter of the Nansemond Wastewater Treatment plant in Suffolk, Virginia, and the final location was chosen jointly with HRSD. The final location was located ~500 ft from the injection test well, has paved road access, and has adequate space for equipment including a large drilling rig, delivery trucks, a generator, work lights, forklift, dumpsters, and backhoe. A roughly 60′ × 100′ area is used to house materials, a work area for geologists and drillers was provided, and a pit was dug for the circulation of drilling fluids.

A permanent structure with a 20′ × 20′ footprint and a 10′ height was built to house the top of the extensometer instrument and the associated monitoring equipment. It is an insulated metal building with a concrete floor and a window and door, but alternative exteriors can be considered later if desired by HRSD. The building will have locked windows and doors.

FIGURE 17.2 USGS drilling rig at Nansemond Treatment Plant Suffolk, VA. (Photograph by F.R. Spellman.)

A cluster of four piezometers is highly recommended to allow monitoring of groundwater levels (pressures) in multiple aquifers. Each piezometer will be housed in a protective casing that protrudes about 3′ above ground. These piezometers are permanent structures and are located near the extensometer building.

PROPOSED SWIFT ADDITIONS

HRSD's SWIFT project proposes to add an advanced treatment process to several of its facilities to produce water that exceeds drinking water standards, and to pump this clean water into the ground and the Potomac Aquifer. This will ensure a sustainable source of water to meet current and future groundwater needs through eastern Virginia while improving water quality in local rivers and the Chesapeake Bay. SWIFT project benefits include (HRSD, 2016a, b):

• Eliminates HRSD discharge to the James, York, and Elizabeth Rivers except during significant storms
• Restores rapidly dwindling groundwater supplies in eastern Virginia upon which hundreds of thousands of Virginia residents and businesses depend
• Creates huge reductions in the discharge of nutrients, suspended solids, and other pollutants to the Chesapeake Bay

- Make available significant allocations of nitrogen and phosphorous to support regional needs
- Protects groundwater from saltwater contamination/intrusion
- Reduces the rate of land subsidence, effectively slowing the rate of sea level rise by up to 25%
- Extends the life of protective wetlands and valuable developed low-lying lands

POTOMAC AQUIFER

In a Nutshell (USGS, 2013)

- The Potomac Aquifer is the largest and most heavily used groundwater source in the Virginia Coastal Plan.
- It spans across parts of Virginia, Maryland, and North Carolina.
- Water-level declines in this aquifer can be substantial, reaching up to 200 ft, which raises concerns about saltwater intrusion.
- The sediments within the Potomac aquifer exhibit regional variations.
- In the central and southern parts of the Virginia Coastal Plain (aka the Norfolk arch depositional subarea), the aquifer consists of coarse-grained sediments that function hydraulically as a single interconnected aquifer.
- Fine-grained intervals, generally avoided for water-supply wells are increasingly thick and widespread northward.
- In far southeastern Virginia and northeastern North Carolina as part of the Albemarle embayment the depositional subarea has a thickness of both coarse and fine-grained intervals.
- The Potomac sediments were deposited approximately 100–145 million years ago during the Cretaceous period.
- Uplifted granite and gneiss source rocks contributed coarse-grained sand and gravel.
- In the Norfolk arch subarea immature high-gradient braided streams deposited sediments.
- Pumping tests conducted at various locations provided information on transmissivity, storativity, hydraulic conductivity, and specific storage of the aquifer sediments.
- The Bottom Line: The Potomac Aquifer plays an essential role in supplying groundwater to the region, and its sediment characteristics vary significantly across separate areas of the Coastal Plain.

Be advised that HRSD is still (as of 2024) in the early stages of operation of its SWIFT initiative and lessons are being learned via this work in progress. Also, before HRSD commenced pumping up to 130 MGD into the Potomac Aquifer System (PAS) or as it will do more in the future into any other aquifer it; HRSD first had to determine the feasibility of aquifer replenishment by recharging clean water, purified from the advanced treatment of wastewater treatment plant (WWTP) effluent. In line with all this, this section provides a description of the essential elements of recharging clean water into the PAS at the following seven HRSD WWTPs: Army Base, Boat Harbor, James River, Nansemond, Virginia Initiative Plant, Williamsburg, and York River. Also, a determination of the capacity of individual injection wells at the seven WWTPS is has been made or is being studied at the present time; moreover, a projection of the injection capacity within the existing site area of the seven WWTPs has or is being determined; and a characterization of the regional beneficial hydraulic response of the PSA to clean water injection has and is being determined.

Note to the reader: The material presented in this section is based on data available in public domain sources, city/country, state and federal databases, reports, scientific papers, and from interviews, operators' findings, personal observations, and current literature to characterize the injection capacity of individual wells at each of the WWTPs. Injection well capacities and analytical mathematical modeling were used to estimate the injection capacity of each WWTP based on the plant's flow rate, property size, and the transmissivity of the underlying PAS aquifer.

THE POTOMAC FORMATION[2]

Given the elevated volume requiring disposal, and the importance of minimizing the number of injection wells, the most suitable aquifer units are those that exhibit the highest production capacity. Furthermore, a thick, confining bed composed of impermeable materials like silt or clay should overlie the aquifer to prevent vertical migration of the injection fluid (injectate) into the surrounding aquifer units. Beneath the HRSD service area, the Cretaceous age, Potomac Formation meets these criteria. The Potomac Formation contains thick sand deposits, forming three discrete aquifer units. Although the modern convention developed by the United States Geological Survey (USGS) and the Virginia Department of Environmental Quality (VDEQ) is to group the three aquifers as one, named the Potomac Aquifer (McFarland and Bruce, 2006), locally and in this book, because they behave hydraulically as three distinct units, they must be and are examined separately. Accordingly, in this presentation, these units are treated as three distinct units referred to as the Upper Potomac Aquifer zone (UPA), Middle Potomac Aquifer zone (MPA), and Lower Potomac Aquifer zone (LPA) (Laczniak and Meng, 1988; Hamilton and Larson, 1988). Each discrete aquifer in the HRSD service area is separated from adjacent aquifers by clay confining beds of measurable thickness, while a cumulative thickness of silt and clay units totaling several hundred feet overlies the top (UPA) aquifer unit of the PAS. Production wells screened in the PAS exhibit significantly greater pumping capacities than wells screened in other aquifers of the Virginia Coastal Plain (Smith, 1999). Some wells can pump at rates approaching 3,000 gallons per minute (gpm) (4.3 MGD). In addition to confinement and production capacity, aquifers in the PAS exhibit deep static water levels, ranging from 80 to 180 feet below grade (fbg), providing available head for injection.

Recent USGS sedimentological studies suggest the HRSD service area, spanning the York-James Peninsula and Southeastern Virginia is well situated regarding the quality of aquifers in the PAS (McFarland, 2013). PAS aquifer sands display greater thickness, coarser grain size, and better sorting in the HRSD service area than units in the northern Virginia Coastal Plain, or to the south in northern North Carolina. As a result, aquifers exhibit excellent hydrologic coefficients (hydraulic conductivity, transmissivity), and thus, more productive well capacity (HRSD, 2016a, b).

The PAS outcrops at the ground surface west of King William WWTP in King William County Virginia (see Figure 17.1). PAS is recharged by infiltrating precipitation. In the recharge area, the aquifers range in thickness from 70 (UPA) to 400 ft (LPA). Further downdip, recharge enters the PAS by leakage through overlying confining beds (Meng and Harsh, 1988).

Individual aquifers thicken and dip to the southeast toward the Atlantic coast line reaching thickness ranging from 170 ft (UPA) to ~1,000 ft (LPA) at the coast (Treifke, 1973). These thicknesses comprise all sediments contained in the vertical section including discrete sand beds representing aquifer materials, and interleaving (i.e., arranged in alternate layers) silt and clay lenses from intra-aquifer contained beds. Individual aquifers exhibit a strongly inter-bedded morphology consisting of thin to thick beds of sands, silts, and clays. Beneath Newport News (refer to Figure 17.1), the MPA consists of six discrete sand intervals. Obtaining maximum production (or injection) capacities from wells installed in layered aquifers such as the PAS requires extending the well screen assembly across the maximum thickness of aquifer sand. Screen assemblies can consist of multiple screens and blank sections (HRSD, 2016a, b).

INJECTION WELLS

Typically, when we think about a well, the image of a hole in the ground and some type of device to lift by bucket or mechanical pumping device to pump water to the surface for use or storage. A well pumping water from the subsurface to the surface functions to provide whatever use it might be intended for. However, wells used for pumping are not what we are concerned within this book. We are concerned with just the opposite; that is, wells that inject-treated injectate and that do not pump fluids to the surface.

Injection wells are known as Class V wells. There are 22 types of Class V injection wells. The types of wells are shown in Table 17.1.

Subsidence Control Wells

It is the last type of Class V well listed in Table 17.1 (subsidence control wells) that is the focus of this presentation. Subsidence control wells are injection wells whose primary objective is to reduce or eliminate the loss of land surface elevation due to the removal of groundwater providing subsurface support. Subsidence control wells are important to HRSD's SWIFT project. The goal is to inject treated wastewater to drinking water quality into the underground PAS to maintain fluid pressure and avoid compaction and to ensure that there is no cross-contamination between infected water and underground sources of drinking water. Thus, the injectate must be of the same quality as or superior in quality to the existing groundwater supply.

TABLE 17.1

Class V Underground Injection Wells

Type of Injection Well	Purpose
Agricultural drainage wells	Receive agricultural runoff
Stormwater drainage wells	Dispose of rainwater and melted snow
Carwashes without undercarriage washing or engine cleaning	Dispose of wash water from car exteriors
Large-capacity septic systems	Dispose of sanitary waste through a septic system
Food processing disposal wells	Dispose of food preparation wastewater
Sewage treatment effluent wells	Used to inject treated or untreated wastewater
Laundromats without dry cleaning facilities	Dispose of fluid from laundromats
Spent brine return flow wells	Dispose of spent brine for mineral extraction
Mine backfill wells	Dispose of mining byproducts
Aquaculture wells	Dispose of water used for aquatic sea life cultivation
Solution mining wells	Dispose of leaching solutions (lixiviants)
In-situ fossil fuel recovery wells	Inject water, air, oxygen solvents, combustibles, or explosives into underground or oil shale beds to free fossil fuels
Special drainage wells	Potable water overflow wells and swimming pool drainage
Experimental wells	Used to test new technologies
Aquifer remediation wells	Use to clean up, treat, or prevent contamination of underground sources of drinking water
Geothermal electrical power wells	Dispose of geothermal fluids
Geothermal direct heat return flow wells	Dispose of spent geothermal fluids.
Heat pump/air conditioning return flow wells	Re-inject groundwater that has passed through a heat exchanger to heat or cool buildings
Saline intrusion barrier wells	Injected fluids to prevent the intrusion of saltwater
Aquifer recharge/recovery wells	Used to recharge an aquifer
Noncontact cooling water wells	Used to inject noncontact cooling water
Subsidence control wells	Used to control land subsidence caused by groundwater withdrawal or over-pumping of oil and gas

INJECTION WELL HYDRAULICS

With regard to aquifer injection hydraulics, a well's injection capacity depends on its specific capacity and the pressure (head) available for injection, a function of the static head (water level) of the aquifer in which the well is screened. Specific capacity describes a well's yield per unit of head decrease (drawdown) in a pumping well, or head increase (drawup) in an injection well. Specific capacity is expressed in units of feet of drawdown/drawup per unit of pumping/injection rate in gpm per foot (gpm/ft), respectively. When injection begins, the water level in the well rises as a function of the transmitting properties of the receiving aquifer and the well's efficiency (Warner and Lehr, 1981). While the transmitting character of the aquifer should remain stable over the service life of an injection well, the available head for injection will decline as injection recharges the aquifer causing the static water level to rise toward the ground surface (HRSD, 2016a, b).

DID YOU KNOW?

Specific capacity is one of the most important concepts in well operation and testing. The calculation should be made frequently in the monitoring of well operation. A sudden drop in specific capacity indicates problems such as pump malfunction, screen plugging, or other problems that can be serious. Such problems should be identified and corrected as soon as possible. *Specific capacity* is the pumping rate per foot of Drawdown (gpm/ft), or

$$\text{Specific capacity} = \text{Well yield} \div \text{drawdown}$$

Problem: If the well yield is 300 gpm and the drawdown is measured to be 20 ft, what is the specific capacity?

Solution:

$$\text{Specific capacity} = 300 \div 20$$

$$\text{Specific capacity} = 15 \, \text{gpm per feet of drawdown}$$

In this discussion of the planning and study phase (and now the processing phase at Nansemond Treatment Plant, NATP) of the SWIFT project it is important to differentiate between production and injection by referring to injection-specific capacity as injectivity. Moreover, for our purposes here, the evaluation of the specific capacity of local production wells and its conversion to injectivity forms an important variable in determining the capacity of individual injection wells and ultimately the total injection capacity across the affected area and the number of wells required at each WWTP.

In determining injectivity for wells screened in specific aquifer units, specific capacity values were evaluated in the three Potomac Aquifer units with respect to their proximity to HRSD's wastewater treatment plants (WWTPs). Virginia Department of Environmental Quality (VDEQ) supplied a database of production wells that contained a total of 98 wells screened in either the UPA, MPA, or LPA, with some wells screened across sands in two of the three aquifers. In these wells, a technique was employed to determine the specific capacity contributed by each aquifer to the well, based on the length of the screen penetrating the aquifer, divided by the total length of the well screen. It proved necessary to separate the aquifers across individual wells for the less represented LPA. Because of its greater depth and water quality, production wells rarely penetrated all the sand units in the LPA, often screening sand intervals in both the MPA and UPA.

Across the study area, the specific capacity of production wells screening the UPA and MPA averaged 35.5 and 32.4 gpm/ft, respectively, essentially equaling each other, while wells screened in the LPA exhibited a 40% lower value of 21 gpm/ft. Wells screening the LPA, as identified by VDEQ, were only located around the Franklin Paper Mill and Franklin City area (see Figure 17.1). Wells screened in the UPA and MPA were better represented across the study area.

As stated previously, the available head for injection represents an important factor in determining injection capacities. Multiplying the well's injectivity by the available head for injection provides the injection capacity of the well. As a driver for this study, local industrial and residential development has resulted in elevated pumpage, drawing water levels in aquifers of the PAS downward at rates averaging 1–2 feet per year (McFarland and Bruce, 2006). The declining water levels represent a greater available head for injection. However, once injection commences, water levels should rebound with a corresponding loss in the available head (HRSD, 2016a, b).

INJECTION OPERATIONS

Because of the deep static water levels at most HRSD wastewater treatment plants and available head for injection in feet at each site for each level of the Potomac Aquifer (see Table 17.2), HRSD could inject water under gravity conditions or pressurized conditions. Under gravity conditions, HRSD would fit a foot valve to the base of the injection piping to reduce head and maintain a positive pressure in the injection and well-header piping. This design facilitates better control of injection rates compared with cascading water down the injection piping, inducing a vacuum through the system. If HRSD decides to inject under gravity conditions, it would not be necessary to seal the injection head, well-header piping, and associated fittings. Instead, the design requires monitoring injection levels rising in the well's annular space to prevent it from topping the ground surface.

An option to inject under gravity conditions could entail sealing the injection head, well-header piping, and associated fittings while allowing the injection head to rise above the elevation of the ground surface by maintaining a positive pressure in the annular space of the well. With a greater available head for injection, HRSD could achieve higher injection rates, while anticipating risking regional water levels inherent in injecting large volumes of water.

DID YOU KNOW?

"No owner or operator shall construct, operate, maintain, convert, plug, abandon, or conduct any other injection activity in a manner that allows the movement of fluid containing any contaminant into underground sources of drinking water if the presence of that contamination may cause a violation of any primary drinking water regulation under 40 CFR part 142 or may otherwise adversely affect the health of persons." (40 CFR 144.121.)

The operator is responsible for ensuring annular pressures stay below an established threshold. Elevated annular pressure scan stresses the sand filter pack surrounding the well screen by lifting it, and initiating the formation of channels that connect the well screen to the surrounding formation materials. Limiting the injection

TABLE 17.2

Summary of Available Head for Injection (ft) in PAS at HRSD's Treatment Plants

Wastewater Treatment Plant	Ground	PAS Aquifer	Available Head for Injection (ft)
		Elevation	
Army base	11	UPA	107.28
		MPA	101.28
		LPA	101.28
Virginia initiative plant	8.5	UPA	109.78
		MPA	106.28
		LPA	101.28
Nansemond	22.5	UPA	123.78
		MPA	106.28
		LPA	106.28
Boat harbor	4	UPA	100.28
		MPA	91.28
		LPA	96.28
James river	18	UPA	107.28
		MPA	81.28
		LPA	86.28
York river	5.5	UPA	86.78
		MPA	76.28
		LPA	76.28
Williamsburg	58.5	UPA	125.78
		MPA	67.28
		LPA	76.28
Rate of water level decline	0.92 ft/yr		

Source: Adapted from CH2M (2016). Sustainable Water Recycling Initiative: Groundwater Injection Hydraulic Feasibility Evaluation, Report No. 1, CH2M, Newport News, VA.

Note: ft = feet.

pressure to ten pounds per square inch (psi) in the well's annular space precludes damage to the filter pack while making available another 23 ft of head for injection.

Regular maintenance is required to operate injection wells to their maximum capacity whenever they are screening fine sandy materials like those found in the PAS. In Aquifer Storage and Recovery (ASR) wells, screening the PAS, of similar aquifers at Chesapeake, Virginia, and in North Carolina, Delaware, and New Jersey, fine sandy aquifers have proven susceptible to clogging from total suspended solids (TSS) entrained in the recharge water in the recharge water (McGill and Lucas, 2009). Even high-quality treated water can contain some amount of TSS that accumulates (clogs) in the screen and pores spaces of sand filter pack and formation. Clogging reduces the permeability around the screen, filter pack, and aquifer proximal to the well (wellbore environment), resulting in higher injection levels while reducing injection capacity.

Aquifer storage and recovery wells operating in these states are equipped with conventional well pumps, allowing periodic well back-flushing during recharge. Back-flushing entails temporarily shutting down recharge and turning on the well pump for a sufficient time to remove fine-grained materials from the well bore environment to the ground surface.

By adopting the approach of maintaining ASR wells by installing a pump in each of HRSD's injection wells for back-flushing, will slow the accumulation of fine-grained materials in the wellbore environment. To generate sufficient energy required to effectively remove solids from the wellbore, the capacities of back-flushing pumps should equal or exceed the injection rates planned for the well.

Even with back-flushing progressive clogging can occur and exceed the ability of back-flushing to maintain injection wells near their maximum capacity. Each WWTP should contain a sufficient number of wells to compensate for removing wells from service for rehabilitation, without compromising the facility's injection capacity.

INJECTION WELL CAPACITY ESTIMATION

In the planning and study stage, the following steps were employed in estimating the capacities of injection wells in the PAS at each HRSD wastewater treatment plant (HRSD, 2016a, b):

- Estimate specific capacity at individual wells (Specific capacity = well yield ÷ drawdown)
- Organize specific capacities by aquifer
- Separate well screens spanning two aquifers and calculate the specific capacity for both
- In short screen assemblies, normalize screen length to 100 ft
- Convert specific capacity to injectivity
- Calculate the available head for injection from the USGS 2005 synoptic study
- Combine injectives of UPA and MPA, or MPA and LPA in a single injection well
- Average available head for injection across UPA and MPA, or MPA and LPA in a single injection well
- Add 23 ft of head to the available injection head to account for maintaining a pressure of 10 psi in the annular space of the well
- Multiply injectivity by available head for pumping to obtain injection well capacity
- Practically limit injection well capacity to 3 MGD (2,100 gpm)
- To estimate the number of injection wells per WWTP, divide the plant's effluent rate by injection well capacity
- To facilitate periodic maintenance, add one injection well for every five, per WWTP

Estimating Specific Capacity and Injectivity

The Virginia Department of Environmental Quality (VDEQ) database of wells and their locations were obtained and evaluated according to their location relative to HRSD's WWTPs at Army Base, Boat Harbor, James River, Nansemond, Virginia Initiative Plant, Williamsburg, and York River. The database contained static water levels, pumping levels, and stable production rates at most wells, enabling the calculation of specific capacity. Moreover, the VDEQ database identified the PAS unit spanned by each screen interval in each well. Several wells featured over five well screen intervals. As previously described, maximizing well screen length increases a well's production capacity. Accordingly, many larger-capacity productized wells were equipped with multiple screen intervals, and in many cases, screening more than one PAS unit (HRSD, 2016a, b).

DID YOU KNOW?

The type and quality of injectate and the geology affect the potential for endangering an underground supply of drinking water. The following examples illustrate potential concerns.

- If injectate is not disinfected pathogens may enter an aquifer. Some states allow injection of raw water and treated effluent. In these states, the fate of microbes and viruses in an aquifer is relevant.
- When water is disinfected prior to injection, disinfection byproducts can form in situ. Soluble organic carbon should be removed from the injectate before disinfection. If not, a chlorinated disinfectant may react with the carbon to form contaminating compounds. These contaminants include trihalomethanes and haloacetic acids.
- Chemical differences between the injectate and the receiving aquifer may create increased health risks when arsenic and radionuclides in the geologic matrix interact with injectate having a high reduction-oxidation potential.
- Carbonate precipitation in carbonate aquifers can clog wells when the injectate is not sufficiently acidic (USEPA, 2016a, b).

Specific capacities were calculated for each well and then grouped according to the aquifer unit(s) in which the well was screened. Specific capacities for wells with screens spanning two aquifers resulted in developing two specific capacities for a single well. Screen assemblies in these wells usually spanned the UPA and MPA, or MPA and LPA. Individual intervals were grouped according to the aquifer which they screened. In wells spanning several aquifers, the specific capacity for a discrete aquifer was then estimated by taking the length of the screen spanning the aquifer and dividing it by the total screen length as follows:

$$\text{Total SC of well} \times \left(\text{SL aquifer 1} / \text{TSL of well} \right) = \text{SC aquifer 1}$$

where
   SC = specific capacity
   SL = screen length
   TSL = total screen length

HRSD should design their injection wells with the screens penetrating an entire aqui-fer's sand thickness to maximize injection capacity. Note, however, that many of the production wells studied in this project used shorter screen assemblies that only partially penetrated the aquifers in which they were installed. Accordingly, to make the estimates compatible with fully penetrating injection wells, the specific capacity for a shortened screen assembly was normalized to a well with 100 feet of screen by applying the following equation.

$$SC\,aquifer\,1 = \left(100\,ft\,of\,screen\,/\,SL\right) = SC\,of\,aquifer\,normalized\,to\,100\,ft\,of\,screen.$$

One hundred feet of screen was recognized as an average total for assemblies fully penetrating the UPA, MPA, and LPA. Specific capacities for wells exhibiting screens extending over 100 ft across a specific aquifer unit of the PAS were accepted as rep-resentative and applied without modification (HRSD, 2016a, b).

Based on observations from aquifer storage and recovery wells, which function in both injection and pumping modes of operation, the specific capacity and the injec-tivity of wells installed in the same aquifer usually vary slightly (Pyne, 2005). In a production well, water migrates toward a potentiometric head lowered by pumping in the wellbore. Water moves through an environment constrained by the size and heterogeneity of the porous media into the well. Upon entering the pumping well, the water is no longer impeded by porous media.

By comparison, injected water is driven down the wellbore by an elevated head against the resistance of the wellbore environment. As a result of the greater resis-tance to flow, the injectivity of an ASR well often falls 10%–50% less than its spe-cific capacity (Pyne, 2005). To re-create this important relationship for the study, converting specific capacity to injectivity involved multiplying specific capacity by a factor of 0.5, an average of the ratio of injectivity to specific capacity in Atlantic Coastal Plain injection-type wells (Pyne, 1995).

AVAILABLE HEAD FOR INJECTION

The available head for injection also played an important role in determining the capacity of an injection well. The available head in each aquifer at individual WWTPs was determined by obtaining the most recent general view of the whole (synoptic) water level information. The US Geodetic Survey last measured synoptic water levels separated by aquifer unit of the PAS, in 2005 (VADEQ, 2006b). More recent work considers all aquifers of the PAS together.

Note that with progressively declining water levels attributed to over-pumping, potentiometric heads measured in 2005 cannot accurately match septic conditions in 2014. To adjust potentiometric levels to 2015 conditions, hydrographs were exam-ined to quantify the annual conditions in water levels (VADEQ, 2006a). The evalu-ation revealed that potentiometric levels have declined an average of 0.9 feet per year since 2005 in the Upper Potomac Aquifer, Middle Potomac Aquifer, and Lower Potomac Aquifer. The annual rate of decline in potentiometric levels was multiplied by 10 years and then applied to the potentiometric head from 2005, projecting elevations to 2014.

In estimating the depth of water, the projected potentiometric level for 2014 in each aquifer unit was subtracted from an elevation of the land surface at each WWTP. A constant head of 23 ft, equal to injecting under a pressure of 10 psi in the well's annual space, was added to the depth of the water level to obtain the available head for injection (HRSD, 2016a, b).

With regard to the flexibility for adjusting injection well capacities maximizing the capacity of individual injection wells required screening two aquifer units, either the Upper Potomac Aquifer and Middle Potomac Aquifer, or the Middle Potomac Aquifer and Lower Potomac Aquifer in each well. Because of potential hydraulic inefficiencies inherent with the difference in screen elevations, wells screening the Upper Potomac Aquifer and Lower Potomac Aquifer were not considered for this evaluation. This is important because it provides HRSD with some flexibility for adjusting injection well capacities while installing the injection well fields. Upon installing the initial injection wells at a WWTP, HRSD could elect to screen all three PAS units in a single well or revert to screening one aquifer if capacities fail to meet, or significantly exceed expectations, respectively.

To obtain the injection well capacity, the final steps in estimating injection well capacities were deterring by adding the injectivities of two aquifer units in a well and averaging the available head for injection between the two aquifers. Then, the resulting injectivity was multiplied by the average available head for injection.

NUMBER OF INJECTION WELLS REQUIRED AT EACH WWTP

Elevated well capacities that exceeded reasonable constructability practices resulted from the contamination of high injectivities from merging screens and large available heads for injection. Constructing the relatively deep wells of sufficient diameter to inject at rates approaching 8 MGD (5,600 gpm) or equipping the well with a back-flush pump capable of the same rate are likely impractical. To reduce well casing, screen, and pumps to dimensions consistent with local well drilling capabilities and operation (such as equipment available and electrical service), a capacity threshold of 3 MGD was set for the injection wells.

The number of injection wells at each WWTP was determined by dividing the plant's effluent rate by the injection well capacities. At large plants, injection wells were split evenly between UPA and MPA, and MPA and LPA combinations. These good totals and aquifer combinations at each WWTP formed the basis for the initial mathematical modeling runs discussed in the modeling section of this work. To accommodate removing wells from service for rehabilitation, one well was added for each five. At plants with smaller effluent flows, one well was added to any total of less than five.

HRSD's Virginia Initiative Plant (VIP) and Nansemond Treatment plant required the largest number of injection wells to replenish the aquifer because of their higher effluent rates and relatively low injectivity in the Middle Potomac Aquifer. Lower injectives also appear in the Upper Potomac Aquifer adjacent to the Virginia Initiative Plant.

Production wells supporting the mapping of specific capacity in the Middle Potomac Aquifer for this project were not present within a 5-mile radius of the Nansemond and Boat Harbor Plants. Regionally, specific capacity values in the Middle Potomac Aquifer appear to increase to the southeast with the exception of two production wells located west of Nansemond and Boar Harbor. To maintain a conservative approach to the project, low-specific capacity values imparted by these wells were maintained in estimating injection well capacity at Nansemond and Board Harbor. By comparison, large-specific capacity values in the Upper Potomac Aquifer and Middle Potomac Aquifer around the James River, York River, and Williamsburg Plants resulted in elevated injectivities, which yielded elevated hypothetical injection capacities despite the relatively shallow available head for injection (HRSD, 2016a, b).

AQUIFER INJECTION MODELING

Hydrologists, hydrogeologists, and groundwater experts in other professional fields soon learn that groundwater flow models are simplified representations of often highly complex hydrogeological flow systems. Modeling tools are well-suited for analyzing aquifer injection experiments. The modeling employed in the SWIFT project helps to project or estimate the capacity of injections of injectate and other important parameters used in this project.

Generally, incorporating as much available hydrogeologic information as possible into the formulation of the conceptual and numerical models of the flow system is advantageous. This is the approach used by HRSD and its consultant in modeling for HRSD's SWIFT project. Hydrogeologic information takes many forms, including maps that show outcropping surfaces of geologic units and faults, cross sections derived from geophysical surveys and well-bore information that show the likely subsurface location of geologic units and faults, maps of water-table levels, independent point well data, maps showing the hydraulic properties of the subsurface materials. HRSD and its consultant used this information to classify the geologic units into hydrogeologic units, which are convenient units with which to define hydrologic properties (Anderman and Hill, 2000).

Estimating the capacity of individual injections at each WWTP along with preliminarily determining the number of wells at each facility comprises one of several key elements of HRSD's Sustainable Water Initiative for Tomorrow (SWIFT); the significance of these determinations can be seen when the goal is to dispose of nearly 130 MGD into the PAS. The modeling executed in this section tests the hydraulic interference between injection wells located within the boundaries of each WWTP property. This evaluation will identify whether individual WWTP properties are sufficiently large to contain the projected number of wells required to dispose of the projected effluent volumes.

MATHEMATICAL MODELING

Estimates of injection well capacity and the appropriate number of wells assigned at each WWTP described earlier did not account for hydraulic interference between wells in the same aquifer. With well screens combining the UPA and MPA or MPA and LPA, hydraulic interference in the MPA will exert the greatest influence on local injection levels. Wastewater treatment plants that are situated on smaller properties will cause an increase in hydraulic interference; they require smaller inter-well spacing for fitting the number of injection wells required to dispose of effluent.

Mathematical modeling techniques were used in quantifying the interference between injection wells located at the WWTPs and rebounding water levels in the aquifers receiving effluent. In this section, analytical groundwater flow modeling is used to evaluate local groundwater mounding at individual WWTPs while injecting effluent (injectate) over 50 years.

DID YOU KNOW?

Hydraulic engineers and others are quite familiar with mathematical modeling. With the continuing advancements in computer technology and the development of advanced computer engineering programs, engineers rely more and more on mathematical modeling. A mathematical model is an abstract model that uses mathematical language to describe the behavior of a system. Eykhoff defined a mathematical model as a representation of the essential aspects of an existing system (or a system to be constructed) which presents knowledge of that system in usable form.

GROUNDWATER FLOW MODELING

Note that it is important to use, that is, to choose an appropriate solution (or model) to the groundwater flow equation using observed data. The factors deemed as fitting, as deemed important include leaky aquitards, delayed yield via unconfined flow, partial penetration of the pumping and monitoring wells, dual porosity, anisotropic aquifers, heterogenous aquifers, and finite aquifers.

In evaluating potentiometric levels in the Upper Potomac Aquifer, Middle Potomac Aquifer, and Lower Potomac Aquifer analytical groundwater flow modeling was applied at each injection well at the seven WWTPs. The modeling study extends the determination of individual injection well capacities by testing the injection rates under the spatial conditions unique to each WWTP property. Although the injection capacities of individual wells may appear feasible at a WWTP based on the head available for injection and the aquifer transmissivities, hydraulic interference between multiple wells can drive injection levels higher in inverse proportion to the available spacing between wells.

At smaller WWTP sites, interfering wells could cause injection levels to exceed 23 ft above the ground surface, the maximum threshold, established for injection heads at individual wells. Mitigating elevated injection levels can entail screening all three PAS in a single well and/or reducing injection rates sufficiently to lower levels below the site injection elevation threshold. Lowering injection rates so that heads fall below site thresholds effectively limits flow injection rates lower than the projected 2040 target flows.

Accordingly, groundwater flow models were customized according to property size and the projected number of wells required at each WWTP. The computer program CAPZONE (Bair, Springer, and Roadcap, 1992) was applied to conduct the analytical groundwater flow modeling at each WWTP.

CAPZONE is an analytical flow model that can be used to construct groundwater flow models of two-dimensional flow systems characterized by isotropic and homogeneous confined, leaky-confined, or unconfined flow conditions. CAPZONE computes drawdowns at the intersections of a regularly-spaced rectangular grid produced by up to 100 wells using either Equation 17.1 for a confined aquifer developed by Charles Vernon Theis (1935) (while working for USGS; aka the Theis solution) or the Hantush-Jacob equation for a leaky-confined aquifer (see Equation 17.2) (Bair et al., 1992). Unlike the numerical mathematical techniques employed by models like MODFLOW (which comprises simple algebraic equations that a computer cycles through multiple iterations to solve the flow equation), CAPZONE directly solves the differential flow equation. Subsequently, CAPZONE provides a more exact and conservative solution that models relying on numerical methods. However, analytical groundwater flow models offer less flexibility in simulating the heterogeneous conditions exhibited in natural systems, including multiple layers, variable grid spacing, spatially varying transmissivity and boundary conditions. At the scale of a single WWTP, where neither the USGS nor DEQ have characterized heterogeneity beyond single wells, CAPZONE offers a reasonable method for simulating the hydraulic response to injection (or pumping) in the PAS.

The Theis Equation is simply

$$s = \frac{Q}{4\pi T} W(u)$$

$$u = \frac{r^2 S}{4Tt}$$

(17.1)

where
s = drawdown (change in hydraulic head at a point since the beginning of the test
u = a dimensionless time parameter
Q = the discharge (pumping) rate of the well
T and S = are the transmissivity and storativity of the aquifer around the well
r = distance from the pumping well to the point where the drawdown was observed
t = the time since pumping began (seconds)
$W(u)$ = well function

The Theis solution is based on the following ideas and assumptions; note that these ideas and assumptions are rarely met but the solution might be useful—to a degree:

• Darcy's law adequately describes the flow in the aquifer (with a Reynolds number (Re) < 10). Reynolds number is a dimensional quantity that is utilized to decide the sort of flow design as laminar or turbulent while flowing through a pipe.

- a confined, homogenous, isotropic aquifer
- the well penetrates the entire thickness of the aquifer
- the well is approximated as a vertical line (has zero radius) meaning no water can be stored in the well
- the well has a constant pumping rate Q
- the head loss is negligible over the well screen
- in radial extent the aquifer is infinite
- the top and bottom of aquifer borders are not slopping (horizontal), flat, and non-leaky (impermeable); the groundwater flow is horizontal
- all changes in the potentiometric surface are the result of the pumping well alone

The Hantush–Jacob well function for leaky confined aquifers is abbreviated w (u, r/B). The Hantush–Jacob equation can be written in compact notation as follows:

$$s = \frac{Q}{4\pi T} w\left(u, r/B\right) \qquad (17.2)$$

where

s = drawdown
Q = pumping rate
T = transmissivity

In practice, CAPZONE produces drawdowns/drawups that are then subtracted from water levels that form either a uniform or non-uniform hydraulic gradient. Thus, the analyst can designate a hypothetical gradient of one based on an observed water-level distribution (non-uniform). The non-uniform options have proven particularly useful for injection well-field analyses at sites where a potentiometric surface exhibits irregularities or deflections that could potentially alter the potentiometric surface geometry.

At HRSD's individual WWTPs, a uniformed hydraulic gradient representing the site's position in the regional potentiometric surface (Figure 17.3) for the PAS developed by USGS was input to CAPZONE. In this approach, the regional gradient was considered locally at each WWTP in estimating ambient groundwater flow direction and hydraulic gradient. In this approach, the regional gradient was considered locally at each WWTP in estimating ambient groundwater flow direction and hydraulic gradient.

In applying CAPZONE, the boundaries of the simulated area were defined, and then the area was divided into a grid. The grid and cell dimensions for CAPZONE were unique to each WWTP, depending on the size of the site. The grid contained up to 75 columns and 75 rows, with a grid node spacing range from 11 to 200 ft at the Boat Harbor and York River WWTPs, respectively.

Other inputs included coefficients of transmissivity from the results of aquifer testing conducted at production wells in the vicinity of each WWTP and a storage coefficient of 0.0001, typical for a confined aquifer. Hydraulic gradients averaged around 0.0001 feet per year foot (ft/ft), with varying directions of groundwater flow based on the site's position within the USGS potentiometric map.

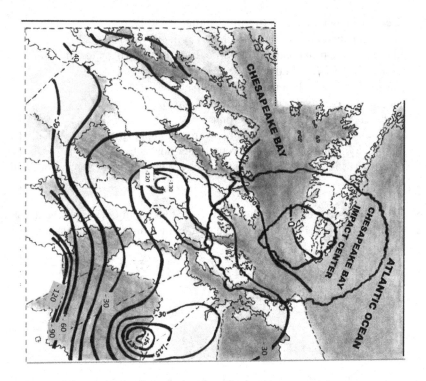

FIGURE 17.3 Potentiometric surface map of the Potomac Aquifer. (Adapted from USGS (2015) *in* HRSD (2016a). Illustration by F.R. Spellman and Kathern Welsh.)

Other inputs included coefficients of transmissivity from the results of aquifer testing conducted at production wells in the vicinity of each WWTP and a storage coefficient of 0.0001, typical for a confined aquifer. Hydraulic gradients averaged around 0.0001 feet per year foot (ft/ft), with varying directions of groundwater flow based on the site's position within the USGS potentiometric map.

A unique static groundwater elevation was entered for each WWTP site and modified slightly depending on whether injection wells were screened across two or three aquifer units of the PAS. As described by the assessment of local-specific capacity, injection wells were first simulated to screen the two adjoining aquifer units of the three aquifers (UPA and MPA; and MPA and LPA) comprising the PAS. This approach entailed adding the transmissivity of the two aquifers together and obtaining an average static water elevation for the two aquifers (USGS, 2016).

For wells screening the UPA and MPA, and the MPA and LPA discrete simulations were conducted. As the MPA received effluent whether it was combined with the UPA or LPA, to effectively simulate well interference in the aquifer, each simulation involved all the wells. As an example, in simulations involving wells screening the UPA and MPA, wells screening the MPA and LPA received one-half of the total effluent flow.

To obtain the number of injection wells used in each simulation, maximum injection rates were held at 3 MGD per well and divided into the effluent flow HRSD projected for 2040. The injection wells were spaced (as much as possible) at roughly equal distances around the perimeter of the WWTP. Care was taken to avoid locating wells on existing structures. Locations were not evaluated for the practicality of positioning wells on lawns, parking lots, along fence lines, or other questionable areas that might host an injection well. Consistent with transient model runs conducted by USGS, CAPZONE simulations were set for a 50-year duration.

Simulated injection elevations were compared against the WWTP's threshold elevation in ft mean sea level, as defined by the ground surface elevation plus 10 psi (23 ft). Simulated injection levels from two aquifers exceeding the threshold elevation, indicated that two aquifers could not facilitate effluent flows for the site. Accordingly, the simulation was run again combining all three aquifers of the PAS in each injection well at the WWTP.

Similar to the approach described previously, the transmissivities were added for each aquifer, and static water level elevations were averaged. In case the simulated injection levels continued to exceed threshold elevations, the effluent flow rates were reduced in each well until a solution was found where injection rates fell below the WWTP's designated threshold elevation. This approach resulted in the determination of the sustainable effluent rate for the site HRSD (2016a, b).

After the model runs that resolved the sustainable number of wells and effluent rates were completed, sensitivity testing was conducted to quantify the uncertainty in input parameters used in obtaining the model solutions. Sensitivity testing was conducted under conditions prevalent at the HRSD's York River Plant. An additional test was conducted at the York River Plant to investigate the relationship between well spacing and interference. This testing proved particularly important to stations with elevated projected effluent flows, or smaller stations that could not support the number of wells to inject effluent at the projected 2040 rates. At these stations, HRSD could potentially locate injection wells at offsite locations at distances sufficient to lower the effects of hydraulic interference.

DID YOU KNOW?

Of the United States more than 140,000 water systems, almost 130,000 rely primarily on groundwater.

After the model grid was set up, injection wells were located around the perimeter of each WWTP site, maximizing the number of wells given the site constraints. Through iterations of well layouts and injection rates, the 2040 projected demands were tested at each WWTP.

ARMY BASE TREATMENT PLANT (NORFOLK)

The Army Base WWTP model was able to meet the 2040 projected demands of 12 MGD using four wells. Two wells were screened in the UPA/MPA, with the other two set in the MPA/LPA, with all four injecting effluent at 3 MGD. The injection head elevation reached a maximum of 9 ft MSL within the UPA/MPA, falling several feet below the threshold elevation of 35 ft MSL. Depending on the aquifer used, the maximum draw-up found at the property boundary is approximately 61–68 ft above static conditions.

BOAT HARBOR TREATMENT PLANT (NEWPORT NEWS)

At Boat Harbor Treatment Plant seven injection wells spaced between 200 and 600 ft apart and screening all three PAS aquifers achieved an injection rate total of 14, falling short of the 16 MGD targeted 2040 flow projections. At 14 MGD, the injection level MGD has a maximum threshold value of 28 ft MSL. The total injection rate was constrained by the location and layout of the plant. The adjacent highway and the harbor limit the space available for wells, placing some wells at distances as close as 200 ft apart. If HRSD can find locations outside the WWTP boundaries, increasing well spacing to >600 ft, six injection wells should prove sufficient to meet the 2040 flow projections. The maximum draw-up found at the property boundary fell ~93 ft above static conditions.

JAMES RIVER TREATMENT PLANT (NEWPORT NEWS)

The James River WWTP model was able to meet the 2040 projected flows of 15 MGD using five wells. Unlike the Army base model, the James River model required that the wells screened all three aquifers in the PAS. Using the three PAS aquifers, the injection head elevation fell below the threshold elevation (42 ft MSL) by almost three feet. The maximum draw-up found at the property boundary totaled ~98 ft above static conditions.

NANSEMOND TREATMENT PLANT (SUFFOLK)

The 2040 projections for the Nansemond WWTP represented the second highest of any of the WWTP sites texted, reaching 28 MGD. The results of the model include using 12 well screening all three PAS aquifers to inject a maximum rate of 24 MGD but fell short of reaching the 2040 projections. At 24 MGD, the injection head elevation remained 2 ft below the threshold elevation (46.5 ft MLS). The large number of wells required and the limited space available led to the 4 MGD shortfall. The adjacent river and marsh limit the space available within the Nansemond WWTP site for locating injection wells. The maximum draw-up found at the property boundary was ~122 ft above static conditions.

Virginia Initiative Plant (Norfolk)

Virginia Initiative Plant's 2040 projections were the highest of all of the WWTP sites tested, requiring 33 MGD. The model included 14 UPA/MPA/LPA wells spread out across parcels north and east of the WWTP, comprising the golf course. The maximum attainable injection rate reached only 21 MGD, falling short of the 2040 projections. Meeting the projected 2040 flow of 33 MGD will require locating wells in offsite locations. The maximum projected draw-up found at the property boundary equaled ~101 ft above static conditions. In this scenario, draw-up was obtained from the boundary of the WWTP with the surrounding open space.

Williamsburg Treatment Plant (Williamsburg)

Williamsburg Treatment Plant, like the Army Base Treatment Plant, was able to meet 2040 projections using five wells split between the UPA/MPA and MPA/LPA. For Williamsburg simulations, two wells were set in the UPA/MPA, with the other three in the MPA/LPA. Given the high threshold elevation at this site (82.5 ft MSL), the WWTP was able to exceed the 2040 demands (13 MGD).

At 15 MGD, injected through five wells, the injection elevation reached 53 ft MSL, well below the threshold elevation for the site. The maximum draw-up found at the property boundary totaled approximately 77–93 ft above static conditions, depending on the aquifer combination (UPA/MPA or MPA/LPA).

York River Treatment Plant (Yorktown)

York River Treatment Plant also successfully met the 2040 projections but required using all three PAS aquifers. The model was able to achieve 15 MGD (over 2040 projections of 14 MGD) using six UPA/MPA/LPA wells, reaching a simulated injection elevation of 29.5 ft MSL, matching the threshold value (29.5 ft MSL). The maximum draw-up found at the property boundary fell ~79 ft above static conditions.

SENSITIVITY OF AQUIFER PARAMETERS

> Thou shall confess in the presence of sensitivity. Corollary: Thou shall anticipate criticism … When reporting a sensitivity analysis, researchers should explain fully their specifications search so that the readers can judge for themselves how that results may have been affected. This is basically an 'honesty is the best policy' approach, advocated by Leamer (1978).

Whenever mathematical models are used as was the case in a phase of HRSD's SWIFT project, there is uncertainty in the inputs applied to get an output. Because of uncertainty in inputs and outputs used in mathematical modeling, a sensitivity analysis is called for; that is, it should be part of the entire process. With regard to the SWIFT project, a sensitivity analysis quantifies the doubt in a calibrated or predicted solution caused by uncertainty in the estimates of the aquifer parameters, injection stresses, and groundwater elevations. Basically what a sensitivity analysis accomplishes is a process of recalculating outcomes under alternative assumptions and has a range of various purposes (Parnell, 1997), including

- In the presence of uncertainty it tests the strength of the results of the model.
- Amplified understanding of the relationships between input and output variables in a system to model.
- Further research can reduce uncertainty by identifying the model inputs that cause significant uncertainty in the output.
- Encountering unexpected relationships between inputs and outputs can be accomplished by searching for errors in the model.
- Model simplification
- Increasing and enhancing the communication and the links from modelers and decision makers via using persuasion and straight talk.
- Employing Monte Carlo filtering to find regions in the space in input factors to optimum criterion.
- Knowing the sensitivity of parameters saves time by ignoring non-sensitive ones (Bahremand and De Smedt, 2008).
- To develop better models, important connections between observations, model inputs, and predictions or forecasts must be identified (Hill et al., 2015; Hill and Tiedeman, 2007).

In HRSD's SWIFT program, the sensitivity of aquifer parameters was performed on the scenario that simulated injecting 14 MGD at the York River WWTP site. The York River site was chosen for selectivity analysis because it represents a site that accommodated the 2040 flows but required using all three PAS aquifers.

Additionally, changes in pumping stresses and groundwater elevations were tested on a single well, eliminating interference from multiple wells. Finally, two wells were simulated to measure interference at varying distances.

For this sensitivity analysis, characterizing the uncertainty of the modeled solution, input values for transmissivity, storativity, injection rate, simulation duration, and the groundwater elevations were systematically adjusted to assess how the changes affected groundwater elevations beneath the WWTP site. To quantify the evaluation, the maximum head generated from a sensitivity run was compared against the head from the original modeled solution.

The sensitivity analysis for aquifer transmissivity was performed by changing one parameter value at a time by −50% and +50% of the original parameter. The storage coefficient, injection rates, and groundwater elevations were tested by increasing and decreasing the values incrementally, not on a percent basis.

Transmissivity

For the purpose of clarity, in this text transmissivity is defined as the capacity of a rock to transmit water under pressure. The coefficient of transmissibility is the rate of flow of water, at the prevailing water temperature, in gallons per day, through a vertical strip of the aquifer one foot wide, extending the full saturated height of the aquifer under a hydraulic gradient of 100%. A hydraulic gradient of 100% means a 1-ft drop in the head in 1 ft of flow distance.

With regard to HRSD's SWIFT project, the transmissivity applied to the combined units of the UPA, MPA, and LPA at the York River Treatment Plant was increased and decreased from the value used in the modeled solution. Values used in the sensitivity analysis ranged from 101,200 to 303,600 gpd/ft. The model was more sensitive to decreasing than increasing transmissivity by a factor approaching three times. Reducing transmissivity by 50% increased the maximum groundwater elevation at York River by 77 ft over the modeled solution of 29.5 ft MSL. Conversely, increasing the transmissivity value by 50% decreased the mounding by only 27 ft from the modeled solution.

STORAGE COEFFICIENT

Storage coefficient is the volume of water released from storage in a unit prism of an aquifer when the head is lowered a unit distance. In the HRSD SWIFT model, the aquifer storage coefficient was increased and decreased from the value used in obtaining the modeled solution (0.0005). The storage coefficient used in the sensitivity analysis ranged from 0.00005 to 0.005. Reducing the storage coefficient by an order of magnitude increased the maximum head across the site by 13.6 ft. Conversely, increasing the storage coefficient by an order of magnitude decreased the maximum injection level by 13.5 ft. As the modeled solution fell close to the threshold elevation, adjusting the storage coefficient can have significant effects on whether the UPA/MPA/LPA can accept the 2040 injection rate.

INJECTION RATES

The sensitivity of changes to injection rates was texted by comparing the maximum injection levels of a single well at different injection rates. Injection rates within the sensitivity analysis ranged from 0 (static conditions) to 4 MGD. The model was almost identically sensitive to increases and decreases in injection rate. Reducing the rate from 3 to 2 MGD reduced the maximum injection head by 8.3 ft while increasing the rate to 4 MGD increased the injection level by 8.2 ft.

SIMULATION DURATION

The duration of injection activity was adjusted to determine the model's sensitivity to changes in this parameter. An increase of 50% (75 years) over the original simulation (50 years) resulted in a rise in the maximum head value of 2.45 ft. Decreasing the simulation duration by 50% (25 years) lowered the maximum head value by 4.04 ft.

STATIC WATER LEVELS

The static water level was set at −56 ft MSL in the model solution. The model solution, with 14 MGD of injection into the UPA/MPA/LPA, reached a maximum groundwater elevation of 29.5 ft MSL. The static water level was raised and lowered to 10 ft for the sensitivity analysis. The model appeared slightly more sensitive to an increase in the groundwater elevation than a decrease. Increasing the static water level to −46 ft MSL increased the maximum head at the well by 10.1 ft, and decreasing the static water level to −66 ft MSL, reduced the maximum head by 9.9 ft.

WELL INTERFERENCE

To measure the effect of multiple wells injecting in proximity to each other, two wells were simulated with injection rates of 3 MGD under the subsurface conditions encountered at the York River WWTP for 50 years. The distance between wells was changed incrementally, while the maximum groundwater elevation was recorded at each well and the mid-point between the wells. The well spacings tested ranged from 500 to 2,000 ft. Injection heads at each well ranged from −30.9 ft MSL with only Well 1 injecting to −12 ft MSL when spaced 500 ft apart. Between 500 and 3,000 ft of spacing, the maximum groundwater elevations in the wells varied about 4 ft. Interferences at Well 1 ranged from almost 19 to 15 ft, with Well 2 spaced from 500 to 3,000 ft away, respectively. At the midpoint between the wells, the head values ranged from −14.8 to −23.1 ft MSL, for distances of 500 to 3,000 ft, respectively. The interference in the PAS changed by 8.3 ft, or slightly <2 ft for every 500 ft of separation between the wells. The large amount of interference (>15 ft) caused by a single nearby injection well even at relatively large spacings (3,000 ft) appears consistent with the results of the modeling at other WWTPs. WWTPs carrying large injection rats require many wells, each well increasing the injection levels at other wells and in the aquifer (HRSD, 2016a, b).

HAMPTON ROADS REGION GROUNDWATER FLOW

The Virginia Coastal Plain Model (VCPM), a SEAWAT groundwater model, was employed to evaluate the hydraulic response of the PAS to injection operations at HRSD's seven WWTPs (Heywood and Pope, 2009; Langevin et al., 2008). This section presents the results of simulating injection at individual WWTPs and in scenarios with all seven WWTPs injected simultaneously.

SEAWAT exemplifies a three-dimensional, variable-density groundwater flow and transport model developed by the USGS based on MODFLOW and MT3DMS (Modular Three-Dimensional Multispecies Transport Model for Simulation). The VCPM groundwater model encompasses all of the coastal plain within Virginia and parts of the coastal plain in northern North Carolina and southern Maryland. The original VCPM was updated and updated for use in the Department of Environmental Quality (DEQ) well-permitting process and is now called VAHydro-GW. The VAHydro-GW model is discretized into 134 rows, 96 columns, and 60 layers. The majority of the model cells are square with the horizontal edges measuring one mile. The upper 48 model layers are 35-ft thick each. Layer thicknesses for the lower model layers increase to 50 ft after layer 48 (top to bottom) and then 100 ft beneath layer 52.

The model simulates potentiometric water levels in 19 coastal plain hydrogeologic units. The water levels are simulated for each year from 1891 through 2012, based on historic pumping records. The VAHydro-GW also simulates water levels for 50 years beyond 2012. These water levels are based on two scenarios: the total permitted scenario and the reported use scenario. The total permitted scenario simulates water levels for 50 years beyond 2012 by using the May 2015 total permitted withdrawal rates established for withdrawal permits issued by the DEQ together with the estimates for non-permitted (domestic wells, wells in Maryland and North Carolina, wells within unregulated portions of Virginia) withdrawals based upon 2012 estimated use. The total permitted scenario represents the estimated water levels 50 years into the future if all permittees within the coastal plain were to pump at their authorized maximum withdrawal rates for the duration of the 50-year period.

The reported use scenario simulates water levels for 50 years using pumping rates reported in 2012, for wells permitted by the DEQ and estimates for non-permitted withdrawals based upon 2012 estimated use. For most large permitted systems (>1 MGD), reported pumping rates fall well below their total permitted diversion. The reported use simulation represents the best available estimate of water levels within the coastal plain aquifers over the next 50 years, if pumping were to continue at the currently reported pumping rates for the permitted wells within the coastal plain.

Virginia regulations have established limits on the amount of drawdown allowed as a result of permitted pumping within the coastal plain. The "critical surface" is defined as the surface that represents 80% of the distance between the land surface and the top of the aquifer. Individual model cells where simulated potentiometric water levels fall below the critical surface are referred to as "critical cells." Both the reported use and total permitted simulations show areas of the coastal plain for the Potomac, Virginia Beach, Aquia, Piney Point, and Yorktown-Eastover aquifers, where the predicted water levels at the end of the 50-year simulation, end below the critical surface for those aquifers.

For any new or renewing permitted withdrawal, DEQ performs a technical evaluation which involves adding the proposed facility to the total permitted simulation. As a major criterion for permit issuance, the facility cannot create new critical cells in any aquifer due to their proposed withdrawal. The critical cells simulated at the end of the reported use simulation are not used for permit evaluation or issuance but represent a more plausible estimate of areas where water levels have lowered to crucial levels.

MODEL INJECTION RATES

VAHydro-GW row and column values were assigned to seven of HRSD's proposed injection WWTPs by using the well locations (latitude and longitude) to plot the position on a Geographic Information System coverage of the VAHydro-GW finite-difference grid. Each facility was simulated as a single point of injection and consequently assigned to only one row and one column. As explained earlier, each model cell is square with each cell edge measuring one mile. As a result, the rates injected through any number of wells are dependent on the individual WWTP. Because of course grid dimensions, the analysis is not intended to evaluate the number of injection wells that are required to dispose of injectate at each facility.

For the initial modeling, the well screen length for each WWTP was assumed to measure between 300 and 350 ft, thus screening across multiple layers of the

VAHydro-GW model. Because the VAHydro-GW module utilizes the HUF package, the model layers are independent of the hydrogeologic units. The HUF package (Hydrogeologic-Unit Flow Package) is an alternative internal flow package that allows the vertical geometry of the system hydrogeology to be defined explicitly within the model using hydrogeologic units that can be different than the definition of the model layers. With regard to the model, a model layer may contain multiple hydrogeologic units. In order to ensure that simulated water levels were not artificially influenced by the Potomac confining unit, each injection well was assigned to the uppermost VAHydro-GW model layer filled by the PAS. The remainder of each injection well screen was assigned to lower, adjacent model layers.

MODELING DURATION

In addition to modeling each WWTP operating individually (with the exception of the York River injection facility), all of the proposed facilities were modeled simultaneously at a combined flow of 114.01 MGD. The York River Treatment Plant was not included in the combined simulations because the facility lies with the outer rim of the Chesapeake Bay Bolide Impact Crater. As simulated in the VAHydro-GW, the horizontal hydraulic conductivity for the PAS at the cells within the bolide impact crater equals 0.0001 feet per day. As a result, simulated heads mounded to unrealistically high values when modeling injection at the York River Treatment Plant.

The individual WWTP scenarios and the combined WWTP scenario were simulated by adding the proposed injection rates to the total permitted and reported use simulations outlined previously, at the beginning of the 50-year predictive portion of those simulations (year 2013). The reported use and total permitted scenarios were also executed before adding the injection facilities to establish "baseline" conditions. This presentation refers to these scenarios as the "reported use baseline" and "total permitted baseline" simulations.

The model runs represent the DEQ's preferred metric for determining the beneficial impacts, if any, of proposed pumping/injection scenarios. The difference between water levels from an injection simulation and water levels from a baseline simulation represents the benefits, or recovery (rebound), resulting from the injection. The results of the injection and baseline simulations were compared at two points, 10 and 50 years into the predictive portion of the simulations.

THE BOTTOM LINE

This section summarizes the findings and conclusions drawn from research, investigation, and modeling procedures conducted by HRSD, USGS, and CH2M for HRSD's SWIFT project. These summarized procedures and findings are based on the evaluation of injection well rates, WWTP injection capacities, and the hydraulic response of the PAS beneath the HRSD service area to injection operations. In addition, these bottom-line conclusions are from HRSD's (2016b) *Sustainable Water Recycling Initiative: Groundwater Injection Geochemical Compatibility Feasibility Evaluation. Report 1.* Virginia Beach, VA. HRSD. Compiled by CH2M Newport News, VA, and are summarized as follows:

- The transmissivities and available head for injection in the PAS beneath each HRSD WWTP appear to support individual injection well capacities ranging between 3 and 8 MGD.
- In adhering to practical well design standards (such as borehole and casing diameter and pumping capacities), injection capacities were capped at 3 MGD for this evaluation.
- To account for maintenance necessary for injection wells screened in sandy aquifers, one additional injection well was added in every five required to meet the effluent disposal rate at each WWTP.
- Accordingly, the number of injection wells ranged from 5 at the Army Base and Williamsburg to 17 at the Virginia Initiative Plant.
- Analytical groundwater flow modeling indicated that of the seven WWTP sites tested,
- Army Base, James River, Williamsburg, and York River were able to meet the 2040 projected demands, within the site boundaries.
- Only Army Base and Williamsburg met the demands using the original two-aquifer approach.
- Conditions at James River and York River required screening all three of the PAS aquifers to meet the 2014 demands.
- Sensitivity testing at York River Treatment Plant revealed that the modeled solution appeared sensitive to all parameters tested (transmissivity, storage coefficient, injection rates, simulation duration, and static water levels).
- The modeled solution exhibited the greatest sensitivity to changes in transmissivity. Changes in static water level resulted in increasing or decreasing the modeled head by the magnitude of the change in the water level.
- An evaluation of hydraulic interference between two wells at the York River WWTP reveals significant (interference >15 ft) between wells spaced even 3,000 ft away.
- The results of the analytical modeling show that hydraulic interference exerts significant influence over the feasibility of replenishing the aquifer at the project 2040 rates.
- Injection was successfully simulated using the VCPM at each WWTP except York River. Because of its location inside the outer rim of the Chesapeake Bay Bolide Impact Crater, the VCPM simulates very low coefficients of hydraulic conductivity for the PAS beneath the York River Treatment Plant.
- Injection at each WWTP resulting in removing most of the critical cells and region-wide, recovering water levels in the PAS.
- Water levels in all injection scenarios resulted in simulated water levels that exceeded the land surface across the HRSD service area.

NOTES

1 Much of the information in this section is based on HRSD's *Proposal to establish and extensometer station at the Nansemond wastewater treatment plant in Suffolk, Va* (2016). Proposed by USGS.
2 Much of the information in this section is based on information from HRSD (2016a) and compiled by CH2M.

REFERENCES

Anderman, E.R., and Hill, M.C. (2000). MODFLOW 2000. Water.USGS.gov/ogw/mudflow/ MODFLOW.html.

Bair, E.S., Springer, A.E., and Roadcap, G.S. (1992). *CAPZONE*. Columbus, OH: Ohio State University.

Bahremand, A., and De Smedt, F. (2008). Distributed hydrological modeling and sensitivity analysis in Torysa Watershed, Slovakia. *Water Resources Management* 22(3): 293–408.

CH2M. (2016). Sustainable water recycling initiative: Groundwater injection geochemical compatibility feasibility evaluation. Report No. 1, Newport News, VA: CH2M.

Eggleston, J., and Pope, J. (2013). Land subsidence and relative sea-level rise in the southern Chesapeake Bay region. U.S. Geological Survey Circular 1392, 30 p. https://dx.doi.or/10.3111/cir1392.

Hamilton, P.A., and Larson, J.D. (1988). Hydrogeology and analysis of the ground-water-flow system in the Coastal Plan of southeastern Virginia. U.S. Geological Survey Water Resources Investigations Report 87-4240, Richmond, Virginia.

Heywood, C.E., and Pope, J.P. (2009). Simulation of groundwater flow in the Coastal Plain Aquifer System of Virginia, Scientific Investigation Report 2009-5039. U.S. Geological Survey. Water.usgs.gov/ogw/seawat.

Hill, M., Kavetski, D., Clark, M., Ye, M., Arabic, M., Lu, D., Foglia, L., and Mehl, S. (2015). Practical use of computationally frugal model analysis methods. *Groundwater* 54(2): 159–170.

Hill, M., and Tiedeman, C. (2007). *Effective Groundwater Model Calibration, with Analysis of Data, Sensitivities, Prediction, and Uncertainty.* New York: John Wiley & Sons.

Holdahl, S.R., and Morrison, N. (1974). Regional investigations of vertical crustal movements in the U.S., using precise relevelings and mareograph data. *Tectonophysics* 23(4): 373–390.

HRSD. (2016a). SWIFT (Sustainable Water Initiative For Tomorrow). Accessed 2/23/24 at www.hrsd.com.

HRSD. (2016b). Sustainable water recycling initiative: Groundwater injection hydraulic feasibility evaluation. Virginia Beach, VA, Hampton Roads Sanitation District. Compiled by CH2M Newport News, VA.

Jones, B.D. (1980). *Service Delivery in the City: Citizens Demand and Bureaucratic Rules.* New York: Longman, p. 2.

Laczniak, R.J., and Meng III, A.A. (1988). Ground-water resources of the York-James Peninsula of Virginia. U.S. Geological Survey Water Resources Investigations Report 88-4059, Richmond, Virginia.

Langevin, C.D., Thorne Jr., D.T., Dausman, A.M., Sukop, M.S., and Guo, W. (2008). SEWAT version. U.S. Geological Survey. https://pubs.usgs.gov/tm/tmbaza//.

Leamer, E. (1978). *Specification Searches: Ad Hoc Inferences with Nonexperimental Data.* New York: John Wiley & Sons.

McFarland, E.R. (2013). Sediment distribution and hydrologic conditions of the Potomac Aquifer in Virginia and parts of Maryland and North Carolina. United States Geological Survey Scientific Investigations report 2013-5116, Reston, Virginia.

McFarland, E.R., and Bruce, T.S. (2006). The Virginian Coastal Plain hydrogeologic framework. United States Geologic Survey Professional Paper 1731, Reston, Virginia.

McGill, K., and Lucas, M.C. (2009). Mitigating specific capacity losses in aquifer storage and recovery wells in the New Jersey Coastal Plain. *New Jersey Chapter of American Water Works Association, Annual Conference,* Atlantic City, NJ.

Meng, A.A., and Harsh, J.F. (1988). Hydrogeologic framework of the Virginia Coastal Plain. United States Geological Survey Professional Paper 1404-C, Washington, DC.

Parnell, D.J. (1997). Sensitivity analysis of normative economic models: Theoretical framework and practical strategies. *Agricultural Economics* 16: 139–152.

Pyne, D.G. (1995). *Groundwater Recharge and Wells*. Ann Arbor, MI: Lewis Publishers.

Pyne, D.G. (2005). *Aquifer Storage and Recovery: A Guide to Groundwater Recharge through Wells*. Gainesville, FL: ASR Press.

Smith, B.S. (1999). The potential for saltwater intrusion in the Potomac aquifers of the York-James Peninsula. U.S. Geological Survey Water Resources Investigations Report 98-4187, Richmond, Virginia.

Spellman, F.R. (2021). *Sustainable Water Initiative for Tomorrow*. Lanham, MD: Bernan Press.

Theis, C.V. (1935). *The Relation between Lowering of the Piezometric Surface and the Rate of Duration of Discharge of a Well Groundwater Storage*. Washington DC: Transactions of the America Geophysical Union, vol. 16, Part 2, pp. 518–524.

Treifke, R.H. (1973). Geologic studies Coastal Plain of Virginia, Bulletin 83, Virginia Division of Mineral Resources, Richmond, Virginia.

Virginia Department of Environmental Quality (VADEQ). (2006a). Status of Virginia's water resources, Richmond, Virginia.

Virginia Department of Environmental Quality (VADEQ). (2006b). Virginia Coastal Palin model 2005 withdrawals simulation, Richmond, Virginia.

USEPA. (2016). Aquifer recharge and aquifer storage and recovery. Accessed at https://epa.gov/uic/aquifer-recharge-and-aquifer-storage-and-recovery.

USGS. (2013). Land subsidence in the United States. Circular 1182. U.S. Department of the Interior, U.S. Geological Survey, Washington, DC.

Warner, D.L., and Lehr, J. (1981). *Subsurface Wastewater Injection, the Technology of Injecting Wastewater into Deep Wells for Disposal*. Berkeley, CA: Premier Press.

18 Mixing Native Groundwater and Injectate

INTRODUCTION

Bear with me as I present a very simplistic view of what this chapter is all about. Envision two 1-L glass beakers. In one of these beakers, we fill it half-full of clean, safe drinking water. In the other beaker, we fill it half-full of salty seawater. A normal person who wants to quench his or her thirst would obviously prefer to drink from the beaker of clean, safe water; he or she would leave the beaker of salty seawater alone. Now, if we take either the beaker of clean, safe drinking water or the beaker of salty seawater and pour one into the other we have obviously mixed the two different solutions. The question now becomes is the new 1-L mixture of clean, safe drinking water and salty seawater something that any of us would want to drink? The truth is some would use this mixture to gargle with. I have seen this done; have you? Anyway, the point I am making here is that by our action of mixing clean, safe drinking water with salty seawater we have changed or adulterated the mixture.

HRSD does not intend to mix clean, safe drinking water with salty seawater or any other contaminant. No, the intent is not to adulterate (i.e., taint, spoil, or pollute) native groundwater in any way. Instead, the intent is to inject, replenish, and recharge the Potomac Aquifer's native groundwater supply with purified, safe water from the advanced treatment of wastewater effluent. In order to do this, to ensure the injection of treated wastewater is of the same quality as the native groundwater contained in the Potomac Aquifer HRSD and its consultant (CH2M) conducted a feasibility study. This feasibility study evaluated the geochemical compatibility of recharging clean water (injectate), native groundwater, and injectate interactions with minerals in the Potomac Aquifer System aquifers. Three-discrete injectate chemistries originate from advanced water treatment processes (AWTP) including reverse osmosis (RO), nanofiltration (NF), and biologically activated carbon (BAC).

The focus of the study was on a single wastewater treatment plant where conditions (such as geography, flow, geology, injectate quality, and groundwater quality) best represent the HRSD system. This was based on a large number of permutations involved with comparing three injectates with native groundwater chemistries from the three PAS aquifers (i.e., Upper, Middle, Lower Potomac Aquifers) beneath HRSD's seven WWTPs, and then applying the injectate chemistries to aquifer minerals in the three aquifers.

DOI: 10.1201/9781003498049-18

HRSD's York River Treatment Plant was selected for this evaluation and the subsequent pilot study. In addition to displaying fairly representative conditions, the property surrounding the York River Treatment Plant is sufficiently spacious to accommodate WWTP upgraded with AWTP and an injection wellfield.

HRSD'S WATER MANAGEMENT CHALLENGES AND VISION[1]

During my first tour of SWIFT's ongoing pilot study at the SWIFT site at Nansemond Treatment Plant, Suffolk, VA, I made a quick survey and inspection of each unit process in the purification train. Later, during a few other visits and observations of the entire process, I spent more time studying each unit process and learning from the operators the 'nuts and bolts' of each component in the treatment train. After checking out the operation and the operator's procedures, I finally decided to sample the treated and tested water. So, I did and I tried to detect anything different in the taste and odor and the only impression I had was how it quenched my thirst on that hot, humid summer day. At that moment, I pretty much felt that HRSD was onto technology that is destined to solve many of Hampton Road's problems and future challenges.

As mentioned earlier, the Hampton Roads region is faced with a variety of future challenges related to the management of the region's water supply and receiving water resources. These challenges involve a combination of technical, financial, and institutional complexities that invite the exploration of using non-traditional approaches that provide benefits on a larger scale beyond what the current wastewater treatment and disposal model can achieve. Accordingly, aquifer replenishment can protect and enhance the region's groundwater supplies, as well as reduce the potential damage caused by discharge (of nutrients) to the lower James River and the Chesapeake Bay and may be slow or arrest relative sea-level rise in the region.

GEOCHEMICAL CHALLENGES

Earlier, it was stated that evaluations and analytical groundwater flow modeling revealed that HRSD will recharge between 77 and 131 million gallons per day (MGD) to the Potomac Aquifer System using over 60 injection wells with maximum capacities approaching 3.0 MGD per well, at seven WWTPs. Groundwater flow modeling revealed that the injection wells will screen combinations of two, or all three PAS aquifer zones, according to the hydrologic conditions found at individual WWTPs. Both physical and geochemical challenges can emerge while recharging clean water into aquifers composed of reactive metal-bearing minerals, and potentially unstable clay minerals, while also containing brackish native groundwater, typical conditions found in the PAS.

Physical and chemical reactions are important relative both to the well facility operation and aquifer water quality. The potentially damaging effects can come from:

- Water-to-water mixing of injectate water with native groundwater—impacts are in the form of physical plugging pore spaces with solids or precipitated metals, reduced permeability local to the wellbore, and eventually lower injectivity.

- Interaction of the injectate/native groundwater mix with the aquifer matrix—impacts are in the form of damage to firewater sensitive clays, precipitation or dissolution of metal-bearing minerals, and potential release of metals troublesome to injection activities (iron and manganese), or water quantity issues (arsenic).

REDUCTION IN INJECTIVITY

The two following factors affect the injectivity of an injection well:

- Physical plugging
- Mineral precipitation

Physical Plugging

A well's injection rate per unit of head buildup (draw-up) in an injection well is known as its injectivity, expressed in gallons per minute per foot (gpm/ft) of draw-up (Warner and Lehr, 1981). When injection begins, the water level in the well rises as a function of the transmitting properties of the receiving aquifer, and the well's efficiency. While the transmitting character of the aquifer should remain stable over the service life of an injection well, the available had for injection will decline as the injectate recharges the aquifer. This causes the static water level to rise toward the ground surface. The well's efficiency will decrease with time, depending on the quality of the injectate, particularly its total suspended solids (TSSs) content. TSSs content, which, in water, is commonly expressed as a concentration in terms of milligrams per liter (mg/L) and typically is described as the amount of filterable solids in a wastewater sample. More specifically, the term *solids* means any material suspended or dissolved in water and wastewater. Although normal domestic wastewater contains a very small amount of solids (usually <0.1%), most treatment processes are designed specifically to remove or convert solids to a form that can be removed or discharged without causing environmental harm. However, 100% removal is unlikely. Even the most purified injectate can contain small amounts of TSS. If left to accumulate in the borehole environment (wellbore), solids can clog the screen, filter pack, and aquifer proximal to the well, which reduces the well's injectivity. Injectivity reduction increases draw-up and eventually lowers the well's injection capacity (Pyne, 2005). TSS can originate from scale or dirt in piping, treatment residuals, and reactions in the injectate that result in solids precipitation. One of the more common reactions occurs when oxygen dissolved in the injectate reacts with dissolved iron or manganese, precipitating ferric or manganese oxides, turning a dissolved component of the injectate into a source of solids.

Mineral Precipitation

In addition to physical plugging, chemical reactions between the injectate and the native groundwater, or injectate and aquifer mineralogy can precipitate metal-bearing oxides and hydroxides. These reactions often come from injectate that contain dissolved oxygen (DO). Considering the relatively small surface areas around the wellbore, precipitating metal-bearing minerals can clog pore spaces, and reduce permeability and well injectivity. An important part of the research and planning involved with HRSD's SWIFT project is to determine precisely the type and composition of injectate from the RO, NF, or BAC process to estimate its potential for plugging the wellbore.

GEOCHEMICAL CONCERNS

Beyond problems associated with physically plugging pore spaces around the borehole, several geochemical reactions can negatively affect injection well operations. These reactions include:

- Clay minerals damage
- Metals precipitation
- Mineral dissolution

Damaging Clay Minerals

The term *clay* is applied to both materials having a particle size of $<2\,\mu m$ ($25,400\,\mu m = 1$ inch) and to a complex group of poorly defined hydrous silicate minerals that contain primarily aluminum, along with other cations (potassium and magnesium) according to the exact mineral species. Displaying a platy or tabular structure, clay minerals exhibit an extremely small grain size and typically adsorb water to their particle surfaces. In aquifer sand, clays occur in trace ($<10\%$) amounts as components of the aquifer's interstitial spaces, coating framework particles like quartz grains, lining or filling pore spaces, or as a weathering product of feldspars.

Damaging clays occur with the disruption of their mineral structure. The damage can arise when injecting water of significantly different ionic strength than the native groundwater, a concern when injecting dilute fresh water into an aquifer containing brackish or saline native groundwater (Drever, 1988). The dilute water contains significantly fewer cations and a weaker charge than brackish native groundwater. When displacing the brackish water in the diffuse-double layer between clay particles, the weaker charge can induce repulsive forces dispersing the particles, fragmenting the clay structure while mobilizing the fragments into flowing pore water. The particles can eventually accumulate in smaller pores physically plugging the pore space and reducing the permeability of the aquifer.

Damage can also arise when injectate displays differing cation chemistry than the native groundwater and the clay minerals (Langmuir, 1997). Exchanging cations can disrupt clay mineral structure particularly when their atomic radius exceeds the radius of the exchanged cation. The larger cation fragments the tabular structure, shearing off the edges of the mineral. Plate-like fragments break off the main mineral particle and migrate with flowing groundwater. Like the damage incurred by water of differing ionic strength, migrating clay fragments will brush pile in pore spaces, physically plugging passageways and reducing aquifer permeability. Unlike the accumulation of TSS in the wellbore, formation damage by migrating clays develops in the aquifer away from the wellbore, making its removal difficult by backflushing or even invasive rehabilitation techniques.

Mineral Precipitation

Metal-bearing minerals can precipitate in the aquifer away from the well. These reactions typically occur when the injectate contains dissolved oxygen (DO) at concentrations exceeding anoxic (DO < 1.0 mg/L levels but can also occur if the pH of the injectate exceeds 9.0. As surface areas in the aquifer increase geometrically away from the well, mineral precipitation does not create as great a concern as the same reactions at the borehole wall.

Mineral Dissolution

Injectate reactions with minerals in the aquifer matrices can dissolve minerals leaching their elemental components (Stuyfzand, 1993). Injectate containing DO above anoxic concentrations will react with common, reduced metal-bearing minerals like pyrite (FeS_2) and siderite ($FeCO_3$), to release iron and other metals like manganese that occupy sites in the mineral structure, iron and manganese can precipitate as oxide and hydroxide minerals if they contact injectate-containing DO. Oxidation of arseniferous pyrite can release arsenic, creating a water quality concern in the migrating injectate (HRSD, 2016a).

WATER QUALITY AND AQUIFER MINERALOGY

Note: Because it would require an unwieldy number of permutations (63) to assess the injection of three injectate chemicals into three discreet aquifers at seven WWTPs, the chemical composition of three injectate types and native groundwater in the three PAS aquifer zones beneath the York River Treatment Plant is chosen for discussion here. The targeting of the York River WWTP for discussion that follows is not an issue and does not skew data because the York River Treatment Plant exhibits effluent and local aquifer characteristics typical of conditions across the Hampton Roads Sanitation District.

Mass–balance relationships between raw water entering the plant and modeling of the advanced water treatment process were used to determine injectate chemistry. As no wells installed in the PAS aquifer zones currently exist at the York River WWTP, water quality data from the area around the site was obtained from the National Water Information System (NWIS) database (maintained by the United States Geological Survey [USGS]). The NWIS database provides samples collected by USGS personnel from local municipal, irrigation, and industrial supply wells, along with designated monitoring wells.

Two methods were used to identify potential minerals in the PAS aquifers that could react with the injectate. First, thermodynamic equilibrium models were applied to identify the potential mineral suite in each aquifer. The models were run using water chemistry analyses obtained from the NWIS database for the PAS zones around the York River WWTP area. The models project potential minerals that occur in equilibrium with water chemistry.

Second, mineralogical analysis of cores collected at the City of Chesapeake's Aquifer Storage and Recovery (ASR) facility was examined to gain information on the mineralogy of the PAS. Cores were collected from the PAS zones at the City of

Chesapeake. The composition of the PAS should remain fairly consistent across the HRSD service area. However, the grain size and sorting (texture) decline proceeding down the stratigraphic dip in the Virginia Coastal Plain (Treifke, 1973). Consistent with the changes in texture, the percentage of fines (texture) increases downdip. As the City of Chesapeake lies over 20 miles downdip from HRSD's York River WWTP, data from cores should portray more conservative aquifer properties than actually occur below the York River.

INJECTATE WATER CHEMISTRY

Injectate chemistry was estimated by modeling water quality entering the York River Treatment Plant and through the Advanced Water Treatment (AWT) processes. Effluent chemistry was estimated for RO, NF, and BAC advanced water treatment processes.

Reverse Osmosis

Injectate modeled for treatment by the RO process featured a pH of approximately 7.8 after adjustment with lime ($CaOH_2$), dilute total dissolved solids (TDSs) (46 mg/L), and, correspondingly, a low ionic strength (0.0015). Cations and anions in the influent were reduced following the treatment process, resulting in concentrations of cations like potassium (<1 mg/L), magnesium (<1 mg/L), and sodium (<10 mg/L) falling below their method detection limits (MDLs), with similarly lower concentrations of anions like phosphate (0.01 mg/L), chloride (<10 mg/L), and sulfate (<1 mg/L). At York River Treatment Plant, RO-treated waste displayed a calcium bicarbonate water type. Metals such as iron, manganese, and arsenic exhibited low concentrations, and all were near MDLs. RO has a limited effect on DO concentrations in the injectate, which ranged around 5 mg/L (HRSD, 2016b).

Nanofiltration

NF is much the same as RO but the device is slightly more permeable, so less energy (pressure) is required for separation. In this particular case, injectate derived from the NF process exhibited a pH of approximately 7.8, moderate TDS (262 mg/L), and corresponding ionic strength (0.005). Cations and anions are reduced following the treatment process, but unlike RO displayed measurable concentrations of major cations like potassium (7.7 mg/L), magnesium (2.5 mg/L), and sodium (58 mg/L), and anions including phosphate (0.03 mg/L), chloride (125 mg/L), and sulfate (1.8 mg/L). NF injectate displayed sodium chloride chemistry. Concentrations of metals including iron, manganese, and arsenic fell below MDLs. NF injectate also displayed near-saturated concentrations of DO around 5 mg/L.

Biologically Activated Carbon

Injectate originating from BAC treatment displayed a pH of approximately 7.8, slightly brackish TDS (615 mg/L), and corresponding ionic strength (0.009). Cations and anions exhibited less reduction following the treatment process compared with NF and RO. Cation concentrations of potassium (13 mg/L), magnesium (10 mg/L), and sodium (103 mg/L) exceeded the concentrations yielded by the membrane (reverse and NF) treatments. Concentrations of anions like phosphate (0.5 mg/L), chloride (212 mg/L), and sulfate (44 mg/L) also appeared correspondingly higher. Unlike RO and NF, DO concentrations fell to 1.3 mg/L after treatment using BAC. BAC injectate featured sodium-chloride water chemistry. Iron and arsenic displayed concentrations higher than the membrane treatment options at 0.002 and 0.73 mg/L, respectively. Iron concentrations above 0.1 mg/L create significantly large amounts of TSS that can quickly clog an injection well.

NATIVE GROUNDWATER

Native groundwater quality from the Upper Potomac Aquifer (UPA), Middle Potomac Aquifer (MPA), and Lower Potomac Aquifer (LPA) zones was obtained from nested observation wells maintained by the USGS and NWIS, located 5 miles west of the York River WWTP. At this location, observation wells installed were screened in the UPA from 527 to 537 feet below grade (fbg), in the MPA from 820 to 830 fbg, and in the LPA from 1,205 to 1,215 fbg.

Upper Potomac Aquifer Zone

The UPA featured a slightly alkaline pH (8.2), brackish TSD (1,280 mg/L, and anoxic water (DO < 1.0 mg/L), displaying an ionic strength of 0.02. The groundwater exhibited low amounts of nutrients with concentrations of ammonia, nitrates, and phosphate falling below 0.01 mg/L. Chloride concentrations approached 500 mg/L, while sodium concentrations appeared similarly elevated (516 mg/L). Concentrations of other cations including calcium (4.5 mg/L), and potassium (13.5 mg/L) were comparatively low. Iron concentrations fell around its MDLs (method detection limit) (0.01–0.04 mg/L), while manganese concentrations approached the drinking water maximum contaminant level (MCL) of 0.05 mg/L. Water from the UPA displayed a sodium chloride chemistry.

Middle Potomac Aquifer Zone

The MPA also featured a slightly alkaline pH (8.0), brackish TDS (2,780 mg/L), and anoxic groundwater (DO < 1.0 mg/L), with an ionic strength of 0.04. Similar to the UPA, concentrations of nutrients fell below 0.01 mg/L. Concentrations of anions, comprising chloride (1,200 mg/L), alkalinity (370 mg/L), and sulfate 73 mg/L), exceeded concentrations encountered in the UPA. Sodium concentrations appeared similarly elevated at 870 mg/L, while concentrations of other cations, such as calcium, magnesium, and potassium fell below 15 mg/L. Iron concentrations were near MDSLs (0.01–0.04 mg/L), while manganese concentrations appeared at 0.02 mg/L. Similar to the UPA, groundwater in the MPA exhibited sodium chloride chemistry.

Lower Potomac Aquifer Zone

The LPA displayed a circum-neutral pH (7.7) brackish TDS (4,580 mg/L), and anoxic groundwater (DO < 1.0 mg/L). Similar to the other PAS aquifers, concentrations of nutrients fell below 0.01 mg/L, although phosphate concentrations, at 0.5 mg/L, appeared notably higher than in the other PAS aquifers. The concentration of anions, comprising chloride (2,950 mg/L) and sulfate (146 mg/L), exceeded concentrations encountered in the UPA and MPA. Sodium concentrations were similarly elevated (1,700 mg/L). Concentrations of other cations such as potassium (25 mg/L) and calcium (51 mg/L) increased over the concentrations encountered in the other PAS zones. Yet, iron and manganese in the LPA mimicked groundwater from the other aquifers with concentrations at MDLs and 0.02 mg/L, respectively. Similar to the other PAS aquifers, groundwater from the LPA displayed sodium-chloride chemistry.

Geochemical Assessment of Injectate and Groundwater Chemistry

In this section, the discussion is about the chemical assessment of the injectate (i.e., the effluent from the advanced water treatment processes, RO, NF, and BAC) that potentially is to be injected as injectate into native groundwater. Obviously, as pointed out with the example of the beakers of pure water and seawater mixing that adulterated the pure water into something not potable HRSD does not want a similar outcome with the injection of its treated wastewater into the Potomac Aquifer's native groundwater. The goal is to determine the appropriate injectate. Keeping the desired outcome and goal in mind, the evaluation of the chemistry of the injectate water and native groundwater from the Potomac Aquifer System revealed the following:

- RO and NF displayed ionic strengths differing by over one order of magnitude from the native groundwater in the Potomac Aquifer System.
- BAC displayed ionic strengths within the same order of magnitude as the Potomac Aquifer System.
- RO exhibits differing cationic chemistry than the groundwater from the Potomac Aquifer System.

Influence of Ionic Strength

The ionic strength of RO diluted, treated water appeared lower than groundwater in the three Potomac Aquifer System zones by at least one order of magnitude. By comparison, the ionic strength displayed by NF differed from the LPA by over none order of magnitude. The ionic strength of biologically treated carbon, although lower than the Potomac Aquifer System aquifers, fell within the same order of magnitude. The low ionic strength of RO compared to the Potomac Aquifer System groundwater represents a concern for injection operations, particularly for RO' potential to disperse clay minerals. Clay dispersion represents an electro-kinetic process (Meade, 1964; Reed, 1972; Gray and Rex, 1966), where an electrostatic attraction between negatively charged clay particles is opposed by the tendency of ions to diffuse uniformly throughout an aqueous solution. One of the most important factors leading to the dispersion of clay minerals involves a change in the double-layer thickness of a clay particle. A double layer of ions lies adjacent to the clay mineral surface or between the mineral's structural layers because a negative charge attracts cations toward the surface. As the fluid must maintain electrical neutrality, a more diffuse layer of anions surrounds the cations.

As in brackish water, when the concentration of ions is large, the double layer around the particle or between the clay's structure layers gets compressed to a smaller thickness. Compressing the double layer causes particles to coalesce, forming larger aggregates. This process is called clay flocculation. When the ionic concentration of a fluid invading the aquifer is significantly lower than the native groundwater, the diffuse double layer expands, forcing clay particles and the structural layers within clay minerals apart. The expansion prevents the clay particles from moving closer together and forming an aggregate. The tendency toward dispersion is measured in clay minerals by their zeta potential—that is, where colloids with high zeta potential are electrically stabilized while colloids with low zeta potentials (aka electrokinetic potential) tend to coagulate or flocculate—according to the following relationship:

$$Z = 4\pi 6q/D \qquad (18.1)$$

where
  Z = zeta potential
  6 = thickness of the zone of influence surrounding the charged particle
  q = charge on the clay particle before attaching cations
  D = dielectric constant of the liquid

For any solution and clay mineral, reducing the zeta potential involves lowering the thickness of the zone of influence. Substituting small, double- or triple-charged cations such as Ca^{2+} or Al^{3+}, respectively, in place of large singly charged and hydrated ions like Na^{2+}, lowers the Zeta potential, permitting clay particles to coalesce. This behavior explains the tendency for sodium to cause clay dispersion, while calcium and aluminum induce its flocculation.

Aquifer sands contain more complex clay minerals like mixed-layer clays and smectite group clays that display small particle sizes, yet large surface charges usually exhibit the greatest sensitivity to fresh water (Brown and Silvey, 1977). As little as 0.4% smectite in a sand body has reduced the aquifer's hydraulic conductivity by 55% after exposure to fresher water (Hewitt, 1963). Mixed-layer clays and smectite encountered in cores from the UPA and MPA at the City of Chesapeake exhibited an abundance equaling trace (<1%–4% of the whole rock composition of the sand).

In the 1970s, the USGS tested an aquifer storage and recovery facility in Norfolk, Virginia; it exhibited >50% reduction in injectivity after only 150 min of starting injection operations (Brown and Silvey, 1977). The aquifer storage and recovery well was installed in the UPA, screening nearly 85 ft of sand in the unit. The USGS employed nuclear, electrical, and mechanical geophysical logging techniques to evaluate the origin of the injectivity losses and discriminate between the causes of physical plugging documented at other sites, like TSS loading. Injectivity losses caused by physical plugging from TSS loading typically occur at discrete zones through the well screen. In contrast, geophysical logging of the aquatic storage and recovery test well at Norfolk showed hydraulic conductivity losses distributed evenly across the entire screen. Also, in comparison to physical plugging by TSS which responds positively to mechanical and chemical rehabilitation techniques, the USGS was unable to restore even a fraction of the well's original injectivity).

USGS used Calcium chloride ($CaCl_2$ > 10,000 mg/L) solution to treat the wellbore and proximal aquifer to arrest the declining injectivity. The double-charged calcium cation forms a stronger particle and inter-layer bond than the monovalent cation, sodium. Using a concentrated solution ensures calcium exchanges for sodium at the maximum number of sites. After applying the treatment at Norfolk, the injectivity of the aquifer storage and recharge test well remained stable over two more test cycles before the project ended. Concentrated solutions containing trivalent aluminum proved to be effective in stabilizing clay minerals prone to dispersion in the presence of dilute injectate (Civan, 2000). Applying a calcium or aluminum chloride treatment to the Potomac Aquifer System before initiating injection operations, offers a viable alternative for stabilizing clay minerals in situ, precluding formation damage, and injectivity loss should regulators select RO as the most viable method for protecting local water users. These treatments could also benefit injection operations using NF or BAC as the preferred injectate.

Cation Exchange

In addition to differing ionic strengths, RO, as calcium carbonate water, differs from the sodium chloride chemistry encountered in the UPA, MPA, and LPA. As previously described, the doubly charged calcium ion should benefit the long-term stability of clay minerals where calcium exchanges for sodium. However, calcium exhibits a large ionic radius that can damage clay minerals when entering the position left by the sodium, fragmenting the edges of the mineral, and mobilizing the fragments in the aquifer environment.

Iron and Manganese

None of the injectates or native groundwater from the Potomac Aquifer Systems aquifers appears to exhibit problematic concentrations of iron or manganese. Iron concentrations in RO and NF effluent typically occurred below method detection levels (MDLs). During injection operation, iron and manganese contained in the injectate or native groundwater can precipitate oxide and hydroxide minerals when exposed to DO. Formation of these minerals presents a problem if they precipitate close to the well bore, which is a zone featuring small surface areas sensitive to physical plugging. Accordingly, the absence of iron and manganese in injectate or native groundwater benefits injection operations.

LITHOLOGY OF THE POTOMAC AQUIFER

The lithology (i.e., the physical characteristics of the rock unit) and minerals comprising the Potomac Aquifer System aquifers are described in this section, starting with the general composition across the study area, and then focusing on cores collected from the UPA and MPA near the City of Chesapeake, VA.

LITHOLOGY

As previously mentioned, the Potomac Aquifer System consists of three discrete aquifer zones (UPA, MPA, and LPA) named for their position in the section. Deposited in river (fluvial) and shallow marine environments, the aquifers consist of coarse to fine sands with occasional gravel, interbedded with thin gray to pale green clays (Treifke, 1973). The aquifers are separated by clay beds of thicknesses exceeding 20 ft. However, thinner clay beds transect the sand units in the MPA and the LPA.

Because of the abundance of clay beds, the MPA and LPA often consist of multiple, stacked units requiring repeated screen and blank combinations for supply wells installed in these aquifers.

Sands comprised primarily of quartz (Meng and Harsh, 1988), often reaching amounts exceeding 90% by weight, forming the predominant framework mineral. Accessory minerals include orthoclase, muscovite, glauconite, and local lignite. Trace minerals mostly occupy the interstitial spaces in the sands and comprise biotite, pyrite, siderite, magnetite, and clays.

CITY OF CHESAPEAKE AQUIFER STORAGE AND RECOVERY FACILITY

In 1989, at the City of Chesapeake's aquifer storage and recovery facility, ten core samples were collected from the UPA and MPA, at depths ranging from 560 to 835 fbg. The cores were submitted to Mineralogy Inc., a laboratory specializing in mineralogical assays, for the following analyses:

- Specific gravity
- Porosity
- Permeability
- X-ray diffraction
- Cation exchange capacity (CEC)
- Grain size distribution
- Energy dispersive chemical analysis
- Scanning electron microscopy

Potomac Aquifer System sediments found at the City of Chesapeake locations, even though they are located ~35 miles apart, should display similar characteristics as those underlying the Yorktown Wastewater Treatment Plant. Because aquifer characteristics like grain size, sorting, textural maturity, porosity, and permeability decline to move downdip, Potomac Aquifer System sediments at Yorktown Wastewater Treatment Plant should display characteristics better suited to injection operations than at the City of Chesapeake.

Core samples from the UPA and MPA were composed of coarse to very coarse-grained sands, in a medium-grained matrix. Aquifer sands appeared conglomeratic and unsorted. However, as unconsolidated sands, they displayed open pore spaces yielding good porosity (21%–34%) and air permeability. Grain size diminished with depth. Samples from the deeper portions of the MPA exhibited a medium-grain size with a larger percentage of fine sands than shallower samples. Most clay minerals were found in interstitial spaces of the aquifers and showed a high degree of crystallinity, which suggests they formed after deposition and burial (i.e., were authigenic).

Sands from the UPA and MPA consisted of 84%–89% quartz with 8%–12% potassium and plagioclase feldspar, classifying the sands as subarkosic, or lithic arkosic. Trace (<10%) amounts of calcite and dolomite were detected in every sample of the aquifer sands. Clay minerals, comprising kaolinite, illite/mica, and smectite made up to 4% of the same samples. The iron carbonate mineral, siderite ($FeCO_3$) was encountered in a confining bed sample (595 fbg) at an amount of up to 19%. Siderite was also encountered in an aquifer core (685.2 fbg), at trace amounts.

Sands from the UPA and MPA consisted of 84%–89% quartz with 8%–12% potassium and plagioclase feldspar, classifying the sands as subarkosic, or lithic arkosic. Trace (<10%) amounts of calcite and dolomite were detected in every sample of the aquifer sands. Clay minerals, comprising kaolinite, illite/mica, and smectite made up to 4% of the same samples. The iron carbonate mineral, siderite ($FeCO_3$) was encountered in a confined bed sample (595 fbg) at amounts up to 19%. Siderite was also encountered in an aquifer core (685.2 fbg) at amounts up to 19%. Siderite was also encountered in an aquifer core (685.2 fbg), at trace amounts.

Permeabilities in air (intrinsic permeability) ranged from 1280 to 5900 millidarcies. Generally, intrinsic permeability and porosity values declined with depth, so the greatest permeabilities were encountered in cores from the UPA. Intrinsic permeability displayed minimal anisotropy with horizontal and vertical values from the same core yielding near equal permeabilities. Cation exchange capacity or CEC refers to the number of exchangeable cations per dry weight that a soil can hold, at a given pH, and are available for exchange with the soil-water solution which is influenced by the amount and type of clay and the amount of organic matter (Drever, 1982). CEC serves as a measure of soil fertility, nutrient retention capacity, and the capacity to protect groundwater from cation contamination. The CEC of minerals contained in confining beds often controls the cation chemistry in the adjacent aquifers by exchanging cations across the contact between the units. For injection purposes, knowing the CEC of the aquifer and confining bed materials can help assess how these materials will react with recharge water displaying a specific cation ionic chemistry. CEC is expressed as milliequivalents of hydrogen per 100 grams (g) of dry soil (meq+/100g). Table 18.1 presents the CEC values of some clay minerals.

All ten samples at the Chesapeake site were analyzed for CEC. Sodium represented the most dominant exchangeable cation followed by magnesium, calcium, and potassium. The confining bed sample at 685.2 fbg, displayed the most elevated CEC at 12.5 meq/100g of core. Aquifer sand samples from the UPA and MPA exhibited CECs for sodium ranging from 0.7 to 3.9 meq/100 g of core. Sodium, a monovalent ion in the exchange position of clays, will not benefit from injection operations.

Despite the dominance of sodium, CEC values from cores from the City of Chesapeake were low, suggesting that the clays should display a minimal tendency to exchange cations. In environments showing more elevated CECs, divalent ions

---

TABLE 18.1

Cation Exchange Capacities of Some Clay Minerals (meq/100 g) (Drever, 1988)

Smectites	80–150
Vermiculites	120–200
Illites	10–40
Kaolinite	1–10
Chlorite	<10

---

like calcium or magnesium in the injectate can exchange with sodium temporarily disrupting the clay's atomic structure. Over the long term replacing sodium with a divalent ion will strengthen the clay mineral's atomic structure eventually transitioning to a stable smectite.

MINERALOGY—GEOCHEMICAL MODELING

The thermodynamic equilibrium model PHREEQC (Parkhurst, 1995) was used to gain a greater understanding of the stability of the clay minerals in the Potomac Aquifer System aquifers beneath the Yorktown Wastewater Treatment Plant, based on the native groundwater and injectate chemistries. As previously described, the stability of clay minerals can control the success of injection operations in sandy aquifers like the Potomac Aquifer System.

Thermodynamic equilibrium models consist of computer programs using a relatively sophisticated set of equations (Davies, 1962; Truesdell and Jones, 1974; Debye and Huckel, 1923) to stimulate the chemical equilibrium of a solution under natural or laboratory conditions and to simulate the effects of chemical reactions. These models perform the following types of calculations:

- Correct all equilibrium constants to the temperature of the specific sample.
- Calculates speciation: the distribution of chemical species by element by solving a matrix of equations.
- Calculation activity coefficients of each chemical species.
- Calculates the state of saturation for potential mineral species that occur in equilibrium with the samples' water chemistry. These calculations identify potential mineral species, and whether they will dissolve or precipitate under the changing conditions consistent with Aquifer Storage and Recovery operations.
- The models perform a wide variety of calculations related to oxidation-reduction processes.

Thermodynamic equilibrium computer models represent a power tool for predicting chemical behavior in a natural system. Manual manipulation of the same equations performed by these programs is time-consuming and prone to calculation errors.

STABILITY OF CLAY MINERALS

PHREEQC was employed to evaluate the stability of clays contained in the $CaO^- Al_2O_3^- SiO_2^- H_2O$; $NaO^- Al_2O_3 SiO_2^- H^2O$; and $K_2O^- Al_2O_3^- SiO_2^- H_2O$ mineral systems. Minerals contained in these systems represent clays and their weathering products (gibbsite, kaolinite) commonly found in sediments of the Potomac Aquifer System. The simulations' objective involved determining how native groundwater chemistries fall into the stability fields of clay minerals and identifying potential; instabilities. Along with ambient clay stabilities, PHREEQC simulates how clay can evolve during the exchange of cations. However, the program does not address instability arising from introducing an injectate of differing ionic strength.

The chemistry of groundwater from the UPA, MPA, and LPA, along with potential injectate waters from the Yorktown Treatment Plant was plotted on stability diagrams for three systems describing common clay minerals ($CaO^- Al_2O_3 . SiO_2^- H_2O$; $NaO^- Al_2O_3 SiO_2^- H^2O$; and $K_2O^- Al_2O_3^- SiO_2^- H_2O$). Common clay minerals including smectite, beidellite, montmorillonite, illite, and the gibbsite (weathering product) were over-saturated in recharge water samples, which suggests a tendency to precipitate over time.

When injecting waters of incompatible ionic strength or differing cations, damage to clay minerals can arise; however, this was not a concern during injection operations in the Potomac Aquifer System because precipitation of clay minerals represented a relatively minimal matter regarding permeability loss. Moreover, the precipitation of clay minerals requires significant amounts of geologic time, rather than the relatively short service life of an injection facility.

DID YOU KNOW?

In modeling used in determining the possible effect of injectate injected into an aquifer various parameters are used including a description of natural groundwater flow; aquifer thickness, porosity, hydraulic conductivity, and groundwater density; degree of plume dispersity in the aquifer; and injection operations—location, flow rate, duration, and injectate density (Johnson et al., 2021).

SIMULATED INJECTATE—WATER INTERACTIONS

Along with characterizing clay mineral stability during injection operations, PHREEQC was also employed in assessing reactions originating from mixing the three injectate types, RO, NF, and BAC injectates, with native groundwater from each of the PAS aquifers, and reactions between the injectate and elected reactive minerals in the PAS aquifers. Mixing reactions occur when injectate interfaces with native groundwater. As injection operations proceed, injectate drives the mixing interface further into the PAS aquifers.

Surface areas in an aquifer undergoing injection are small around the injection wellbore but increase geometrically with distance away from the well. The larger surface areas away from the injection wells help buffer reactions that cause permeability losses. Thus, these reactions diminish in importance as injection operations progress. Reactions between injectate and reactive minerals cause the following concerns

- Permeability losses with the precipitation of iron or manganese oxide minerals
- Leaching of environmentally problematic constituents like iron, manganese, and arsenic along with other metals, depending on the ambient mineralogy

Mixing

Mixing could arise during injection operations, including:

- Mixing between the injectate and native groundwater
- Mixing between groundwater from the UPA, MPA, and LPA in the injection wellbore.

As mentioned, mixing reactions prove most troublesome around the injection wellbore. One common reaction involves the precipitation of oxide minerals when injectate-containing dissolved oxygen contacts dissolved iron or manganese entrain in the injector or native groundwater. Although each injectate from the advanced water treatment processes contained measurable concentrations of dissolved oxygen, dissolved iron, and manganese concentrations, with the exception of BAC, were absent in the injectate and native groundwater. Despite the absence of iron and manages other minerals can also precipitate during mixing.

PHREEQC modeling was employed to simulate mixing between the injectate and native groundwater chemistries at a 1:1 ratio in order to evaluate the mixing between the differing water types. The modeling was also used to evaluate reactions between the native groundwater chemistries in the UPA, MPA, and LPA as they are mixed in an injection wellbore, simulating an injection well screening the three Potomac Aquifer System aquifers.

During the mixing simulations, important reactions were tracked including the potential precipitation of metal oxide, hydroxide, sulfate, and carbonate minerals along with the dissolution of silicates, including clays. Because of the similar bulk chemistry of the three injectates and native groundwater from the UPA, MPA, and LPA, potential mineral suites identified by PHREEQC in the mixed water and their saturation indices were repeated across the nine mixtures. The *saturation index* (SI) of a mineral determines whether the mineral occurs in equilibrium (Langmuir, 1997) with mixed water chemistry (SI = 0.0); is undersaturated, and if present, should dissolve (SI < 0.0); or is supersaturated (SI > 0.0) and should precipitate. Estimation of saturation indices is usually not exact, often varying over ±0.3 units, depending on the composition of the solution. The SI provides a guideline on how minerals will behave in a water sample.

Of the common mineral and their weathering products identified in nine combinations of mixed water quartz the most common mineral in the sands of the Potomac Aquifer System returned a slightly over-saturated SI (SI = 0.66–0.87) for all simulations. Oversaturation of quartz and near-equilibrium saturation indices for less crystalline forms of silica such as chalcedony (SI = −0.06 to 0.29) and cristobalite (SI = 0.02–0.37) indicates feldspars are dissolving, releasing silica. Gibbsite ($Al(OH)_3$) appeared oversaturated in the mixed water consistent with the dissolution of feldspar, and precipitation of residual byproducts in a weathered environment.

Other minerals potentially reacting in the mixed water included the carbonates, comprising calcite, aragonite, and siderite. Calcite ($CaCo_3$) appeared uniformly undersaturated (SI = −1.42 to −0.24) in all the mixed waters suggesting it will not precipitate. Calcite, in an undersaturated state, benefits injection operations by not precipitating, blocking pore spaces, and reducing the permeability of the aquifer. Aragonite, an isomorph of calcite shows similar indices, ranging from −1.56 to −0.39. The iron carbonate mineral ($FeCO_3$), siderite displayed SI values similar to calcite and aragonite with strongly undersaturated indices ranging from −8.09 to −11.73.

Other important minerals include gypsum and jarosite, a weathering product of iron-bearing mineral sand sulfides. Similar to the carbonates, with one exception, gypsum and jarosite exhibited undersaturated indices ranging from −2.21 to −3.37 and −5.62 to 0.33, respectively. As a single exception, BAC mixing with groundwater from the Lower Potomac Aquifer resulted in near-equilibrium SI for jarosite.

MIXING IN THE INJECTION WELLBORE

With an open conduit extending between the three Potomac Aquifer System aquifers at York River Treatment Plant, groundwater will mix in the injection wellbore before the start of injection operations. Once injection operations start, injectate will displace the groundwater away from the wellbore so mixing groundwater will no longer present an issue.

An examination of static water level elevations at the nested NWIS wells near the York River Wastewater Treatment suggests a vertically downward hydraulic gradient of 0.085 feet/foot (ft/ft) occurring between the UPA and MPA, while an upward gradient of 0.031 ft/ft appears between the LPA and MPA. The differing gradient directions impose converging flow in the wellbore. Accordingly, groundwater will flow in through intervals screening the UPA and LPA and out through the screen against the MPA, promoting the mixing of the three water types in the wellbore, rather than stratification.

In the geochemical simulation, groundwater from the UPA, MPA, and LPA was mixed at even proportions between the three units. To maintain a conservative approach to the simulations, ferrous iron (Fe II) concentrations at UPA, MPA, and LPA were assumed to occur at 0.1 mg/L, the MDL for iron. Concentrations in the mixed water oxidized from Fe II to ferric iron (Fe III), remain at a concentration of 0.1 mg/L.

No deleterious reactions associated with mixing in the Aquifer Storage and Recovery wellbore were detected through modeling. The pH of the mixed water declined slightly to 7.83. SI for calcite and siderite appeared near equilibrium at −0.08 and −0.04, respectively. This suggests that these minerals should neither dissolve nor precipitate in the mixed water. The mixed water displayed sodium chloride chemistry similar to groundwater from the three Potomac Aquifer System aquifers. Aragonite, gypsum, and jarosite displayed unsaturated SIs indicating that the mineral should not precipitate in the mixed water. The weathering products of clay minerals, gibbsite remained saturated roughly the same as the individual groundwater chemistries.

INJECTATE AND AQUIFER MINERAL REACTIONS

In addition to reactions between dilute injectate and clay minerals, dissolved oxygen in the injectate can react with reduced metal-bearing minerals releasing metals and other constituents, which can compromise the quality of water disposed in the Potomac Aquifer System aquifers. These reactions should not affect injection operations but can result in environmental concerns, prompting the attention of regulators.

Analysis of cores from the Aquifer Storage and Recovery project at the City of Chesapeake encountered microcrystalline siderite in the interstices of aquifer sands and as larger crystalline forms in adjoining confining beds.

Pyrite is another reduced, metal-bearing mineral common to the Virginia Coastal Plain Aquifers, including the Potomac Aquifer System (Meng and Harsh, 1988; McFarland and Bruce, 2006). Although not detected in core samples from the City of Chesapeake, pyrite and siderite typically occur together in sediments subject to flooding by marine and freshwater systems (Postma, 1982), typical of the near coastal environment in which the formation bearing the Potomac Aquifer System aquifers was deposited. Accordingly, pyrite was considered in the geochemical modeling evaluation.

The primary metal in siderite and pyrite is FE II but both minerals can also contain cadmium and manganese. Additionally, pyrite occasionally contains varying amounts of arsenic (Evangelou, 1995). Dissolved oxygen in the three injectates should range between 1.3 and 5 mg/L, providing a source for oxidizing reactions.

SIDERITE DISSOLUTION

Dissolved oxygen reacts with siderite to release Fe II and CO_3^{2-} (carbon trioxide). Upon encountering dissolved oxygen, Fe II oxidizes to Fe III, which acts as a strong oxidant, continuing the dissolution of siderite. At equilibrium, a small amount of siderite can release large amounts of Fe II into the surrounding pore water.

PHREEQC was employed to simulate potential reactions between the injectates and siderite. In this reaction, one mole of siderite was reacted with RO, NF, and BAC each containing dissolved oxygen concentrations ranging from 1 to 7 mg/L in 1.0 mg/L increments. Resulting in Fe II at dissolved concentrations of 7 mg/L ranging from 90 to 130 mg/L for RO and BAC, respectively, with NF exhibiting concentrations between RO and BAC. Even acting with only 1 mg/L dissolved with siderite produces Fe II concentrations ranging from 50 to 90 mg/L for RO and BAC, respectively.

Bicarbonate concentrations increased from 95 mg/L for RO at 1 mg/L dissolved oxygen to over 200 at 7 mg/L dissolved oxygen for BAC. A portion of the total bicarbonate comprised carbonic acid, lowering the simulated pH of the pore water from 7.8 to 6.91 for RO at 7 mg/L dissolved oxygen. The pH of RO, NF, and BAC dropped below 7.15 when reacting siderite with dissolved oxygen of only 1 mg/L RO, the most dilute injectate, exhibited the lowest capacity to buffer the pH during the reaction between siderite and dissolved oxygen.

PYRITE OXIDATION

Although pyrite was not encountered in the cores collected at the City of Chesapeake, its appearance elsewhere in the Potomac Aquifer System, and its deleterious reactions when encountering dissolved oxygen make evaluating the mineral an important part of any injection feasibility study. Reacting dissolved oxygen with pyrite releases Fe II and the bisulfide ion (Evangelou, 1995). Upon encountering dissolved oxygen, Fe II oxidized to Fe III, which also acts as a strong oxidant, continuing the oxidation of pyrite. The bisulfide ion $\left(S_2^2\right)$ further reacts with dissolved oxygen to form sulfuric acid (H_2SO_4), lowering the pH of the surrounding pore water.

PHREEQC was used to simulate an operational injection scenario to predict the chemistry of effluent containing varying amounts of dissolved oxygen, exposed to pyrite in the Potomac Aquifer System aquifer matrices. Pyrite was equilibrated with RO, NF, and BAC containing concentrations ranging from 1.0 to 7.0 mg/L in 1.0 mg/L increments.

Similar to the siderite simulations, the injectate chemistries were equilibrated with one mole of pyrite. Thus, simulations provide conservative results, overestimating the concentrations of iron, sulfate, and arsenic in the recovered water. Presently, in the Atlantic Coastal Plain aquifers, pyrite comprises <1% of the whole rock composition or 0.05–0.1 moles.

Modeling results showed that Fe II concentrations increased from 3 to over 10 mg/L, while sulfate increased at twice this rate. Sulfate concentrations simulated with BAC were elevated above the other effluent types by its initial concentration of 44 mg/L. Similar to the simulations with siderite, RO exhibited the greatest decline in pH after reacting with pyrite, with the pH declining from 7.8 to <6.8.

The iron concentrations simulated from the modeling are considered conservative as a large portion of Fe II released by the oxidation of pyrite will precipitate as hydrous ferric oxides (HFO). The HFO typically precipitates on the pyrite mineral surface, progressively reducing its reactivity (passivate). Moreover, these surfaces can adsorb Fe II migrating in the aquifer environment. In the absence of pyrite, these simulations illustrate potential groundwater quality problems that can emerge from effluent containing dissolved oxygen in the presence of this reactive mineral.

Arsenic

The release of arsenic from pyrite was simulated with PHREEQC. As a substitution for sulfur, arsenic concentrations were estimated at 1% by weight of the mass of pyrite. Similar to previous simulations, one mole of pyrite was equilibrated with RO, NF, and BAC containing dissolved oxygen concentrations ranging from 1 to 7 mg/L. Applying this approach, arsenic concentrations increased from 31 to nearly 58 µg/L in reactions with RO, and from 58 to 85 µg/L during reactions with BAC.

Similar to relationships between pyrite and iron, arsenic concentrations simulated with PHREEQC, represent conservative conditions. HFO surfaces in aquifer settings display a strong affinity for adsorbing the oxyanions of arsenic and lowering its concentrations in groundwater. Moreover, continuing oxidation of pyrite precipitates HFO on the mineral surface, passivating the mineral, also diminishing the concentration of constituent released during oxidation reactions.

MITIGATING PYRITE OXIDATION

At Aquifer Storage and Recovery facilities recharging beneath the Atlantic Coastal Plain, siderite and pyrite dissolution is addressed by increasing the pH of the injectate by adding potassium and sodium hydroxide (caustic). Increasing the injectate pH raises it above the solubility limit of iron, buffering the dissolution of iron-bearing minerals. Hydroxyl ions in sodium and potassium hydroxide will react with Fe II released from siderite or pyrite, oxidizing Fe II and Fe III. It precipitates HFO on the surface of these minerals, which then passivates the minerals to future reactions in the aquifer environment. In addition to isolating the minerals, HFO surfaces display excellent adsorption properties, adsorbing metals migrating in the aquifer environment including arsenic and Fe II. Iron adsorbs as a surface precipitate on HFO, while these surfaces exhibit an affinity for adsorbing arsenic at pH values encountered in groundwater environments (Dzombak and Morel, 1990).

PHREEQC was employed to simulate adjusting the pH of RO, NF, and BAC injectate containing dissolved oxygen from 7.8 to 8.5 in the presence of pyrite. Consistent with other simulations, dissolved oxygen was varied between 1.0 and 7.0 mg/L. Fe II concentrations approaching 10 mg/L during reactions with only dissolved oxygen, fell to $<1.0 \times 10^{-7}$ mg/L in simulations with injectate pH adjusted to 8.5. Fe II was nearly completely oxidized to Fe III, which precipitated as $Fe(OH)_3$. The modeling results illustrate how well adjusting the pH of the injectate can control Fe II concentrations. As equilibrium simulations, the modeling did not account for the reactions passivating siderite and pyrite over time, which also reduces Fe II concentrations in groundwater (HRSD, 2016b).

THE BOTTOM LINE

Modeling and geochemical evaluation resulting from mixing native groundwater and injectate and the reactions between injectate and aquifer minerals in the Potomac Aquifer System beneath HRSD's York River Treatment have generated several conclusions. These bottom-line conclusions are from *Sustainable Water Recycling Initiative: Groundwater Injection Geochemical Compatibility Feasibility Evaluation. Report 2.0* by Hampton Roads Sanitation District, Virginia Beach, VA, and its primary consultant CH2M Newport News, VA, and are summarized as follows:

- The chemistry of RO, a potential injectate, differed significantly from the chemistry of native groundwater exhibited by the UPA, MPA, and LPA.
  - RO displayed a dilute ionic strength that differed by over one order of magnitude from the chemistry encountered in the UPA and MPA while approaching two orders of magnitude when compared against the Lower Potomac.
  - RO is calcium-bicarbonate water chemistry, while groundwater from the three Potomac Aquifer System aquifers uniformly exhibited sodium chloride chemistry.
- The low ionic strength of RO compared to groundwater from the Potomac Aquifer System represents a concern for injection operations, particularly for its potential to disperse clay minerals. Once dispersed, clay particles migrate through connected pores in the aquifer until accumulating and blocking narrowed pores, reducing aquifer permeability and ultimately injection well capacity.
- A USGS-sponsored Aquifer Storage and Recovery facility tested at Norfolk in the 1970s used an injectate similar in ionic strength and cation chemistry

to RO. The injection capacity of the Aquifer Storage and Recovery well declined by 50% after only 4 hours of operation, dropping 75% over several days. USGS was not able to restore the capacity of the Aquifer Storage and Recovery well, despite applying several, for the time, state-of-the-art rehabilitation techniques.

- Cores collected at the City of Chesapeake's Aquifer Storage and Recovery Facility exhibited trace concentrations of smectitic clays dispersed throughout the interstices of every sample collected in aquifer sands. Smectites possess a complex lattice-expanding structure vulnerable to dispersion or swelling when exposed to dilute water.

- In considering the varied cation chemistry between RO and groundwater in the Potomac Aquifer System aquifers, the doubly charged calcium ion should benefit the long-term stability of clay minerals where calcium exchanges for sodium. However, calcium, when hydrated, exhibits a large ionic radius that can damage clay minerals upon entering the position vacated by sodium, fragmenting the edges of the mineral, and mobilizing the fragments in the aquifer environment.

- Conversely, cores from the City of Chesapeake project, analyzed for cation exchange capacity displayed little tendency to exchange, which is a benefit of injection operations.

- Geochemical modeling of potential clay minerals in the Potomac Aquifer Systems aquifers produced a similar result with the stability of clay minerals improving over time during injection operations.

- Given the concerns with RO as a source of injectate and the problems experienced at the City of Norfolk's Aquifer Storage and Recovery project, HRSD should consider eliminating RO from further evaluation of the SWIFT project.

- The ionic strength of NF and BAC injectate fell within one order of magnitude of the groundwater chemistries originating from the Potomac Aquifer System aquifers. NF and biological activated carbon as a source of injectate, represent significantly less of a concern of dispersing water-sensitive clays in the Potomac Aquifer System aquifers during injection operations.

- Applying a calcium or aluminum chloride treatment to the Potomac Aquifer System aquifers before initiating injection operations, offers a viable alternative for stabilizing clay minerals in situ, precluding formation damage, and injectivity losses should regulators select RO as the most viable method for protecting local water users. These treatments could also benefit injection operations using NF of BAC as the preferred injectate.

- BAC exhibits 0.7 mg/L iron, which presents a considerable source of TSS in the injectate, and a strong physical plugging agent in injection wells. HRSD will need to remove iron from BAC effluent before employing it as an injectate.

- Geochemical modeling runs, simulating mixing between RO, NF, and BAC and groundwater from the Potomac Aquifer System aquifers showed no evidence of deleterious reactions that might clog the injection wells or surrounding aquifer such as precipitating oxide, hydroxide, carbonate, or sulfate minerals, or the dissolution of silicate minerals.

- Geochemical modeling, simulating the mixing between the three ground-waters in an injection well screening the UPA, MPA, and LPA displayed no evidence of deleterious reactions that might clog the injection well or surrounding aquifers.
- Cores collected at the City of Chesapeake contained the iron carbonate mineral siderite at amounts ranging from 0.5% to 19% of the whole rock composition. In reactions with injectate containing dissolved oxygen, siderite released up to 130 mg/L Fe II.
- Although not a concern for injection operations, dissolving siderite can compromise the quality of the disposed water, prompting attention from state and federal regulators.
- Adjusting the pH of the injectate water with a source of hydroxyl like sodium or potassium hydroxide can help lower Fe II concentrations. During model runs simulating reactions between injectate containing varying amounts of dissolved oxygen and pH of 8.5 with pyrite. Fe II concentrations fell below 10E-7 mg/L. Fe II oxidized to Fe III, which precipitated as Fe III-oxide and Fe III-hydroxide minerals.

NOTE

1 HRSD (2016b). *Sustainable Water Recycling Initiative: Groundwater Injection Geochemical Compatibility Feasibility Evaluation.* Virginia Beach, VA. Hampton Roads Sanitation District. Compiled by CH2M Newport News, VA.

REFERENCES

Brown, D.L., and Silvey, W.D. (1977). Artificial recharge to a freshwater-sensitive brackish-water sand aquifer, Norfolk, Virginia. U.S. Geological Survey Professional Paper 939, 53 p.

Civan, F. (2000). *Reservoir Formation Damage: Fundamentals, Modeling, Assessment, and Mitigation.* Houston, TX: Gulf Publishing Company.

Debye, P., and Huckel, E. (1923). On the theory of electrolytes, I. Freezing point depression and related phenomena. *Physikalische Zeitschrift* 24(9): 185–206.

Drever, J.M. (1982). *The Geochemistry of Natural Waters.* New York: Prentice-Hall Englewood Cliffs.

Davies, C.W. (1962). *Ion Association.* Washington, DC: Butterworths, 190 p.

Drever, J.M. (1988). *The Geochemistry of Natural Waters,* 2nd ed. New York: Prentice-Hall Englewood Cliffs.

Dzombak, D.D., and Morel, F.M.M. (1990). *Surface Complex Modeling.* New York: John Wiley & Sons.

Evangelou, V.P. (1995). *Pyrite Oxidation and Its Control.* New York: CRC Press, 275 pp.

Gray, D.H., and Rex, R.W. (1966). *Formation Damage in Sandstones Caused by Clay Dispersion and Migration.* La Habra, CA: Chevron Research Company.

Hewitt, E.J. (1963). Mineral nutrition of plants in culture media. In F.C. Steward (ed.), *Plant Physiology,* Vol. III. New York: Academic Press, pp. 97–133.

HRSD. (2016a). *Sustainable Water Recycling Initiative: Groundwater Injection Hydraulic Feasibility Evaluation.* Report 1. Virginia Beach, VA: Hampton Roads Sanitation District. Compiled by CH2M Newport News, VA.

HRSD. (2016b). *Sustainable Water Recycling Initiative: Groundwater Injection Geochemical Compatibility Feasibility Evaluation.* Report 2. Virginia Beach, VA: Hampton Roads Sanitation District. Compiled by CH2M Newport News, VA.

Johnson, C.D., Franklin, T., Molina, B., Wassing, J.Q., Tran, P.K., Kump, M.S., Demirkanli, I., Yonkofski, C.M., and Zhang, F. (2021). *Aquifer Injection Modeling (AIM) Toolbox User Guide.* Washington, DC: United States Environmental Protection Agency.

Langmuir, D. (1997). *Aqueous Environmental Geochemistry.* New York: Prentice-Hall.

McFarland, E.R., and Bruce, T.S. (2006). The Virginian Coastal Plain hydrogeologic framework. United States Geologic Survey Professional Paper 1731, Reston, Virginia.

Meade, R.H. (1964). Removal of water and rearrangement of particles during the compaction of clayey sediments. Geological Survey Prof. Paper 497-B.

Meng, A.A., and Harsh, J.F. (1988). Hydrogeologic framework of the Virginia Coastal Plain. United States Geological Survey Professional Paper 1404-C, Washington, DC.

Parkhurst, D.D. (1995). User's guide to PHREEQC-A computer program for speciation, reaction-path, advective-transport, and inverse geochemical calculations. U.S. Geological Survey Water-Resources Investigations Report 95–4227, 143 p.

Postma, D. (1982). Pyrite and siderite formation in brackish and freshwater swamp sediments. *American Journal of Science* 282: 1154–1183.

Pyne, D.G. (2005). *Aquifer Storage and Recovery: A Guide to Groundwater Recharge through Wells.* Gainesville, FL: ASR Press.

Reed, M.G. (1972). Stabilization of formation clays with hydroxy-aluminum solutions. Society of Petroleum Engineers. https://dx.doi.org/10.2118/3694-PA.

Stuyfzand, P.J. (1993). Hydrochemistry and hydrology of the coastal dune area of the Western Netherlands, PH.D. Thesis, 366 pp. Free University, Amsterdam, Netherlands.

Treifke, R.H. (1973). Geologic studies Coastal Plain of Virginia, Bulletin 83, Virginia Division of Mineral Resources, Richmond, Virginia.

Truesdell, A.H., and Jones, B.F. (1974). WATEQ, a computer program for calculating chemical equilibria of natural waters. *U.S. Geological Survey Journal of Research* 2: 233–274.

Warner, D.L., and Lehr, J. (1981). *Subsurface Wastewater Injection,* the Technology of Injecting *Wastewater into Deep Wells for Disposal.* Berkeley, CA: Premier Press.

19 Advanced Water Purification

It is common practice when treating wastewater to churn out water cleaner than the local waterways it is ultimately outfalled into. It is ultimately sent on to the ocean with no downstream use—in other words "one and done" usage. Why? Why waste such a valuable resource? Why not use cutting edge technology to purify wastewater to drinking water quality? Why not reuse it? Don't we already use it … in de facto water recycling?

—**F.R. Spellman (2017)**

INTRODUCTION

Hampton Roads Sanitation District (HRSD) is faced with a variety of future challenges related to the treatment and disposition of wastewater in its region of responsibility; this region includes much of the southern Chesapeake Bay, with its many major tributaries and surrounding communities. HRSD envisions that it can protect and enhance the region's groundwater supplies by reusing highly purified wastewater through advanced treatment and subsequent injection into the region's groundwater aquifers. For those of us who understand the natural water cycle, humankind's urban water cycle, the proper use of various advanced wastewater treatment processes to purify the water, and HRSD's commitment to absolute excellence to its ratepayers and all those who live in the Hampton Roads Region, as well as to the restoring and sustaining the Chesapeake Bay—we understand that HRSD's vision not only has merit but is also necessary. It is necessary because the groundwater supply within the Potomac Aquifer is dwindling; it is in danger of contamination from saltwater intrusion; treated wastewater from HRSD's wastewater treatment plants outfall nutrients into the Bay which contribute to dead zones and other environmental issues; and, finally, HRSD understands that as native groundwater is withdrawn from the underlying aquifers this contributes to land subsidence in the region. Land subsidence plus global sea-level rise is the contributor to relative sea-level rise and if not abated will soon (in <150 years) inundate many of the Hampton Roads major cities and other low-lying areas in the region.

In the previous chapter, it was pointed out that HRSD and its contractor along with the U.S. Geological Survey (USGS) are working in unison to implement steps and modeling to ensure the compatibility of treated injectate with native groundwater. It is also important to make sure the chemical match between injectate and native groundwater is safe for consumption. This is where advanced water treatment (AWT) comes into play. Wastewater treated only to conventional standards is probably safe enough for pipe-to-pipe connection but suffers from the public perception

 DOI: 10.1201/9781003498049-22

of the "Yuck Factor." That is, the old "your toilet water at my tap, no way, Jose and Maria," syndrome. It should be pointed out that when HRSD's plans were published in the surrounding Hampton Roads area and that the overwhelming majority of the populace voiced no objection to HRSD's plans. However, one of the few but common complaints was from those with wells drawing water from the Potomac Aquifer; they were worried that the HRSD SWIFT project might contaminate their water source with toilet water. This is exactly what HRSD is working hard to prevent by implementing advanced water treatment. Understand that the advanced water treatment process is in addition to normal wastewater treatment and filtration. In this chapter, we describe the three treatment processes, reverse osmosis (RO), nanofiltration (NF), and biologically activated carbon, and the pilot studies that were used to determine which process or processes is best suited to facilitate HRSD's goal of producing and injecting the safest water possible.

IN ACCORDANCE WITH ESTABLISHED PROCEDURES

The goal of most public service entities is to perform their functions in accordance with established procedures—by the book. The book in most cases is the written volume that contains the applicable regulations—the so-called laws of the land—that apply to their activities. For operations that can directly or indirectly affect the environment the applicable regulations are generally federal-based and enforced (by the Environmental Protection Agency, for example). However, it is interesting to note that the injection of reclaimed water into an aquifer that is used as a potable water supply is referred to as indirect potable reuse (IPR) and regulations have not been developed by the U.S. Environmental Protection Agency (EPA) for potable reuse projects; therefore, states in which IPR is being practiced (or is being actively considered) have developed state-specific potable reuse regulations. With regard to indirect potable water reuse compliance, it is the state regulator knocking at the door, not the Feds.

THOSE PLAYING BY THE BOOK IN INDIRECT POTABLE REUSE[1]

California and Florida have developed regulations governing the practice of indirect potable reuse (IPR). Other states allow IPR but establish project-specific requirements on a case-by-case basis (e.g., Virginia, Texas). Because Virginia has not developed IPR regulations but does allow IPR, the state will likely look to successful full-scale IPR projects within Virginia (e.g., Upper Occoquan Service Authority (USOA), Loudon Water) and other states' IPR regulations for guidance in regulating HRSD's proposed direct injection IPR project. Table 19.1 presents the treated water quality requirements for other IPR projects and associated regulations that the State of Virginia could reference with respect to an HRSD IPR project.

What does Table 19.1 indicate?

- **Total Organic Carbon (TOC)**: TOC is the amount of carbon found in an organic compound and is often used as a non-specific indicator of water quality. California's strict TOC limit of 0.5 mg/L will most likely require the use of RO, which could increase HRSD's project costs significantly.

TABLE 19.1

Example Treated Water Quality Requirements for Indirect Potable Reuse

Parameter	Virginia's Occoquan and Dulles Area Watershed Policies (Surface Water Augmentation)	TCEQ Policy for El Paso, TX (Direct Injection)	Florida IPR Regulations for Direct Injection	California's IPR Regulations for Direct Injection	EPA's IPR Guidelines for Direct Injection
Relevant IPR projects	Upper Occoquan Service Authority, Centreville, Va; Broad Run WRF, Loudoun County, VA	Hueco Bolson Recharge Project El Paso, TX	N/A	West Basin Water Recycling Plant, Los Angeles, CA, Los Alamitos Seawater Intrusion Barrier, Long Beach, CA; Groundwater Replenishment System, Orange County, CA	N/A
TOC	COD <10 mg/L (~3 mg/L TOC)	None	<3 mg/L, TOX <0.2 mg/L	0.5 mg/L	≤2 mg/L (of wastewater origin)
Pathogens	Multiple barriers required, $E.\ coli$ <2 cfu/100 mL	None, but multiple barriers required	Multiple barriers required; total coliform <4 cfu/100 mL	LRV from raw wastewater to finished water: 12-log for enteric viruses; 10-log for Giardia and Cryptosporidium	Multiple barriers are required; total coliform below the detection limit
Nitrogen	TKN <1 mg/L; TN <4 mg/L (Broad Run WRF only)	NOx-N <10 mg/L	TN <10 mg/L	TN <10 mg/L	None
TDS	None	<1,000 mg/L	None	RO-treatment required	None
Misc.	TSS <1 mg/L turbidity <0.5 ntu; TP <0.1 mg/L	Turbidity <1 NTU	Turbidity <2–2.5 NTU	RO and AOP treatment required for CECs	Turbidity ≤2 NTU

Source: HRSD/CH2M (2016). *Sustainable Water Recycling Initiative: Advanced Water Purification Process Feasibility Evaluation*, Report No. 3. Newport News, VA.

Note: Not all parameters are listed; for example, other requirements such as compliance with all drinking water MCLs, travel time, disinfection residual, and such are required in some states and locations.

Conversely, the application of the TOC and chemical oxygen demands (COD) limits (used to measure the amount of organic compounds in water) used with other IPR facilities (~2–3 mg/L TOC) may allow the implementation of more sustainable alternative treatment technologies. For example, the use of ozone and activated carbon operating in biological and adsorption modes has been studied for use in potable reuse projects and can often produce water with TOC <3 mg/L.

- **Pathogens**: California's strict Log Reduction Values (LRVs) are more challenging to meet than requirements at other locations, which may require additional disinfection-based treatment technologies to achieve and increase project costs significantly if adopted by Virginia.
- **Nitrogen**: USEPA's Maximum Contaminant Level (MCL)—that is, the legal threshold limit on the amount of substance that is allowed in public water systems—for nitrate in drinking water is 10 mg/L (as nitrogen [N]). Therefore, total nitrogen (TN) or nitrite/nitrate (if the predominant nitrogen species) is typically limited in IPR applications to 10 mg/L to achieve MCL compliance to meet the nitrate limit or to prevent the conversion of ammonia (NH_3) or nitrite or nitrate in the aquifer.
- **Total dissolved solids**: USEPA's secondary MCL for total dissolved solids (TDSs)—that is, the measure of the combined content of all inorganic and organic substances contained in a liquid in molecular or colloidal suspended form—is 500 mg/L. Secondary MCLs are not enforceable, and therefore compliance is not required; but drinking water customers may complain of objectionable taste when TDS levels exceed 500 mg/L, depending on the ionic makeup of the TDS. However, the TDS concentration in the Potomac Aquifer is suspected to be high (>750 mg/L, see Table 19.2), so TDS removal may not be necessary, although further investigation is warranted (HRSD/CH2M, 2016).

TABLE 19.2

Estimated TDS Concentrations (mg/L) in the Potomac Aquifer at HRSD's WWTP Locations

WWTP	Upper Potomac	Middle Potomac	Lower Potomac
Boat Harbor	750	1,000	3,500
Army Base	1,500	2,500	15,000
Virginia Initiative Plant	1,000	2,600	15,000
Nansemond	750	800	5,000
James River	1,000	1,500	10,000
York River	5,000	5,000	10,000
Williamsburg	1,000	1,500	3,000

Source: TDS data from Focazio, M.J., Speiran, G.K., and Rowan, M.E. (1993). Quality of groundwater in the Coastal Plain physiographic Province of Virginia. U.S. Geological Survey Water-Resources Investigations Report 92-4175, 20 p.

ADDITIONAL DRINKING WATER CONSIDERATIONS

All IPR projects are required to comply with drinking water MCLs, although this requirement is typically not difficult to meet because most modern wastewater treatment plants comply with drinking water MCLs, except for Tennessee, where nitrification and denitrification are not practiced. Contaminants of emerging concern (CEC), such as Personal Care Products and Pharmaceuticals (PCPPs) are any synthetic or naturally occurring chemical or any microorganism that is not commonly monitored in the environment but has the potential to enter the environment and cause known or suspected adverse ecological and/or human health effects. Pharmaceuticals and Personal Care Products comprise a very broad, diverse collection of thousands of chemical substances, including prescription and over-the-counter therapeutic drugs, fragrances, cosmetics, sun-screen agents, diagnostic agents, nutrapharmaceuticals, biopharmaceuticals, and many others. These emerging contaminants have been and are gathering significant media attention in recent years because of improvements in analytical techniques allowing measurement of these chemicals at part per trillion levels (ppt). Although some impact on the ecology has been noted at a few WWTP discharge locations due to endocrine-disrupting compounds, no impact on human health has been observed. However, at this point, the best we can say about PCPPs and their effect on the environment and/or human health is that we do not know what we do not know about them.

With regard to the impact of endocrine disruptors, what we do know is that there is a growing body of evidence that suggests that humans and wildlife species have suffered adverse health effects after exposure to endocrine-disrupting chemicals (aka environmental endocrine disruptors). In this book, environmental endocrine disruptors are defined as exogenous agents that interfere with the production, release, transport, metabolism binding, action, or elimination of natural hormones in the body responsible for the maintenance of homeostasis and the regulation of developmental processes. The definition reflects a growing awareness that the issue of endocrine disruptors in the environment extends considerably beyond that of exogenous estrogens and includes anti-androgens and agents that act on other components of the endocrine system such as the thyroid and pituitary glands (Kavlock et al., 1996). Disrupting the endocrine system can occur in various ways. Some chemicals can mimic a natural hormone, fooling the body into over-responding to the stimulus (e.g., a growth hormone that results in increased muscle mass) or responding at inappropriate times (e.g., producing insulin when it is not needed). Other endocrine-disrupting chemicals can block the effects of a hormone from certain receptors. Still others can directly stimulate or inhibit the endocrine system, causing overproduction or underproduction of hormones. Certain drugs are used to intentionally cause some of these effects, such as birth control pills. In many situations involving environmental chemicals, an endocrine effect may not be desirable.

In recent years, some scientists have proposed that chemicals might inadvertently be disrupting the endocrine system of humans and wildlife. Reported adverse effects include declines in populations, increases in cancers, and reduced reproductive function. To date, these health problems have been identified primarily in domestic or wildlife species with relatively high exposures to organo-chlorine compounds, including DDT and its metabolites, polychlorinated biphenyls (PCBs) and dioxides, or naturally occurring plant estrogens (phytoestrogens). However, the relationship between human diseases of the endocrine system and exposure to environmental contaminants is poorly understood and scientifically controversial.

Although domestic and wildlife species have demonstrated adverse health consequences from exposure to environmental contaminants that interact with the endocrine system, it is not known if similar effects are occurring in the general human population, but again there is evidence of adverse effects in populations with relatively high exposures. Several reports of declines in the quality and decrease in the quantity of sperm production in humans over the last five decades and the reported increase in incidences of certain cancers (breast, prostate, testicular) that may have an endocrine-related basis have led to speculation about environmental etiologies (Kavlock et al., 1996). For example, Carlson et al. (1992) point to the increasing concern about the impact of the environment on public health, including reproductive ability. They also point out that controversy has arisen from some reviews that have claimed that the quality of human semen has declined. However, only little notice has been paid to these warnings, possibly because the suggestions were based on data on selected groups of men recruited from infertility clinics, from among semen donors, or from candidates for vasectomy. Furthermore, the sampling of publications used for review is not systematic, thus implying a risk of bias. As a decline in semen quality may have serious implications for human reproductive health, it is of great importance to elucidate whether the reported decrease in sperm count reflects a biological phenomenon or, rather, is due to methodological errors.

Data on semen quality collected systematically from reports published worldwide indicate clearly that sperm density had declined appreciably during 1938–1990, although we cannot conclude whether or not this decline is continuing. Concomitantly, the incidence of some genitourinary abnormalities including testicular cancer and possibly also maldescent (faulty descent of the testicle into the scrotum) and hypospadias (abnormally placed urinary meatus opening in the male penis) have increased. Such remarkable changes in semen quality and the occurrence of genitourinary abnormalities over a relatively short period are more probably due to environmental rather than genetic factors. Some common prenatal influences could be responsible both for the decline in sperm density and for the increase in cancer of the testis, hypospadias, and cryptorchidism (one or both testicles fail to move to the scrotum). Whether estrogens or compounds with estrogen-like activity or other environmental or endogenous factors damage testicular function remains to be determined (Carlson et al., 1992). Even though we stated that we do not know what we do not know about endocrine disruptors it is known that the normal functions of all organ systems are regulated by endocrine factors, and small disturbances in endocrine function, especially during certain stages of life cycle such as development, pregnancy, and lactation, can lead to profound and lasting effects. The critical issue is whether sufficiently high levels of endocrine-disrupting chemicals exist in the ambient environment to exert adverse health effects on the general population.

Current methodologies for assessing, measuring, and demonstrating human and wildlife health effects (e.g., the generation of data in accordance with testing guidelines) are in their infancy. USEPA has developed testing guidelines and the

Endocrine Disruption Screening Program (EDSP; discussed later in this text), which is mandated to use validated methods for screening the testing chemicals to identify potential endocrine disruptors, determine adverse effects, dose-response, assess risk, and ultimately manage risk under current laws. The best way to end this brief discussion is to provide a statement by someone who really knows and understands endocrine disruptors and their potential impact on humans, wildlife, and the environment in general.

> Large numbers and large quantities of endocrine-disrupting chemicals have been released in the environment since World War II. Many of these chemicals can disturb development of the endocrine system and of the organs that respond to endocrine signals in organisms indirectly exposed during prenatal and/or early postnatal life; effects of exposure during development are permanent and irreversible.

> **—Theo Colborn et al. (1993)**

Regardless of our current lack of definitive evidence on the impact of PCPPs on human health, multiple barriers and relatively low TOC limits (0.5–3 mg/L) have been established for most IPR projects to limit the presence and concentration of CECs. In addition, some states require specific treatment (e.g., advanced oxidation) to ensure oxidation of a large portion of CECs.

Other nonregulated water quality parameters are often considered when implementing an IPR project, including the following:

- **Nitrosamines**: Nitrosamines are chemical compounds of the chemical structure $R^1N(-R^2)-N=O$; they are suspected carcinogens and have recently been shown to form during wastewater treatment, primarily through the disinfection process when chloramines react with n-nitrosodimethylamine (NDMA) precursors such as secondary, tertiary, and quaternary amines present in the water. California has established a notification level for NDMA of 10 ng/L, and USEPA is considering regulating some of the nitrosamines under the Safety Drinking Water Act as part of the Contaminant Candidate List 3 process, which may result in limitations applied to IPR projects.
- **Total hardness**: Total hardness in WWTP secondary effluent can often exceed levels generally deemed aesthetically acceptable by the public unless treatment or blending is specifically provided.

Note: Compatibility between the aquifer and the injected water is also required to avoid scale formation that may plug injection wells or release undesirable compounds from the soil matrix (e.g., arsenic).

ADVANCED WATER TREATMENT PROCESSES

Advanced treatment provided at indirect potable reuse plants varies but is typically focused on providing multiple barriers for the removal of pathogens and organics. Nitrogen and TDS removal is provided at some locations where necessary. Water extracted from direct injection and surface spreading projects that recharge ground-water is not typically treated again prior to distribution into the potable water system; however, water from surface augmentation projects is typically treated again at water treatment plants because of water treatment requirements stipulated by USEPA's Surface Water Treatment Rule. For example, Fairfax County's Griffith Water Treatment Plant provides coagulation, sedimentation, ozone oxidation, biologically activated carbon filtration, and chlorine disinfection for water extracted from that Occoquan Reservoir that is augmented by Upper Occoquan Service Authority's (UOSA's) indirect potable reuse plant.

Treatment provided for indirect potable reuse projects is typically a combination of multiple barriers for the removal of pathogens and organics. Multiple barriers for pathogens are typically provided through a combination of coagulation, flocculation, sediment, lime clarification, filtration (granular or membrane), and disinfection (chlorine, ultraviolet [UV], or ozone). Multiple barriers for organics removal are typically provided through a combination of advanced treatment processes (e.g., RO, granular activated carbon [GAC], ozone in combination GAC [biologically activated carbon]), although conventional treatment processes (e.g., coagulation, softening) also provide removal at some locations. All potable reuse plants and/or processes discussed in this presentation include a robust organics removal process of either GAC, RO or either GAC, RO, or soil aquifer treatment (SAT), which are effective barriers to bulk and trace organics, and represent the backbone of the potable treatment process. SAT land treatment is the controlled application of wastewater to earthen basins in permeable soils at a rate typically measured in terms of meters of liquid per week. The purpose of a SAT system is to provide a receiver aquifer capable of accepting liquid intended to recharge shallow groundwater. System design and operating criteria are developed to achieve that goal. However, there are several alternatives with respect to the utilization or final fate of the treated water (USEPA, 2006):

- Groundwater recharge.
- Recovery of treated water for subsequent reuse or discharge.
- Recharge of adjacent surface streams.
- Seasonal storage of treated water beneath the site with seasonal recovery for agriculture.

The SAT process typically includes the application of the reclaimed water using spreading basins and subsequent percolation through the vadose zone. SAT provides significant removal of both pathogens and organics through biological activity and natural filtration. However, because the Potomac Aquifer is confined, providing SAT for treatment through the vadose zone to recharge the aquifer is not possible for the HRSD SWIFT project. On the other hand, movement of reclaimed water through the aquifer after direct injection will provide significant treatment benefits, including excellent removal of pathogens, RO- and GAC-based advanced water treatment plants are often provided at locations where SAT treatment through the vadose zone is not feasible because these processes can be implemented at most locations.

RO- and GAC-based advanced treatment trains were developed for the HRSD groundwater recharge project using the historical WWTP effluent water quality data

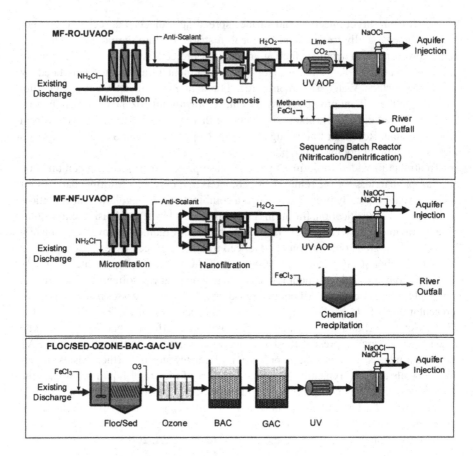

FIGURE 19.1 Process flow diagrams for three advanced wastewater treatment processes. (HRSD/CH2M (2016), CH2M Report No. 3.)

and the preliminary aquifer recharge water quality goals discussed previously. Three treatment trains were developed from this analysis and include a RO-based train, a NF-based train, and a GAC-based train (Figure 19.1). Consideration for each of these treatment trains includes the following (HRSD, 2016):

- **RO-based train**: RO has become common for potable water reuse projects in California and many international locations (e.g., Singapore, Australia) because of its effective removal of TDS, TOC, and trace organics. California regulations require the use of RO for direct injection reuse projects or a comparable alternative with regulatory approval. RO creates a waste (concentrate) stream that can be difficult and costly to dispose of, especially at inland locations. Most locations where RO has been implemented are located near the ocean where disposal of RO concentrate is convenient and much less costly than at inland locations.

- **NF-based train**: The NF-based train is similar to the RO-based train but operates at significantly lower pressure and generates a less saline concentrate, which results in significant cost savings. This process does not meet California's IPR regulatory requirements but provides excellent treatment with significant removal of pathogens and organics. This train provides partial TDS removal by providing a high level of removal of divalent ions (e.g., calcium, magnesium) and moderate removal of monovalent ions (e.g., sodium chloride). NH_3 and NOx-N removal is much lower with NF compared to RO, which results in lower TN concentration in the concentrate.
- **GAC-based train**: This train is a modernized version of full-scale operational IPR plants that have successfully been in operation for decades in Virginia (1978, 2008), Texas (1985), and more recently, Georgia (2000). GAC adsorption is used as the backbone process for organics removal, and other treatments have been added for multiple barriers to pathogens and organics. Flocculation and sedimentation provide removal of solids, pathogens, organics, and phosphorous. Ozone provides disinfection of pathogens and oxidation of organics, including oxidation of CECs and high molecular weight organic matter to smaller organic fractions that can be assimilated by biological activity present on GAC media, which is referred to as biologically active carbon (BAC) filtration. This treatment train does not provide any TDS removal and, therefore, does not generate TDS-enriched waste that might require further treatment prior to discharge.

TREATMENT PLANT EFFLUENT WATER QUALITY

HRSD provided historic effluent water quality data for seven WWTPs to identify specific water quality challenges requiring treatment. The WWTPs analyzed include Army Base, Boat Harbor, James River, Nansemond, Virginia Initiative Plant (VIP), and York River.

DATA SOURCES FOR EVALUATION

The following three primary data sources were used in the evaluation:

- **2013/2014 water quality data**: Detailed water quality data were provided for each WWTP effluent from October 2013 through September 2014. The data were provided as raw data in tables. With few exceptions, these data included the following parameters at the state frequency: chloride (1×/week), calcium (2×/week), magnesium (2×/week), potassium (2×/week), sodium (2×/week), total alkalinity (3×/week), 5-day biochemical oxygen demand (BOD_5; 5×/week), pH (1×/day), turbidity (1×/day), ammonia-nitrogen (NH_3-N; 2×/week), NOx-N (4×/week), Orthophosphate (3×/week), TKN (4×/week), total phosphorus (TP; 5×/week), and TSS (5×/week).
- **2011, 2012, and 2013 water quality data**: Detailed influent and effluent water quality data were provided for each WWTP from the 2011, 2012, and 2013 HRSD Wastewater Characteristics Studies (no James River data were

provided for 2013). The data provided were presented as minimum, average, and maximum values. The data included flow (continuous), temperature (3x/day), pH (12x/day), total alkalinity (1x/week), BOD$_5$ (5x/week), TSS (5x/week), turbidity (5x/week), fecal coliform (5x/week), TKN (frequency not reported), NOx-N (frequency not reported), and TP (frequency not reported). Data were also provided for influent chloride (1x/week) and influent sulfate (4x/week), selected heavy metals (1x/year), and a variety of organics (volatile, base/acid, pesticides; total trihalomethanes [TTHMs] 1x/year).

- **2014 total dissolved solids data**: Effluent TDS data were provided for each WWTP for January through September 2014. The data were provided as raw data in tables, and data points were provided once per week.

Effluent COD and TOC data collected by HRSD on a weekly basis from February 2015 through April 2015 were also used in the evaluation.

DATA EVALUATION

Evaluation of effluent quality from each of the seven treatment plants involved in the HRSD's SWIFT project included identification of the strength of each data source and was qualitatively documented as excellent, good, and limited. Excellent data included detailed 2013/2014 raw data and minimum/average/maximum annual data from 2011 to 2013. Good data included a full data set from only one of the sources. Limited data included data that were only collected once per year.

Total Dissolved Solids

The drinking water secondary MCL for TDS is 500 mg/L. The average effluent TDS from each WWTP except James River exceeds 500 mg/L; Army Base (1,292 mg/L) and VIP (853 mg/L) have notably high TDS concentrations.

Ammonia

Army Base and Boat Harbor plants have an average effluent NH$_3$ of 25.2 and 16.0 mg-N/L, respectively, although the Army Base plant was recently upgraded to biological nutrient removal (BNR) and now produces effluent with low ammonia and total nitrogen concentrations that are comparable to HRSD's other biological nutrient removal (BNR) plants. Total nitrogen (TN) concentrations in excess of 10 mg/L are typically not allowed for groundwater recharge into potable aquifers; therefore, additional nitrogen treatment would likely be required at the Boat Harbor WWTP.

Total trihalomethanes

Elevated TTHM levels were recorded at Nansemond (82.4 micrograms per liter [μg/L]) and Williamsburg (64.7 μg/L). TTHM levels at the other plants ranged from 3 to 50 μg/L. More TTHM data should be collected as the data sources used were limited. The drinking water Primary MCL for TTHMs is 80 μg/L.

Total hardness

Hardness in water is caused by the presence of certain positively charged metallic ions in solution in the water. The most common of these hardness-causing ions are calcium and magnesium; others include iron, strontium, and barium. The two primary constituents of water that determine the hardness of water are calcium and magnesium. If the concentration of these elements in the water is known, the total hardness of the water can be calculated. To make this calculation, the equivalent weights of calcium, magnesium, and calcium carbonate must be known; the equivalent weights are given below.

Equivalent Weights	
Calcium, Ca	20.04
Magnesium, Mg	12.15
Calcium Carbonate, CaCO$_3$	50.045

Calcium and magnesium ions are the two constituents that are the primary cause of hardness in water. To find total hardness, we simply add the concentrations of calcium and magnesium ions, expressed in terms of calcium carbonate, CaCO$_3$, using Equation (19.1).

$$\text{Total hardness (mg/L) as CaCO}_3 = \text{Cal. hardness (mg/L) as CaCO}_3$$

$$+ \text{Mg. hardness (mg/L) as CaCO}_3 \tag{19.1}$$

As mentioned, total hardness is comprised of calcium and magnesium hardness. Once total hardness has been calculated, it is sometimes used to determine another expression of hardness—carbonate and noncarbonate. When hardness is numerically greater than the sum of bicarbonate and carbonate alkalinity, that amount of hardness equivalent to the total alkalinity (both in units of mg CaCO$_3$/L) is called the **carbonate hardness**; the amount of hardness in excess of this is the **noncarbonate hardness**. When the hardness is numerically equal to or less than the sum of carbonate and noncarbonate alkalinity, all hardness is carbonate hardness and noncarbonate hardness is absent. Again, the total hardness is comprised of carbonate hardness and noncarbonate hardness:

$$\text{Total hardness = Carbonate hardness + Noncarbonate hardness} \tag{19.2}$$

During the evaluation, total hardness data were not specifically provided, but it was calculated using the detailed calcium and magnesium data. Total hardness in drinking water systems is often limited to 150 mg/L (as CaCO$_3$) or less to avoid customer complaints; and Boat Harbor (161 mg/L), VIP (181 mg/L), and York River (194 mg/L) all show average effluent data above this value. Total hardness concentrations in potable water in the surrounding area are less than the secondary MCL of 500 mg/L.

Dissolved Organic Carbon and Soluble Chemical Oxygen Demand

DOC and sCOD are important parameters to measure for potable reuse plants because advanced treatment goals and regulatory requirements are often developed for these constituents. The average DOC and sCOD concentrations in the effluent from the seven WWTPs range from 8.6 to 11.2 mg/L and 25 to 49 mg/L, respectively, which are within the typical range for WWTPs practicing biological nutrient removal. TOC concentrations above 10 mg/L become increasingly difficult to treat to recommended levels for certain advanced treatment trains so additional DOC sampling and bench and pilot testing are recommended to confirm adequate treatment performance.

Heavy Metals, VOCs, Synthetic Organic Chemicals, and Other Organics:

The 2011, 2012, and 2013 data set included minimum, average, and maximum data for heavy metals, volatile organic chemicals, synthetic organic chemicals, and other organics. Although the data sets were limited (typical frequency of once per year), the data were compared to the applicable drinking water MCLs in order to identify any potential contaminants of concern. The recorded data that were compared to drinking water MCLSs include antimony, arsenic, beryllium, cadmium, chromium, copper, lead, mercury, selenium, thallium, benzene, TTHMs, carbon tetrachloride, chlorobenzene, 1,2-dichloroethane, 1,1-dichloroethylene, 1,2-dichloropropne, ethylbenzene, toluene, 1,1,1-trichlorotehtan, 1,2,2-trichloroethand, trichloroethane, and vinyl chloride.

There are additional parameters with regulated MCLs that did not have any data provided for these WWTPs. Consequently, additional regular sampling for all MCLs, EPA Candidate Contaminant List three parameters, and other chemicals of concern at each WWTP is recommended. In addition, a comprehensive study identifying industrial and commercial facilities that discharge to the wastewater collection system is advisable to identify potential locations where chemical contamination could occur that could negatively affect finished water quality or treatment processes if not already known. An outreach program aimed at limiting the discharge of contaminants of concern and/or additional water quality monitoring at specific locations may need to be implemented if aquifer replenishment is ultimately pursued and implemented.

In addition to the average WWTP effluent data collected, selected 99th percentile effluent data from the 2013/2014 data set were analyzed to determine peak loadings from the WWTP that could be problematic for various treatment processes. Peak loadings can either be accounted for by a selected I treatment process that is designed for the maximum values or by providing a large enough equalization volume of primary or secondary effluent to attenuate the loading. The following selected parameters of concern based on the 99th percentile data include the following:

- **Nitrate/nitrite-nitrogen**: Average effluent NOx-N levels were well under the nitrate MCL of 10 mg-N/L; however, 99th percentile data at VIP (10.5 mg/L) and Williamsburg (0.3 mg/L) show that NOx-N levels could periodically exceed the nitrate MCL, which could require NOx-N specific treatment or additional storage. The 99th percentile NOx-N concentration at Boat Harbor is also high, but biological nutrient removal (BNR) is not currently practiced at this plant. When BNR is implemented at Boat Harbor, the variability of the NOx-N data should be re-evaluated.

- **Biochemical oxygen demand and total suspended solids**: High 99th percentile BOD_5 and TSS levels suggest occasional plant upsets that could be problematic for filtration (granular or membrane). This could require increased storage or treatment or automated monitoring to divert flow away from the AWTP during high biochemical oxygen demand (BOD) and TSS loadings.
- **Total dissolved solids**: WWTP effluent TDS values are not expected to fluctuate significantly, yet Army Base, VIP, and Boat Harbor each show 99th percentile values that are significantly higher than the average. Periodically high TDS values could violate treatment goals if RO is not selected and would require additional storage or provision for divisions.

ADVANCED TREATMENT PRODUCT WATER QUALITY

INORGANIC WATER QUALITY

Using the historical water quality data presented previously and the expected performance of each unit process based on professional judgment mass balance calculation for key inorganic parameters was performed for each treatment train at seven of HRSD's WWTPs. Summary tables for each treatment train are provided in Tables 19.3–19.5. Detailed mass balances reveal the following:

- The RO-based treatment process provides the lowest concentration of all water quality parameters. However, treatment to this level may not be necessary in all cases. For example, the finished water TDS concentration is about 50 mg/L, which is well below the secondary MCL (500 mg/L) and the

---

TABLE 19.3

Projected Average Finished Water from the RO-Based Treatment Train—Inorganics

Parameter	Unit	Recommended Range	AB	BH	JR	NP	VIP	WB	YR
Avg. flow	mgd	N/A	9.26	11.90	11.30	14.02	25.49	7.56	10.28
Alkalinity	mg/L	40–150	40	40	40	40	40	40	40
TDS	mg/L	0–750	87	46	37	52	61	47	46
Hardness	mg/L $CaCO_3$	50–150	43	43	42	43	43	42	44
TN	mg/L	0–10	4.84	3.95	0.91	1.01	1.31	1.27	0.89
NOx-N	mg/L	0–8	0.02	0.06	0.47	0.57	1.02	0.81	0.48
TP	mg/L	0–1	0.005	0.0035	0.004	0.01	0.00	0.01	0.00

Source: HRSD/CH2M (2016).

TABLE 19.4
Projected Average Finished Water from the NF-Based Treatment Train—Inorganics

Parameter	Unit	Recommended Range	AB	BH	JR	NP	VIP	WB	YR
Avg. flow	mgd	N/A	9.26	11.90	11.30	14.02	25.49	7.56	10.28
Alkalinity	mg/L	40–150	67	77	45	85	31	87	57
TDS	mg/L	0–750	642	280	191	333	415	283	262
Hardness	mg/L $CaCO_3$	50–150	65	76	68	57	68	67	47
TN	mg/L	0–10	<10	15.7	4.8	5.5	7.9	7.2	48
NOx-N	mg/L	0–8	0.10	0.40	3.1	3.8	6.7	5.3	3.2
TP	mg/L	0–1	0.05	0.04	0.04	0.06	0.03	0.06	0.03

Source: HRSD/CH2M (2016).

Note: Highlighted cells indicated areas of concern; modifications to WWTP operations may be required and/or additional treatment at the AWTP.

TABLE 19.5
Projected Average Finished Water from the GAC-Based Treatment Train—Inorganics

Parameter	Unit	Recommended Range	AB	BH	JR	NP	VIP	WB	YR
Avg. flow	mgd	N/A	10.90	14.00	13.30	16.50	30.00	8.89	12.09
Alkalinity	mg/L	40–150	80	99	38	132	13	118	79
TDS	mg/L	0–750	1,422	616	420	734	918	623	615
Hardness	mg/L $CaCO_3$	50–150	143	161	99	66	181	97	194
TN	mg/L	0–10	<10	23	5.7	6.4	8.5	8.1	5.6
NOx-N	mg/L	0–8	0.10	0.40	3.1	3.8	<5	5.4	3.2
TP	mg/L	0–1	0.50	0.50	0.50	0.50	0.50	0.50	0.50

Source: HRSD/CH2M (2016).

Note: Highlighted cells indicate areas of concern; modifications to WWTP operations may be required and/or additional treatment at the AWTP.

minimum background TDS in the Potomac aquifer (~750 mg/L). The very low TDS RO permeate may increase the mobilization of trace metals in the aquifer, which is undesirable.

- The NF-based treatment process provides excellent water quality as shown in Table 19.3. The NF process removes very little nitrogen; therefore, the TN concentration in the finished water exceeds the recommended upper range (10 mg/L) at Boat Harbor and is approaching the 10-mg/L limit at two other WWTPs (VIP and Williamsburg). Nitrification and denitrification improvements at these WWTPs may be necessary to ensure regular compliance with the recommended TN limit. Alternatively, NF membranes that have higher nitrogen removal can be considered; however, their use will result in a higher TDS concentrate stream.

- The GAC-based treatment process provides excellent water quality as shown in Table 19.5. Specific considerations related to this process include the following:
  - Although some incidental nitrification may occur in the biologically activated carbon filters, the process is not intended, not typically designed, to remove nitrogen. Therefore, nitrogen removal should be consulted at the upstream WWTPs which will require nitrification and denitrification improvements at several plants to ensure regular compliance with the recommended TN limit.
  - No TDS are removed through this process. The Army Base and VIP plants have elevated TDS that regularly exceed 750 mg/L. Upstream mitigation, such as reducing infiltration and inflow in areas with high TDS or eliminating industrial discharge high in TDS, may be required at these locations if a TDS limit of 750 mg/L is established.
  - Hardness removal with chemical precipitation may be required at three plants (Boat Harbor, VIP, and York River), although more investigation is necessary to determine if the total harness at these plants is acceptable (161–194 mg/L as $CaCO_3$) from aesthetic and aquifer geochemistry perspectives. If not, the proposed flocculation-sedimentation process shown for the GAC-based treatment train could be modified to a chemical softening process for those plants with elevated hardness.

Organic Water Quality

Bulk Organics

The application of robust treatment barriers for the removal of organics has historically been a center tenant in the implementation of full-scale potable reuse projects to address the presence of unknown organic compounds of chronic health concern that may be present in the secondary effluent—a significant and pressing part of the old we do not know what we do not know syndrome. As presented in Table 7.1 regulations and permits for potable reuse projects have been developed by establishing limits on bulk organic parameters, such as COD and TOC, which act as surrogates for organic compounds of wastewater origin. Virginia established a COD limit of 10 mg/L for the Occoquan and Dulles areas Watershed Policies, which apply to the UOSA IPR project (constructed in 1978) and the Broad Run WRF project (constructed in 2008), respectively. California and Florida have established TOC limits of 0.5 and 3 mg/L, respectively, in their IPR regulations.

The advanced water treatment plant's finished water COD and TOC concentration that would need to comply with the established permit limit is dependent on the initial concentration in the WWTP effluent and the specific treatment processes employed at the advanced water treatment plant. Table 19.6 shows the estimated finished water TOC concentration from each of the three proposed advanced water

TABLE 19.6

Estimated TOC Concentration in WWTP Effluent and AWTP Finished Water

Location	AB	BH	JR	NP	VIP	WB	YR
WWTP Effluent DOC	9.9	9.8	9.0	11.2	8.6	10.1	9.5
RO-AWTP	0.3	0.3	0.3	0.3	0.3	0.3	0.3
NF-AWTP	1.0	1.0	0.9	1.1	0.9	1.0	1.0
GAC-AWTP	2.5	2.5	2.3	2.8	2.2	2.6	2.4

Source: HRSD/CH2M (2016).

Notes: Preliminary TOC Goa: 0.5–3 mg/L. The following DOC removal percentages were used in the table based on full-scale treatment performance data: 97% for RO; 9-0% for NF; 35% for flocculation + sedimentation; 40% for ozone + BAC; and 40% for GAC adsorption.

treatment plant treatment trains (i.e., RO, NA, and BAC) when treating effluent from each of HRSD's WWTPs. The calculations use full-scale advanced water treatment plant effluent TOC and DOC sampling and treatment process pilot testing, the following can be concluded from the information in the table

- Compliance with a California-based TOC limit of 0.5 mg/L could only be achieved by implementing an RO-based treatment train.
- Compliance with a Florida-based UOSA-type permit (3 mg/L TOC and 10 mg/L COD, respectively) could likely be achieved at most WWTPs by any of the three proposed treatment trains, and by a hybrid treatment train that combined partial RO treatment with GAC-based treatment.
- The GAC-based AWTP will require regular replacement for regeneration of the GAC media to provide consistent TOC removal. Pilot testing is necessary to determine the GAC regeneration frequency requirements.
- Measurement of the TOC and DOC in the final effluent from each WWTP is recommended on a regular basis to accurately determine TOC removal requirements.

The following DOC removal percentages were used in this based on full-scale treatment performance data: 97% for RO; 90% for NF; 35% for flocculation + sedimentation; 40% for ozone + BAC; and 40% for GAC adsorption.

Trace Organics

Earlier, personal care products and pharmaceuticals (PCPPs) were mentioned along with other contaminants of emerging concern (CECs). Additional concerns about CECs continue to be raised about the potential for, yet generally unknown, chronic human health effects related to the thousands of organic chemicals that may end up in wastewater effluent at trace levels (mg/L). Furthermore, the efficacy of conventional water treatment processes that may end up treating source waters that have some effluent contribution is typically low. Each advanced treatment process considered in this discussion differs in its effectiveness at removing CECs. Research has shown that RO-based, NF-based, and GAC-based potable reuse treatment trains provide multiple unit processes that are effective barriers to a wide range of CECs. The RO- and NF-Based treatment rains provide substantial removal through membranes (RO/NF) and advanced oxidation (UVAOP), while the GAC-based treatment train provides significant removal through ozone-biologically active granular activated carbon (BAC) and GAC. Representative removals by advanced treatment processes for a variety of CECs were determined through recent research and monitoring of full-scale treatment facilities. These processes are redundant in the removal of some CECs (provide multiple barriers to their passage) and are complementary in the removal of others. For example, both ozone and GAC are effective barriers to the anticonvulsant drug carbamazepine, but only GAC acts as an effective barrier to the flame-retarded TCEP. No one process provides complete removal of all compounds, but RO generally provides the best removal of a wide range of compounds. However, these compounds are not destroyed or transformed by RO, but transferred to the RO concentrate (at high concentration); thus, their presence in the concentrate must be considered, particularly where the concentrate is discharged to a receiving water body.

At present, treatment for all CECs does not appear to be a differentiator among potable reuse treatment trains. Although the health effects of many CECs—either alone or as a mixture—have not been demonstrated at the ng/L concentrations typically detected in wastewater effluent, the proposed treatment trains do reduce the concentrations of many of these chemicals to a significant degree. Meanwhile, USEPA is prioritizing and studying a number of chemicals through its candidate contaminant list program.

RO CONCENTRATE DISPOSAL[2]

To this point in the book, we have discussed the benefits of RO operating systems. It is important to point out, however, that along with the good, there is the not-so-good; that is, RO systems have their advantages but they also have a few disadvantages. The one disadvantage pointed out and discussed here is the major one; that is, concentrate disposal. Where is the concentrated waste stream to be disposed of?

Mass Balance

To gain a better understanding of RO concentrate disposal issues and techniques we begin with a discussion of mass balance. The simplest way to express the fundamental engineering principle of *mass balance* is to say, "Everything has to go somewhere." More precisely, the *law of conservation of mass* says that when chemical reactions take place, matter is neither created nor destroyed. What this important concept allows us to do is track materials (concentrates), that is, pollutants, microorganisms, chemicals, and other materials from one place to another. The concept of mass balance plays an important role in RO system operations (especially in desalination) where we assume a balance exists between the material entering and leaving the RO system: "what comes in must equal what goes out." The concept is very helpful in evaluating biological systems, sampling and testing procedures, and many other unit processes within any treatment or processing system.

All desalination processes have two outgoing process streams—the product water which is lower in salt than the feed water, and a concentrated stream that contains the salts removed from the product water. Even distillation has a "bottom" solution that contains salt from the vaporized water. The higher concentrated stream is called the "concentrate." The nature of the concentrate stream depends on the salinity of the feed water, the amount of product water recovered, and the purity of the product water. To determine the volume and concentration of the two outgoing streams, a mass balance is constructed. The recovery rate of water, the rejection rate of salt, and the input flow and concentration are needed to solve equations for the flow and concentration of the product and concentrate.

RO Concentrate Disposal Practices[3]

RO concentrate is disposed of by several methods, including surface water discharge, sewer discharge, deep well injection, evaporation ponds, spray irrigation, and zero liquid discharge.

Surface Water and Sewer Disposal

Disposal of concentrate to surface water and sewer are the two most widely used disposal options for both desalting membrane processes. Data from the present survey (post-1992 data only) provide the following statistics:

Disposal Option	Desalting Plants (%)
Surface water disposal	45
Disposal to sewer	42
Total	87

This disposal option, though not always available, is the simplest option in terms of equipment involved and frequently the lowest cost options. As will be seen, however, the design of an outfall structure for surface water disposal can be complex.

Disposal to surface water involves the conveyance of the concentrate or backwash to the site of disposal and an outfall structure that typically involves a diffuser and outlet ports or valves mounted on the diffuser pipe. Factors involved in the outfall design are discussed in this section, and cost factors are presented. However, due to the large number of cost factors and the large variability in design conditions associated with surface water disposal, a relatively simple cost model cannot be developed. Disposal to surface waters requires an NPDES permit.

Disposal to the sewer involves conveyance to the sewer site and typically it negotiated fee to be paid to the WWTP. Because the negotiated fees can range from zero to substantial, there is no model that can be presented. No disposal permits are required for this disposal option. Disposal of concentrate or backwash to the sewer, however, affects the WWTP's effluent and requires an NPDES permit.

With regard to design considerations for disposal to surface water, a brief discussion of ambient conditions, discharge conditions, regulations, and the outfall structure are discussed herein.

Because receiving waters can include rivers, lakes, estuaries, canals, oceans, and other bodies of water, the range of ambient conditions can vary greatly. *Ambient conditions* include the geometry of the receiving water bottom, and the receiving water salinity, density, and velocity. Receiving water salinity, density, and velocity may vary with water depth, distance from the discharge point, and time of day and year.

Discharge conditions include the discharge geometry and the discharge flow conditions. The discharge geometry can vary from the end of the pipe to a lengthy multi-port diffuser. The discharge can be at the water's surface or submerged. The submerged outfall can be buried (except for ports) or not. Much of the historical outfall design work deals with discharges from WWTPs. These discharges can be very large—up to several hundred MGDs in flow. In ocean outfalls and many inland outfalls, these discharges are of lower salinity than the receiving water, and the discharge has positive buoyancy. The less-dense effluent rises in the more dense receiving water after it is discharged.

The volume of flow of membrane concentrates is on the lower side of the range of WWTP effluent volumes, extending up to perhaps 15 MGD at present. Membrane concentrate, as opposed to WWTP effluent, tends to be of higher salinity than most receiving waters, resulting in a condition of negative buoyancy where the effluent sinks after it is discharged. This presents a concern about the potential impact of the concentration on the benthic community at the receiving water bottom. Any possible effect on the benthic community is a function of the local ecosystem, the composition of the discharge, and the degree of dilution present at the point of contact. The chance of an adverse impact is reduced by increasing the amount of dilution at the point of bottom contact through diffuser design.

With regard to concentrate discharge regulations, it is important to note that receiving waters can differ substantially in their volume, flow, depth, temperature, composition, and degree of variability in these parameters. The effect of discharge of a concentrate or backwash to a receiving water can vary widely depending on these factors. The regulation of effluent disposal to receiving water involves several considerations, some of which are the end-of-pipe characteristics of the concentrate or backwash. Comparison is made between receiving water quality standards (dependent on the classification of the receiving water) and the water quality of the effluent to determine disposal feasibility. In addition, in States such as Florida, the effluent must also pass a test where test species, chosen based on the receiving water characteristics, are exposed to various dilutions of the effluent. Because the nature of the concentrate or backwash is different than that of the receiving water, there is a region near the discharge area where mixing and subsequent dilution of the concentrate or backwash occurs.

Where conditions cannot be met at the end of the discharge pipe, a mixing zone may be granted by the regulatory agency. The mixing zone is an administrative construct that defines a limited area or volume of the receiving water where this initial dilution of the discharge is allowed to occur. The definition of an allowable mixing zone is based on receiving water modeling. The regulations require that certain conditions be met at the edge of the mixing zone in terms of concentration and toxicity.

Once the mixing zone conditions are met, then the outfall structure can be properly designed and installed. Actually, the purpose of the outfall structure is to ensure that mixing conditions can be met and that discharge of the effluent, in general, will not produce any damaging effect on the receiving water, its lifeforms, wildlife, and the surrounding area.

In highly turbulent and moving receiving water with a large volume relative to the effluent discharge, simple discharge from the end of a pipe may be sufficient to ensure rapid dilution and mixing of the effluent. For most situations, however, the mixing can be improved substantially through the use of a carefully designed outfall structure. Such design may be necessary to meet regulatory constraints. The most typical outfall structure for this purpose consists of a pipe of limited length mounted perpendicular to the end of the delivery pipe. This pipe, called a diffuser, has one or more discharge ports along its length.

Disposal to the Sewer

Where possible, this means of disposal is simple and usually cost-effective. Disposal to sewer does not require a permit but does require permission from the wastewater treatment plant. The impact of both the flow volume and composition of the concentrate will be considered by the WWTP, as it will affect their capacity buffer and their NPDES permit. The high volume of some concentrates prohibits their discharge to the local WWTP. In other cases, concerns are focused on the increased TDS level of the WWTP effluent that results from the concentrate discharge. The possibility of disposal to sewer is highly site-dependent. In addition to the factors mentioned, the possibility is influenced by the distance between the two facilities, by whether the two facilities are owned by the same entry, and by future capacity increases anticipated. Where disposal to the sewer is allowed, the WTP may be required to pay fees based on volume and/or composition.

Deep Well Disposal

As mentioned earlier, Injection wells are a disposal option in which liquid wastes are injected into porous subsurface rock formations. The depths of the wells typically range from 1,000 to 8,000 ft. The rock formation receiving the waste must possess the natural ability to contain and isolate it. Paramount in the design and operation of an injection well is the ability to prevent the movement of wastes into or between underground sources of drinking water. Historically, this disposal option has been referred to as deep well injection or disposal to waste disposal wells. Because of the very slow fluid movement in the injection zone, injection wells may be considered a storage method rather than a disposal method; the wastes remain there indefinitely if the injection program has been properly planned and carried out.

Because of their ability to isolate hazardous wastes from the environment, injection wells have evolved as the predominant form of hazardous waste disposal in the United States. According to a 1984 study by EPA, almost 60% of all hazardous waste disposed of in 1981, or ~10 billion gallons, was injected into deep wells. By contrast, only 35% of this waste was disposed of in surface impoundments and <5%

in landfills. The EPA study also found that a still smaller volume of hazardous waste, under 500 million gallons, was incinerated in 1981 (Gordon, 1984). Although RO concentrate is not classified as hazardous, injection wells are widely used for concentrate disposal in the state of Florida.

A study prepared for the Underground Injection Practices Council showed that relatively few injection well malfunctions have resulted in the contamination of water supplies (Strycker and Collins, 1987). However, other studies document instances of injection well failure resulting in contamination of drinking water supplies and groundwater resources (Gordon, 1984).

Injection of hazardous waste can be considered safe if the waste never migrates out of the injection zone. However, there are at least five ways water may migrate and contaminate potable groundwater (Strycker and Collins, 1987). Wastes may:

- Escape through the wellbore into an underground source of drinking water because of insufficient casing or failure of the injection well casing due to corrosion or excessive injection pressure
- Escape vertically outside of the well casing from the injection zone into an underground source of drinking water (USDW) aquifer
- Escape vertically from the injection zone through confining beds that are inadequate because of high primary permeability, solution channels, joints, faults, or induced fractures
- Escape vertically from the injection zone through nearby wells that are improperly cemented or plugged or that have inadequate or leaky casing
- Contaminate groundwater directly by lateral travel of the injected wastewater from a region of saline water to a region of freshwater in the same aquifer

Evaporation Pond Disposal

Solar evaporation, a well-established method for removing water from a concentrated solution, has been used for centuries to recover salt (sodium chloride) from seawater. There are also installations that are used for the recovery of sodium chloride and other chemicals from strong brines, such as the Great Salt Lake and the Dead Sea, and for the disposal of brines resulting from oil well operation (USEPA, 2016).

Evaporation ponds for membrane concentrate disposal are most appropriate for smaller volume flows and regions having a relatively warm, dry climate with high evaporation rates, level terrain, and low land costs. These criteria apply predominantly in the western half of the United States—in particular, the southwestern portion. Advantages associated with evaporation ponds are described in the following list:

- They are relatively easy and straightforward to construct.
- Properly constructed evaporation ponds are low maintenance and require little operator attention compared to mechanical equipment.
- Except for pumps to convey the wastewater to the pond, no mechanical equipment is required.

- For smaller volume flows, evaporation ponds are frequently the least costly means of disposal, especially in areas with high evaporation rates and low land costs.

Despite the inherent advances of evaporation ponds, they are not without disadvantages that can limit their application, as described in the following list:

- They can require large tracts of land if they are located where the evaporation rate is low or the disposal rate is high.
- Most States require impervious liners of clay or synthetic membranes such as polyvinylchloride (PVC) or Hypalon. This requirement substantially increases the costs of evaporation ponds.
- Seepage from poorly constructed evaporation ponds can contaminate underlying potable water aquifers.
- There is little economy of scale (i.e., no cost reduction resulting from increased production) for this land-intensive disposal option. Consequently, disposal costs can be large for all but small-size membrane plants.

In addition to the potential for contamination of groundwater, evaporation ponds have been criticized because they do not recover the water evaporated from the pond. However, the water evaporated is not "lost"; it remains in the atmosphere for about 10 days and then returns to the surface of the earth as rain or snow. This hydrologic cycle of evaporation and condensation is essential to life on land and is largely responsible for weather and climate.

With regard to evaporation pond design considerations, sizing of the ponds, determination of the evaporation rate, and pond depth are important. Again, evaporation ponds function by transferring liquid water in the pond to water vapor in the atmosphere about the pond. The rate at which an evaporation pond can transfer this water governs the size of the pond. The selection of pond size requires the determination of both the surface area and the depth needed. The surface area required is dependent primarily on the evaporation rate. The pond must have adequate depth for surge capacity and water storage, storage capacity for precipitated salts, and freeboard for precipitation (rainfall) and wave action.

Proper sizing of an evaporation pond depends on accurate calculation of the annual evaporation rate. Evaporation from a freshwater body, such as a lake, is dependent on local climatological conditions, which are very site-specific. To develop accurate evaporation data throughout the United States, meteorological stations have been established at which special pans simulate evaporation from large bodies of water such as lakes, reservoirs, and evaporation ponds. The pans are fabricated to standard dimensions and are situated to be as representative of a natural body of water as possible. A standard evaporation pan is referred to as a Class A pan. The standardized dimensions of the pass and the consistent methods for collecting the evaporation data allow comparatively and reasonably accurate data to be developed for the United States. The data collection must cover several years to be reasonably accurate and representative of site-specific variations in climatic conditions.

Published evaporation rate databases typically cover 10-year or more periods and are expressed in inches per year.

The pan evaporation data from each site can be compiled into a map of pan evaporation rates. Because of the small heat capacity of evaporation ponds, they tend to heat and cool more rapidly than adjacent lakes and to evaporate at a higher rate than an adjacent natural pond of water. In general, experience has shown the evaporation rate from large bodies of water to be ~70% of that measured in a Class A pan (Spellman, 2017). This percentage is referred to as the Class A pan coefficient and must be applied to measured pan evaporation to arrive at actual lake evaporation. Over the years, site-specific Class A pan coefficients have been developed for the entire United States. Multiplying the pan evaporation rate by the pan coefficient results in a mean annual lake evaporation rate for a specific area.

Maps depicting annual average precipitation across the United States also are available. Subtracting the mean annual evaporation from the mean annual precipitation gives the net lake surface evaporation in inches per year. This is the amount of water that will evaporate from a freshwater pond (or the amount the surface level will drop) over a year if no water other than natural precipitation enters the pond. All these maps assume an impervious pond that allows no seepage. Note that for some parts of the country, the results of this calculation give a negative number; and in other parts of the country, it is a positive number. A negative number indicates a net loss of water from a pond over a year or a drop in the pond surface level. A possible number indicates more precipitation than evaporation at a particular site. A freshwater pond at one of these sites would gain water over a year, even if no water other than natural precipitation was added. Thus, such a site would not be a candidate for an evaporation pond.

DID YOU KNOW?

RO concentrate streams are not easily disposed of in inland areas, as surface water and sanitary sewer discharges would not be allowed, and deep well injection may not be feasible depending on geologic features.

It is important to realize that data of this type are representative only of the particular sites of the individual meteorological stations, which may be separated by many miles. Climatic data specific to the exact site should be obtained if at all possible before the actual construction of an evaporation pond.

The evaporation data described above are for freshwater pond evaporation. However, brine density has a marked effect on the rate of solar evaporation. Most procedures for calculating evaporation rate indicate evaporation is directly proportional to vapor pressure. Salinity reduces evaporation primarily because the vapor pressure of the saline water is lower than that of freshwater and because dissolved salts lower the free energy of the water molecules. Cohesive forces acting between the dissolved ions and the water molecules may also be responsible for inhibiting evaporation, making it more difficult for the water to escape as vapor (Miller, 1989).

The lower vapor pressure and lower evaporation rate of saline water result in a lower energy loss and, thus, a higher equilibrium temperature than that of freshwater under the same exposure conditions. The increase in temperature of the saline water would tend to increase evaporation, but the water is less efficient in converting radiant energy into latent heat due to the exchange of sensible heat and long-wave radiation with the atmosphere. The net result is that, with the same input of energy, the evaporation rate of saline water is lower than that of freshwater. To provide the non-engineer or non-scientist with an idea of how (for illustrative reasons only) the evaporation pond evaporation rate is mathematically determined, the following is provided.

ESTIMATING RATE OF EVAPORATION—PONDS

In lake, reservoir, and pond management, knowledge of evaporative processes is important to the environmental professional in understanding how water losses through evaporation are determined. Evaporation increases the storage requirement and decreases the yield of lakes and reservoirs. Several models and empirical methods are used for calculating lake and reservoir evaporative processes. In the following, applications used for the water budget and energy budget models, along with four empirical methods: The Priestly-Taylor, Penman, DeBruin-Keijman, and Papadakis equations are made.

Water Budget Model

The water budget model for lake evaporation is used to make estimations of lake evaporation in some areas. It depends on an accurate measurement of the inflow and outflow of the lake. It is expressed as

$$\Delta S = P + R + \text{GI} - \text{GO} - E - T - O \qquad (19.3)$$

where
ΔS = change in lake storage, mm
P = precipitation, mm
R = surface runoff or inflow, mm
GI = groundwater inflow, mm
GO = groundwater outflow, mm
E = évaporation, mm
T = transpiration, mm
O = surface water release, mm

If a lake has little vegetation and negligible groundwater inflow and outflow, lake evaporation can be estimated by:

$$E = P + R - O \pm \Delta S \qquad (19.4)$$

Note: Much of the following information is adapted from Mosner and Aulenbach (2003).

Energy Budget Model

According to Rosenberry et al. (1993), the energy budget (Lee and Swancar, 1997) is recognized as the most accurate method for determining lake evaporation. Mosner and Aulenbach (2003), point out that it is also the most costly and time-consuming method. The evaporation rate, E_{EB}, is given by Lee and Swancar (1997)

$$E_{EB}\,(\text{cm/day}) = \frac{Q_s - Q_r + Q_a + Q_{ar} - Q_{bs} + Q_v - Q_x}{L(1 + BR) + T_0} \qquad (19.5)$$

where

E_{EB} = evaporation, in centimeters per day (cm/day)
Q_s = incident shortwave radiation, in cal/cm^2/day
Q_r = reflected shortwave radiation, in cal/cm^2/day
Q_a = incident longwave radiation from the atmosphere, in cal/cm^2/day
Q_{ar} = reflected longwave radiation, in cal/cm^2/day
Q_{bs} = longwave radiation emitted by the lake, in cal/cm^2/day
Q_v = net energy advected by streamflow, groundwater, and precipitation, in cal/cm^2/day
Q_x = change in heat stored in a water body, in cal/cm^2/day
L = latent heat of vaporization, in cal/g
BR = Bowen Ratio, dimensionless
T_0 = water-surface temperature (°C)

Priestly-Taylor Equation

Winter and Colleagues (1995) point out that the Priestly-Taylor equation is used to calculate potential evapotranspiration (PET) which is a measure of the maximum possible water loss from an area under a specified set of weather conditions or evaporation as a function of latent heat of vaporization and heat flux in a water body, and is defined by the equation:

$$\text{PET}\,(\text{cm/day}) = \alpha\left(s/s + \gamma\right)\left[\left(Q_n - Q_x\right)/L\right] \qquad (19.6)$$

where

PET = potential evapotranspiration, cm/day
$\alpha = 1.26$, Priestly-Taylor empirically derived constant, dimensionless
$(s/s + \gamma)$ = parameters derived from the slope of saturated vapor pressure-temperature curve at the mean air temperature; γ is the psychrometric constant, s is the slope of the saturated vapor pressure gradient, dimensionless
Q_n = net radiation, cal/cm^2/day
Q_x = change in heat stored in water body, cal/cm^2/day
L = latent heat of vaporization, cal/g

Penman Equation

Winter et al. (1995) point out that the Penman equation for estimating potential evapotranspiration, E_0, can be written as:

$$E_0 = \frac{(\Delta/\gamma)H_e + E_a}{(\Delta/\gamma) + 1} \tag{19.7}$$

where
  Δ = slope of the saturation absolute humidity curve at the air temperature
  γ = the psychrometric constant
  H_e = evaporation equivalent of the net radiation
  E_a = aerodynamic expression for evaporation

DeBruin-Keijman Equation

The DeBruin-Keijman equation (Winter et al., 1995) determines evaporation rates as a function of the moisture content of the air above the water body, the heat stored in the still water body, and the psychrometric constant, which is a function of atmospheric pressure and latent heat of vaporization.

$$\text{PET}(\text{cm/day}) = \left[\text{SVP}/(0.95SVP + 0.63\gamma) \right] * (Q_n - Q_x) \tag{19.8}$$

where
  SVP = saturated vapor pressure at mean air temperature, millibars/K
  All other terms have been defined previously

Papadakis Equation

The Papadakis equation (Winter et al., 1995) does not account for the heat flux that occurs in the still water body to determine evaporation. Instead, the equation depends on the difference in the saturated vapor pressure above the water body at maximum and minimum air temperatures, and evaporation is defined by the equation

$$\text{PET}(\text{cm/day}) = 0.5625 \left[E_0 \max - (E_0 \min - 2) \right] \tag{19.9}$$

where all terms have been defined previously.

++

For water saturated with sodium chloride salt (26.4%), the solar evaporation rate is generally about 70% of the rate for freshwater (USEPA, 2006). Studies have shown that the evaporation rate from the Great Salt Lake, which has a TDS level of between 240,000 and 280,000 mg/L, is about 80%–82% of the rate for freshwater. Other studies indicate that evaporation rates of 2%, 5%, 10%, and 20% sodium chloride solutions are 97%, 98%, 93%, and 78%, respectively, of the rates of freshwater (USEPA, 2006). These ratios are determined from both experiment and theory. However, there is no simple relationship between salinity and evaporation, for there are always complex interactions among site-specific variables such as air temperature, wind velocity, relative humidity, barometric pressure, water surface temperature, heat exchange rate with the atmosphere, including incident solar absorption and reflection, thermal currents in the pond, and depth of the pond. As a result, these ratios should be used only as guidelines and with discretion. It is important to recognize that salinity can significantly reduce the evaporation rate and to allow for this effect in sizing the evaporation pond surface area. In lieu of site-specific data, an evaporation ratio of 0.70 is a reasonable allowance for long-term evaporation reduction. This ratio is also considered to be an appropriate factor for evaporation ponds that are expected to reach salt saturation over their anticipated service life.

DID YOU KNOW?

To gain an understanding of what is meant by incident solar absorption, the following definitions are provided:

- **Incident ray**: a ray of light that strikes (impinges upon) a surface. The angle between this ray and the perpendicular or normal to the surface is the angle of incidence.
- **Reflected ray**: a ray that has rebounded from a surface.
- **Angle of incidence**: the angle between the incident ray and a normal line.
- **Angle of reflection**: the angle between the reflected ray and the normal line.
- **Angle of refraction**: the angle between the refracted ray and the normal line.
- **Index of refraction (n)**: is the ratio speed of light (c) in a vacuum to its speed (v) in a given material; always greater than one.

As mentioned earlier, pond depth is an important parameter in determining pond evaporation rate. Studies indicated that pond depths ranging from 1 to 18 inches are optimal for maximizing the evaporation rate. However, similar studies indicate only a 4% reduction in the evaporation rate as the pond depth is increased from 1 to 40 inches (USEPA, 2006). Very shallow evaporation ponds are subject to drying and cracking of the liners and are not functional in long-term service for concentrate disposal. From a practical operating standpoint, an evaporation pond must not only evaporate wastewater but also provide

- Surge capacity or contingency water storage
- Storage capacity for precipitated salts
- Freeboard for precipitation and wave action

For an evaporation pond to be a viable disposal alternative for membrane concentrate it must be able to accept concentrate at all times and under all conditions so as not to restrict operation of the desalination plant. The pond must be able to accommodate variations in the weather and upsets in the desalination plant. The desalination plant cannot be shut down because the evaporation pond level is rising faster than anticipated.

DID YOU KNOW?

Current concentrate disposal of membrane concentrate using evaporation
ponds accounts for 5% of total disposal practices.

To allow for unpredictable circumstances, it is important that design contingencies be applied to the calculated pond area and depth. Experience from the design of industrial evaporation ponds has shown that discharges are largest during the first year of plant operation, are reduced during the second year, and are relatively constant thereafter. A long-term, 20% contingency may be applied to the surface areas of the pond or its capacity to continuously evaporate water. The additional contingencies above 20% (up to 50%) during the first and second years of operation are applied to the depth holding capacity of the pond.

The freeboard for precipitation should be estimated on the basis of precipitation intensity and duration for the specific site. There may also be local codes governing freeboard requirements. In lieu of site-specific data, an allowance of 6 inches for precipitation is generally adequate where evaporation ponds are most likely to be located in the United States (USEPA, 2016).

Freeboard for wave action can be estimated as follows:

$$H_w = 0.047 \times W \times \sqrt{(F)} \tag{19.10}$$

where:
H_w = wave height (ft)
W = wind velocity (mph)
F = fetch, or straight-line distance the wind can blow without obstruction (mile)

The run-up of waves on the face of the dike approaches the velocity head of the waves and can be approximated as $1.5 \times$ wave height (H_w). H_w is the freeboard allowance for wave action and typically ranges from 2 to 4 ft. The minimum recommended combined freeboard (for precipitation and wave action) is 2 ft. This minimum applies primarily to small ponds.

Over the life of the pond (which should be sized for the same duration as the projected life of the desalination facility), the water will likely reach saturation and precipitate salts. The type and quantity of salts are highly variable and very site-specific.

Spray Irrigation Disposal

Land application methods include irrigation systems, rapid infiltration, and overland flow systems (Crites et al., 2000). These methods, and in particular irrigation, were originally used to take advantage of sewage effluent as a nutrient or fertilizer source as well as to reuse the water. Membrane concentrate has been used for land application in the spray irrigation mode. Using the concentrate in lieu of fresh irrigation water helps conserve natural resources, and in areas where water conservation is of great importance, spray irrigation is especially attractive. Because of the higher TDS concentration of RO concentrate, unless it is diluted, concentrate is less likely to be used for spray irrigation purposes.

The concentrate can be applied to cropland or vegetation by sprinkling or surface techniques for water conservation by the exchange when lawns, parks, or golf courses are irrigated and for preservation and enlargement of green belts and open spaces.

Where the nutrient concentration of the wastewater for irrigation is of little value, hydraulic loading can be maximized to the extent possible, and system costs can be minimized. Crops such as water-tolerant grasses with low potential for economic return but with high salinity tolerance are generally chosen for this type of requirement.

Fundamental considerations in land application systems include knowledge of wastewater characteristics, vegetation, and public health requirements for successful design and operation. Environmental regulations at each site must be closely examined to determine if spray irrigation is feasible. Contamination of the groundwater and runoff into surface water are key concerns. Also, the quality of the concentrate—its salinity, toxicity, and soil permeability—must be acceptable.

The principal objective in spray irrigation systems for concentrate discharge is the ultimate disposal of the applied wastewater. With this objective, the hydraulic loading is usually limited by the infiltration capacity of the soil. If the site has a relatively impermeable subsurface layer or a high groundwater table, underdrains can be installed to increase the allowable loading. Grasses are usually selected for the vegetation because of their high nutrient requirements and water tolerance.

Other conditions must be met before concentrate irrigation can be considered as a practical disposal option. First, there must be a need for irrigation water in the vicinity of the membrane plant. If the need exists, a contract between the operating plant and the irrigation user would be required. Second, a backup disposal or storage method must be available during periods of heavy rainfall. Third, monitor wells must be drilled before an operating permit is obtained (Conlon, 1989).

With regard to design factors, the following considerations apply to spray irrigation of concentrate for ultimate disposal:

- Salt, trace metals, and salinity
- Site selection
- preapplication treatment
- Hydraulic loading rates
- Land requirements
- Vegetation selection
- Distribution techniques
- Surface runoff control

Salt, Trace Metals, and Salinity

Three factors that affect an irrigation source's long-term influence on soil permeability are the sodium content relative to calcium and magnesium, the carbonate and bicarbonate content, and the total salt concentration of the irrigation water. Sodium salts remain in the soil and may adversely affect its structure. High sodium concentrations in clay-bearing soils disperse soil particles and decrease soil permeability, thus reducing the rate at which water moves into the soil and reducing aeration. If the soil permeability, or infiltration rate, is greatly reduced, then the vegetation on the irrigation site cannot survive. The hardness level (calcium and magnesium) will form insoluble precipitates with carbonates when the water is concentrated. This buildup of solids can eventually block the migration of water through the soil.

The U.S. Department of Agriculture's Salinity Laboratory developed a sodium adsorption ratio (SAR) to determine the sodium limit. It is defined as follows

$$SAR = Na \Big/ \big[(Ca + Mg)/2 \big]^{1/2} \qquad (19.11)$$

Where
   Na = sodium, milliequivalent per liter (meq/L)
   Ca = calcium, meq/L
   Mg = magnesium, meq/L

High SAR values (>9) may adversely affect the permeability of fine-textured soils and can sometimes be toxic to plants.

Trace elements are essential for plant growth; however, at higher levels, some become toxic to both plants and microorganisms. The retention capacity for most metals in most soils is generally high, especially for pH above 7. Under low pH conditions, some metals can leach out of soils and may adversely affect the surface waters in the area.

Salinity is the most important parameter in determining the impact of the concentrate on the soil. High concentrations of salts whose accumulation is potentially harmful will be continually added to the soil with irrigation water. The rate of salt accumulation depends upon the quantity applied and the rate at which it is removed from the soil by leaching. The salt levels in many brackish RO concentrates can be between 5,000 and 10,000 parts per million, a range that normally rules out spray irrigation.

DID YOU KNOW?

Soluble salts in a water solution will conduct an electric current; thus, changes in electrical conductivity (EC) can be used to measure the water's salt content in electrical resistance units (decisiemens per meter, or dS/m).

In addition to the effects of total salinity on vegetation and soil, individual ions can cause a reduction in plant growth. Toxicity occurs when a specific ion is taken up and accumulated by the vegetation, ultimately resulting in damage to it. The ions of most concern in wastewater effluent irrigation are sodium, chloride, and boron. Other heavy metals can be very harmful, even if present only in small quantities. These include copper, iron, barium, lead, and manganese. These all have strict environmental regulations in many States.

In addition to the influence on the soil, the effect of the salt concentrations on the groundwater must be considered. The possible impact on groundwater sources may be a difficult obstacle where soil saturation is high and the water table is close to the surface. The chance of increasing background TDS levels of the groundwater is high with the concentrate. Due to this consideration, spray irrigation requires a runoff control system. An underdrain or piping distribution system may have to be installed under the full areas of irrigation to collect excess seepage through the soil and, thus, to protect the groundwater sources. If high salinity concentrate is being used, scaling of the underdrain may become a problem. The piping perforations used to collect the water can be easily scaled because the openings are generally small. Vulnerability to scaling must be carefully evaluated before a project is undertaken.

Site Selection

Site selection factors and criteria for effluent irrigation are presented in Table 19.7. A moderately permeable soil capable of infiltration up to 2 inches per day on an intermittent basis is preferable. The total amount of land required for land application is highly variable but primarily depends on application rates.

TABLE 19.7
Site Selection Factors and Criteria

Factor	Criterion
Soil	
Type	Loamy soils are preferred, but most soils from sands to clays are acceptable
Drainability	Well-drained soil is preferred
Depth	Uniformly 5–6 ft or more throughout sites is preferred
Groundwater	
Depth to groundwater	A minimum of 5 ft is preferred
Groundwater control	Control may be necessary to ensure renovation if the water table is <10 ft from the surface
Groundwater movement	Velocity and direction of movement must be determined
Slopes	Slopes of up to 20% are acceptable with or without terracing
Underground formations	Formations should be mapped and analyzed with respect to isolation depending on wastewater characteristics, methods of application, and crop
Distance from the source of wastewater	An appropriate distance is a matter of economics

Preapplication Treatment

Factors that should be considered in assessing the need for preapplication treatment include whether the concentrate is mixed with additional wastewater before application, the type of vegetation grown, the degree of contact with the wastewater by the public, and the method of application. In four Florida sites, concentrate is aerated before discharge, because each plant discharges to a retention pond or ponds before irrigation. Aeration by increasing DO prevents stagnation and algae growth in the ponds and also supports fish populations. The ponds are required for flow equalization and mixing. Typically, concentrate is blended with biologically treated wastewater.

Hydraulic Loading Rates

Determining the hydraulic loading rate is the most critical step in designing a spray irrigation system. The loading rate is used to calculate the required irrigation area and is a function of precipitation, evapotranspiration, and percolation. The following equation represents the general water balance for hydraulic loading based upon a monthly time period and assuming zero runoff

$$HLR = ET + PER - PPT \tag{19.12}$$

where:
  HLR = hydraulic loading rate
  ET = evapotranspiration
  PER = percolation
  PPT = precipitation

In most cases, surface runoff from fields irrigated with wastewater is not allowed without a permit or, at least, must be controlled; it is usually controlled just so that a permit does not have to be obtained.

Seasonal variations in each of these values would be considered by evaluating the water balance for each month as well as the annual balance. For precipitation, the wettest year in 10 is suggested as reasonable in most cases. Evapotranspiration will also vary from month to month, but the total for the year should be relatively constant. Percolation includes that portion of the water that, after infiltration into the soil, flows through the root zone and eventually becomes part of the groundwater. The percolation rate used in the calculation should be determined on the basis of a number of factors, including soil characteristics underlying geologic conditions, groundwater conditions, and the length of drying period required for satisfactory vegetation growth. The principal factor is the permeability of hydraulic conductivity of the least permeable layer in the soil profile.

Resting periods, standard in most irrigation techniques, allow the water to drain from the top few inches of soil. Aerobic conditions are thus restored, and air penetrates the soil. Resting periods may range from a portion of each day to 14 days and depend on the vegetation, the number of individual plots in the rotation cycle, and the availability of backup storage capacity.

To properly calculate an annual hydraulic loading rate, monthly evapotranspiration, precipitation, and percolation rates must be obtained. The annual hydraulic loading rate represents the sum of the monthly loading rates. Recommended loading rates range from 2 to 20 feet per year (Goigel, 1991).

Land Requirements

Once a hydraulic loading rate has been determined, the required irrigation area can be calculated using the following equation:

$$A = Q \times \text{Kl}/\text{ALR} \qquad (19.13)$$

where:

A = irrigation area (acre)
Q = concentrate flow (gpd)
ALR = annual hydraulic loading rate (ft/yr)
Kl = 0.00112 $d \times \text{ft}^3 \times \text{acres}/(\text{hr} \times \text{gal} \times \text{ft}^2)$

The total land area required for spray irrigation includes allowances for buffer zones and storage and, if necessary, land for emergencies or future expansion.

For loadings of constituents such as nitrogen, which may be of interest to golf course managers who need fertilizer for the grasses, the field area requirement is calculated as follows:

$$\text{Field area (acres)} = 3{,}040 \times C \times Q/Lc \qquad (19.14)$$

where:

C = concentration of constituent (mg/L)
Q = flow rate (mgd)
Lc = loading rate of constituent (pounds per acre-year [lb/acre-yr])

Vegetation Selection

The important aspects of vegetation for irrigation systems are water needs and tolerances, sensitivity to wastewater constituents, public health regulations, and vegetation management considerations. The vegetation selection depends highly on the location of the irrigation site and natural conditions such as temperature, precipitation, and topsoil conditions. Automated watering alone cannot always ensure vegetation propagation. Vegetation selection is the responsibility of the property owners. Woodland irrigation for growing trees is being conducted in some areas. The principal limitations on this use of wastewater include low water tolerances of certain trees and the necessity to use fixed sprinklers, which are expensive.

Membrane concentrate disposal will generally be to landscape vegetation. Such application, for example to highway median and border strips, airport strips, golf courses, parks and recreational areas, and wildlife areas, has several advantages. Problems associated with crops for consumption are avoided, and the irrigated land is already owned, so land acquisition costs are saved.

Distribution Techniques

Many different distribution techniques are available for engineered wastewater efflu-ent applications. For irrigation, two main groups, sprinkling and surface application, are used. Sprinkling systems used for spray irrigation are of two types—fixed and moving. Fixed systems, often called solid set systems, may be either on the ground surface or buried. Both types usually consist of impact sprinklers mounted on risers that are spaced along lateral pipelines, which are, in turn, connected to main pipe-lines. These systems are adaptable to a wide variety of terrains and may be used for irrigation of either cultivated land or woodlands. Portable aluminum pipe is normally used for aboveground systems. This pipe has the advantage of relatively low capital cost but is easily damaged, has a short, expected life because of corrosion, and must be removed during cultivation and harvesting operations. Pipes used for buried sys-tems may be buried as deep and 1.5 ft below the ground surface. Buried systems usu-ally have the greatest capital cost; however, they are probably the most dependable and are well-suited to automated control.

There are a number of different moving sprinkle systems, including center-pivot, side-roll, wheel-move, rotating-boom, and which-propelled systems.

Surface Runoff Control

Surface runoff control depends mainly on the proximity of surface water. If runoff drains to surface water, an NPDES permit may be required. This situation should be avoided if possible due to the complication of quantifying overland runoff. Berms can be built around the irrigation field to prevent runoff. Another alternative, although expensive, is a surrounding collection system. It is best to use precautions and backup systems to ensure that overwatering and subsequent runoff does not occur in the first place.

Zero Liquid Discharge Disposal

In this approach, evaporation is used to further concentrate the membrane con-centrate. In the extreme limit of processing concentrate to dry salts, the method becomes a zero-discharge option. Evaporation requires major capital investment, and the high energy consumption together with the final salt or brine disposal can result in significant disposal costs. Because of this, disposal of municipal mem-brane concentrate by mechanical evaporation would typically be considered as a last resort; that is when no other disposal option is feasible. Cost aside, however, there are some advantages:

- It may avoid a lengthy and tedious permitting process.
- It may gain quick community acceptance.
- It can be located virtually anywhere.
- It represents a positive extreme in recycling by efficiently using the water source.

When this thermal process is used following an RO system, for example, it produces additional product water by recovering high-purity distillate from the concentrate wastewater stream. The distillate can be used to help meet the system product water volume requirement. This reduces the size of the membrane system and, thus, the size of the membrane concentrate to be treated by the thermal process. In addition, because the product purity of the thermal process is so high (TDS in the range of 10 mg/L), some of the product water volume reduction of the system may be met by blending the thermal product with untreated source water. The usual concerns and considerations about using untreated water for blending need to be addressed. The end result may be a system where the system product requirement is met by three streams: (1) membrane product, (2) thermal process product, and (3) bypass water.

Single- and Multiple-Effect Evaporators

Using steam as the energy source, it takes about 1,000 British thermal units (BTI) to evaporate a pound of water. In a single-effect evaporator, the heat released by the condensing steam is transferred across a heat exchange surface to an aqueous solution boiling at a temperature lower than that of the condensing stream. The solution absorbs heat; and part of the solution water vaporizes, causing the remaining solution to become richer in solution. The water vapor flows to a barometric or surface condenser, where it condenses as its latent heat is released to cooling water at a lower temperature. The finite temperature differences between the steam, the boiling liquid, and the condenser are the driving forces required for the heat transfer surface area to be less than infinite. Practically all the heat removed from the condensing stream (which had been generated initially by burning fuel) is rejected to cooling water and is often dissipated to the environment without being of further use.

The water vapor that flows to the condenser in a single-effect evaporator is at a lower temperature and pressure than the heating stream but has almost as much enthalpy. Instead of releasing its latent heat to cooling water, the water vapor may be used as heating steam in another evaporator effect operating at a lower temperature and pressure than the first effect. Additional effects may be added in a similar manner, each generating additional vapor, which may be used to heat a lower-temperature effect. The vapor generated in the lowest-temperature effect is finally condensed by releasing its latent heat to cooling water in a condenser. The economy of a single- or multiple-effect evaporator may be expressed as the ratio of kilograms of total evaporation to kilograms of heating steam. As effects are added, the economy increases representing more efficient energy utilization. Eventually, added effects result in marginal added benefits, and the number of effects is thus limited by both practical and economic considerations. Multiple-effect evaporators increase the efficiency (economy) but add capital cost in additional evaporator bodies.

More specifically, the number of effects, and thus the economy achieved, is limited by the total temperature difference between the saturation temperature of the heating steam (or about heat source) and the temperature of the cooling water (or other heat sink). The available temperature difference may also be constrained by the temperature sensitivity of the solution to be evaporated. The total temperature difference, less any losses, becomes allocated between effects in proportion to their resistance to heat transfer, the effects being thermal resistances in series.

The heat transfer surface area for each effect is inversely proportional to the net temperature difference available for that effect. Increasing the number of effects reduces the temperature difference and evaporation duty per effect, which increases the total area of the evaporator in rough proportion to the number of effects. The temperature difference available to each effect is reduced by boiling point elevation and by the decrease in vapor saturation temperature due to pressure drop. The boiling point elevation of a solution is the increase in boiling point of the solution compared to the boiling point of pure water at the same pressure; it depends on the nature of the solute and increases with increasing solute concentration. In a multiple-effect evaporator, the boiling point elevation and vapor pressure drop losses for all the effects must be summed and subtracted from the overall temperature difference between the heat source and sink to determine the net driving force available for heat transfer.

Vapor-Compression Evaporator Systems (Brine Concentrators)

A vapor compression evaporator system, or brine concentrator, is similar to a conventional single-effect evaporator, except that the vapor released from the boiling solution is compressed in a compressor. Compression raises the pressure and saturation temperature of the vapor so that it may be returned to the evaporator steam chest to be used as heating steam. The latent heat of the vapor is used to evaporate more water instead of being rejected to cooling water. The compressor adds energy to the vapor to raise its saturation temperature above the boiling temperature of the solution by whatever net temperature difference is desired. The compressor is not completely efficient, having small losses due to mechanical friction and larger losses due to nonisentropic compression. However, the additional energy required because of nonisentropic compression is not lost from the evaporator system; it serves to superheat the compressed vapor. The compression energy added to the vapor is of the same magnitude as the energy required to raise feed to the boiling point and make up for radiation and venting losses. By exchanging heat between the condensed vapors (distillate) and the product with the feed, it is usually possible to operate with little or no makeup heat in addition to the energy necessary to drive the compressor. The compressor power is proportional to the increase in saturation temperature produced by the compressor. The evaporator design must trade off compressor power consumption versus heat transfer surface area. Using the vapor compression approach to evaporate water requires only about 100 BTU to evaporate a pound of water. Thus, one evaporator body driven by mechanical vapor compression is equivalent to ten effects or a ten-body system driven by steam.

DID YOU KNOW?

In the British system of units, the unit of heat is the British thermal unit or Btu. One *Btu* is the amount of heat required to raise one pound of water one degree Fahrenheit at normal atmospheric pressure (1 atm).

Although most brine concentrators have been used to process cooling water, concentrators have also been used to concentrate reject from RO plants. Approximately 90% of these concentrators operate with a seeded slurry process that allows the reject to be concentrated as much as 40 to 1 without scaling problems developing in the evaporator. Brine concentrators also produce distilled product water that can be used for high-purity purposes or for blending with other water supplies. Because of the ability to achieve such high levels of concentration, brine concentrators can reduce or eliminate the need for alternative disposal methods such as deep well injection or solar evaporation ponds. When operated in conjunction with crystallizers or spray dryers, brine concentrators can achieve zero liquid discharge of RO concentrate under all climatic conditions.

Individual brine concentrator units range in capacity from approximately 10 to 700 gpm of feedwater flow. Units below 150 gpm of capacity are usually skid mounted, and larger units are field fabricated. A majority of operating brine concentrators are single-effect, vertical tube, falling film evaporators that use a calcium sulfate-seeded slurry process. The energy input to the brine concentrator can be provided by an electric-driven vapor compressor or by process steam from a host industrial facility. Steam-driven systems can be configured with multiple effects to minimize energy consumption.

Product water quality is normally <10 mg/L TDS. Brine rejection from the concentrator typically ranges between 2% and 10% of the feedwater flow, with TDS concentrations as high as 250,000 mg/L.

DID YOU KNOW?

Solids in water occur either in solution or in suspension and are distinguished by passing the water sample through a glass-fiber filter. By definition, the *suspended solids* are retained on top of the filter, and the *dissolved solids* pass through the filter with the water. When the filtered portion of the water sample is placed in a small dish and then evaporated, the solids in the water remain as residue in the evaporating dish. This material is called *total dissolved solids*, or TDS. Dissolved solids may be organic or inorganic. Water may come into contact with these substances within the soil, on surfaces, and in the atmosphere. The organic dissolved constituents of water are from the decay products of vegetation, from organic chemicals, and organic gases. Removing these dissolved minerals, gases, and organic constituents is desirable because they may cause physiological effects and produce aesthetically displeasing colors, taste, and odors.

Because of the corrosive nature of many wastewater brines, brine concentrators are usually constructed of high-quality materials, including titanium evaporator tubes and stainless-steel vessels suitable for 30-year evaporator life. For conditions of high chloride concentrations or other more corrosive environments, brine concentrators can be constructed of materials such as AL6XN, Inconal 825, or other exotic metals to meet performance and reliability requirements.

DID YOU KNOW?

The method of evaporation selected is based on the characteristics of the RO membrane concentrate and the type of energy source to be used.

Crystallizers

Crystallizer technology has been used for many years to concentrate feed streams in industrial processes. More recently, as the need to concentrate wastewater has increased, this technology has been applied to reject desalination processes, such as brine concentrate evaporators, to reduce wastewater to a transportable solid. Crystallizer technology is especially applicable in areas where solar evaporation pond construction costs are high, solar evaporation rates are negative, or deep well disposal is costly, geologically not feasible, or not permitted.

Crystallizers used for wastewater disposal range in capacity from about 2 to 50 gpm. These units have vertical cylindrical vessels with heat input from vapor compressors or an available steam supply. For small systems ranging from 2 to 6 gpm, steam-driven crystallizers are more economical. Steam can be supplied by a package boiler or a process source if one is available. For larger systems, electrically driven vapor compressors are normally used to supply heat for evaporation.

Typically, the crystallizer requires a purge stream of about 2% of the feed to the crystallizer. This is necessary to prevent extremely soluble species (such as calcium chloride) from building up in the vapor body and to prevent the production of dry cake solids. The suggested disposal of this stream is to a small evaporation pond. The crystallizer produces considerable solids that can be disposed of in commercial landfills.

The first crystallizers, applied to power plant wastewater disposal, experienced problems related to materials selection and process stability; but subsequent design changes and operating experience have produced reliable technology.

For RO concentrate disposal, crystallizers would normally be operated with a brine concentrator evaporator to reduce brine concentrator blowdown to a transportable solid. Crystallizers can be used to concentrate RO reject directly, but their capital cost and energy usage are much higher than for a brine concentrator of equivalent capacity.

Spray Dryers

Spray dryers provide an alternative to crystallizers for the concentration of wastewater brines to dryness. Spray dryers are generally more cost-effective for smaller feed flows of <10 gpm.

Pathogen Removal

As shown in Table 19.1, various states have developed different approaches to regulating pathogen removal by indirect potable reuse plants. For example, Virginia permitted the UOSA indirect potable reuse plant and the Broad Run WRF based on achieving a nondetected concentration of E. coli (<2 cfu/100 mL). Other states have taken a different approach. For example, California requires a 12-log reduction of viruses and a 10-log reduction of Cryptosporidium and Giardia from the raw wastewater to the advanced water treatment plant finished water.

Pathogen removal by each of the three proposed treatment trains is significant and would result in nondetectable concentrations for all indicator organisms typically used in wastewater treatment (e.g., *E. coli*, total coliform, fecal coliform) assuming proper operation. Therefore, compliance with a UOSA-type permit or Florida's pathogen-related indirect potable reuse regulations would be met by all three proposed treatment trains. Compliance with the California regulations may be more challenging, especially for GAC-based treatment, because of the high log reduction requirements. Discussion with the Virginia regulators is necessary to determine how the proposed HRSD groundwater recharge project would be regulated with respect to pathogen removal.

DISINFECTION BY-PRODUCTS

Excessive formation of trihalomethanes (THMs) at WWTPs is fairly common, especially for plants that provide good nitrification. Low effluent NH_3 concentrations at these plants lead to the formation of free chlorine (rather than chloramines) during chlorine disinfection, which, when reacted with bulk organics, has a propensity to form high levels of THMs. NDMA, which is another disinfection by-product, can also form in significant concentrations during the disinfection process at WWTPs depending on the precursors in the water and the type of chlorination practiced. Little NDMA forms in the presence of free chlorine, but significant concentrations typically form in the presence of chloramines, with dichloramine resulting in more rapid formation kinetics than monochloramine. Both THMs and NDMA can be removed by AWT processes through specialized design, but a more cost-efficient approach is to prevent their formation by withdrawing the water from the WWTP prior to disinfection (upstream of the chlorine contact basin). Specific withdrawal points sat each WWTP and the potential treatment required for TTHMs and NDMA removal should be considered in the next stage of this project.

ANTICIPATED IMPROVEMENTS TO HRSD'S WWTPS

Some operational and capital improvements to the existing WWTPs may be required depending on the AWT train selected for implementation and the final effluent water quality produced at each WWTP. Table 19.8 shows the improvement that will likely be required. Analysis regarding WWTP improvements should be ongoing.

From the data and evaluations presented to this point, it is obvious that the bottom line is that any of the three advanced water treatment trains—RO-based, NF-based, and GAC-based—are likely viable for groundwater recharge of effluent generated from HRSD's WWTPs. The finished water quality produced by each train will be excellent with respect to pathogen and organics removal, but the use of the RO-based or NF-based treatment train is necessary if TDS reduction is required. Partial RO or NF treatment could be used depending on the degree of TDS reduction required. BNR improvements will be required at some of the WWTPs to reduce the TN concentration and the propensity for organic fouling membranes in the RO- and NF-based trains.

Selection of the advanced water treatment train to be implemented at each WWTP should be based on numerous factors such as finished water quality, wastewater discharge requirements, operability, sustainability, site-specific factors (e.g., space

TABLE 19.8

Required WWTP Improvements to Address AWT Operational Impacts and/or Finished Water Quality Deficiency

WWTP	MF-RO-UVAOP	MF-NF-UVAOP	FS-O3-BAC-GAC-UV
Army Base	None	None	AWT finished water deficiency: TDS > 750 mg/L Required WWTP: if possible, reduce influent TDS to WWTP
Boar Harbor	AWT Operational Impact: Greater Membrane fouling requiring more design due to lack due to lack of BNR AWT finished water deficiency: None Required WWTP Improvements: Add nit/denit at WWTP	AWT Operational Impact: Greater membrane fouling requiring more conservative design due to lack of BNR AWT finished water deficiency: TN > 10 mg/L Required WWTP Improvements: Add nit/denit at WWTP	AWT Finished Water Deficiency: TN > 10 mg/L; total hardness > 150 mg/L Required WWTP Improvements: Add nit/denit at WWTO; if determined Necessary, add softening to AWT
James River	None	None	None
Nansemond	None	None	None
VIP and Williamsburg	None	AWT Finished Water Deficiency: TN approaching 10 mg/L Required WWTP Improvements: Improve nit/denit at WWTP	AWT Finished Water Deficiency: TN approaching 10 mg/L; TDS > 750 mg/L; total hardness > 150 mg/L (VIP only) Required WWTP Improvements: Improve nit/denit at WWTP and reduce influent TDS to WWTP (VIP only), if possible; if determined necessary, add softening to AWTP (VIP only)
York River	None	None	AWT Finished Water Deficiency: Hardness > 150 mg/L Required WWTP or AWT Improvements: If determined necessary, add softening to AWT

Source: HRSD/CH2M (2016).
Note: nit/denit = nitrification/denitrification.

existing infrastructure, hydraulic), and capital and operating costs (discussed later). The ultimate selection of the advanced water treatment train will also be dictated by regulatory requirements related to treatment, finished water quality, and wastewater discharge requirements that have not yet been established; therefore, engaging the appropriate regulatory agency(ies) is important during the next phase of this project. Treatment selection may also be influenced by public perception.

Because HRSD's SWIFT project is presently operational at its Nansemond Treatment but is still a work in progress, with time for adjustment here, there, and almost anywhere, other action items that will influence advanced water treatment train selection should be performed including the following:

- Regularly sample at each WWTP for COD, sCOD, TOC, DOC, all contaminants regulated by primary MFLs, selected CECs, and parameters specific to the design of RO and NF treatment (i.e., barium, strontium, fluoride, silica, alkalinity, pH).
- Regularly measure water quality (e.g., pH, alkalinity, TDS, hardness) in numerous Potomac Aquifer product wells.
- Evaluate site-specific conditions at each WWTP that may influence AWT train selection, including site space, hydraulics, geotechnical conditions, electrical service, and use of existing infrastructure for AWT treatment.
- Conduct an industrial and commercial water quality discharge study to characterize risk and to identify chemicals of concern that may be discharged into the collection system.
- Determine potential causes for high TDS concentrations at WWTPs where effluent TDS is >500 mg/L.

NOTES

1 Much of the material in this section is from HRSD/CH2M (2016). *Sustainable Water Recycling Initiative: Advanced Water Purification Process Feasibility Evaluation. Report 3.* Hampton Roads Sanitation District. Compiled by CH2M Newport News, VA.
2 Adapted from Spellman, F.R. (2015). *Reverse Osmosis: A Guide for the Non-Engineering Professional.* Boca Raton, FL: CRC Press.
3 Material in this section is from US Department of Interior Bureau of Reclamation (Mickley & Associates). (2006). *Membrane Concentrate Disposal: Practices and Regulation.* Washington, DC.

REFERENCES

Carlson, E., Giwercam, A., Keiding, N., and Skakkebaek, N.E. (1992). Evidence for decreasing quality of semen during past 50 years. *BMJ* 305: 609–612.
Colborn, T., vom Saal, F.S., and Soto, A.M. (1993). Developmental effects of endocrine-disrupting chemicals in wildlife and humans. *Environ Health Perspect* 101(5): 378–384.
Conlon, W.J. (1989). Disposal of concentrate from membrane process plants. *Waterworld News* January/February 18–19.
Crites, S.L.J. et al. (2000). *Land Treatment Systems for Municipal and Industrial Wastes.* New York: McGraw.

Goigel, J.F. (1991). Regulatory investigation and cost analysis of concentrate disposal from membrane plants. M.Sc thesis, University of Central Florida, Orlando, FL.

Gordon, W. (1984). *A Citizen's Handbook on Ground Water Protection*. New York: Natural Resources Defense Council.

HRSD/CH2M. (2016). *Sustainable Water Recycling Initiative: Groundwater Injection Hydraulic Feasibility Evaluation, Report 1*. Virginia Beach, VA. Hampton Roads Sanitation District. Compiled by CH2M Newport News, VA. *Sustainable Water Recycling Initiative: Groundwater Injection Geochemical Compatibility Feasibility Evaluation, Report 2*. Virginia Beach, VA. Hampton Roads Sanitation District. Compiled by CH2M Newport News, VA. *Sustainable Water Recycling Initiative: Advanced Water Purification Process Feasibility Evaluation, Report 3*. Hampton Roads Sanitation District. Compiled by CH2M Newport News, VA.

Kavlock, R.J., et al. (1996). Research Needs for the risk assessment of health and environmental effects of endocrine disruptors: A report of the U.S. EPA-sponsored workshop. *Environmental Health Perspectives* 104(4): 715–740.

Lee, T.M., and Swancar, A. (1997). *Influence of Evaporation, Groundwater, and an Uncertainty in the Hydrologic Budget of Lake Lucerne, and Seepage Lake in Polk County*. Florida: USGS.

Miller, W. (1989). Estimating evaporation from Utah's Great Salt Lake using thermal infrared satellite imagery. *Water Resources Bulletin*25: 541–542.

Mosner, M.S., and Aulenbach, B.T. (2003). *Comparison of Methods used to Estimate Lake Evaporation for a Water Budget of Lake Seminole, Southwestern Georgia and Northwestern Florida*. Atlanta, GA: U.S. Geological Survey.

Rosenberry, D.O., Sturrock, A.M., and Winter, T.C. (1993). Evaluation of the energy budget method of determining evaporation at Williams Lake, Minnesota, using alternative instrumentation and study approaches. *Water Resources Research* 29(8): 77–82.

Spellman, F.R. (2017). *The Science of Water*, 3rd ed. Boca Raton, FL. CRC Press.

Strycker, A., and Collins, A.G. (1987). *State-of-the-Art Report: Injection of Hazardous Wastes into Deep Wells*. Ada, OK: Robert S. Kerr Environmental Research Laboratory.

USEPA. (2006). *Land Treatment of Municipal Wastewater Treatment*. Cincinnati, OH: United States Environmental Protection Agency.

USEPA. (2016). Aquifer Recharge and Aquifer Storage and Recovery. Accessed at https://epa.gov/uic/aquifer-recharge-and-aquifer-storage-and-recovery.

Winter, T.C., Rosenberry, D.O., and Sturrock, A.M. (1995). Evaluation of eleven equations for determining evaporation for a small lake in the north central United States. *Water Resources Research* 31(4): 983–993.

20 SWIFT
The Process

Note to reader: At the beginning of this presentation, it was pointed out that in order to understand the basics of HRSD's SWIFT initiative, it would be necessary to connect the dots. In this chapter the final dot is put in place—HRSD's process of converting sewage (wastewater) into drinking water, then injecting it into the Potomac Aquifer. Right now, it is happening at the Nansemond Treatment Plant, which HRSD is using as a research center to gather data about the effects on water quality and geology. The goal is to inject around 1 million gallons per day, but the process has had a few stops and starts as adjustments are made. The information presented in the following is operational data as we know it today and continue to learn and make adjustments.

SUSTAINABLE WATER INITIATIVE FOR TOMORROW

Figure 20.1 presents a process flow block diagram of eight-unit processes (eight-step process) connected in a treatment train for the SWIFT Demonstration Facility at HRSD's Nansemond Treatment Plant in Suffolk, VA. In addition, a very brief fundamental explanation of each of the eight-unit processes is provided below.

SWIFT UNIT PROCESS DESCRIPTION

In this section each step in the SWIFT process is described—we are connecting the dots.

> **Step 1: Influent pump station** (see Figure 20.2): Wastewater effluent from the Nansemond Treatment Plant is directed to the SWIFT processing site at the plant location. Thus, the highly treated water from the Nansemond Treatment plant is pumped to the Research Center's advanced treatment facility (SWIFT) where it undergoes advanced treatment within the unit processes housed within; the highly treated SWIFT effluent (of better drinking water quality than that contained within the Potomac Aquifer) is out-falled (injected) into the Upper, Middle, and Lower Potomac Aquifer layers.
>
> **Step 2: Mixing, flocculation, and sedimentation**: This unit process removes suspended solids by settling large particles at the bottom of the water column.
>
> **Step 3: Ozone contact**: This unit process breaks down organic material and provides disinfection.

DOI: 10.1201/9781003498049-23

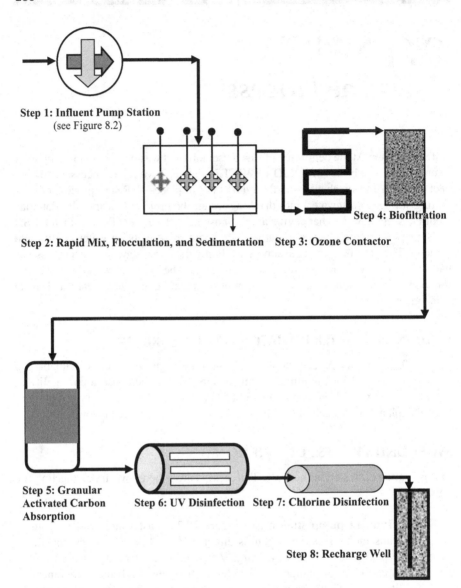

Step 1: Influent Pump Station
(see Figure 8.2)

Step 4: Biofiltration

Step 2: Rapid Mix, Flocculation, and Sedimentation   Step 3: Ozone Contactor

Step 5: Granular
Activated Carbon   Step 6: UV Disinfection Step 7: Chlorine Disinfection
Absorption

Step 8: Recharge Well

FIGURE 20.1   Process flow diagram for SWIFT demonstration facility.

Step 4: Biologically active filtration: This unit process filters out suspended
   particles, and pathogens, and removes dissolved organic compounds through
   microbiological activity.
Step 5: Granulated activated carbon contactors: This unit removes trace
   organic compounds and prepares the water for ultraviolet disinfection.
Step 6: Ultraviolet (UV) disinfection: This unit process provides a barrier
   to pathogens by disinfecting the water with high-intensity ultraviolet light.

FIGURE 20.2   SWIFT influent pump station (Photo by F.R. Spellman).

Step 7: Chlorine contact and chemical addition: This unit process provides disinfection of finished water using chlorine and serves as an additional barrier to pathogens. Chemical addition is used on the disinfected water and is adjusted by small doses to match the geochemistry of the water already more closely in the aquifer.

THE ULTIMATE BOTTOM LINE

HRSD's SWIFT initiative is a far-thinking innovative initiative that potentially offers enormous benefits not only for the Hampton Roads Region but also for any region with similar needs, issues, and/or problems. To conclude with an ultimate bottom line on the SWIFT process and its benefits presented in this account, we must recognize the water challenges faced by those in the Hampton Roads Region and other locations. In the Hampton Roads Region specifically, these water challenges consist of questions that only operational time, operational adjustments, and results can and will eventually answer. These questions are:

- Will SWIFT restore the Chesapeake Bay?
- Will SWIFT mitigate groundwater depletion?
- Will SWIFT prevent saltwater intrusion?
- Will SWIFT counter relative sea level rise?

FIGURE 20.3 The hand pump used by site visitors to sample SWIFT water (Photo by F.R. Spellman).

- Will SWIFT prevent recurrent flooding?
- Will SWIFT prevent sanitary sewer overflows?
- Will SWIFT be affordable?

The jury is still out on whether these questions will be answered positively. We do not know what we do not know at this precise moment. However, early, very early observations, measurements, and results indicate promise.

By the way, if you ever visit the SWIFT research center at the Nansemond Treatment Plant in Suffolk, Virginia make sure to ask to sample the final product at the sample tasting location shown in Figure 20.3. Those who have sampled the recycled water have had nothing but positive comments and have how surprised they were that the SWIFT water tasted better than any water they have ever consumed.

And the positive comments just related are the *real ultimate bottom line.*

Index

Note: **Bold** page numbers refer to tables and *italic* page numbers refer to figures.

Printed in the United States
by Baker & Taylor Publisher Services